Environmental Engineering Dictionary and Directory

Thomas M. Pankratz

LEWIS PUBLISHERS
Boca Raton London New York Washington, D.C.

Library of Congress Cataloging-in-Publication Data

Pankratz, Tom M.
    Environmental engineering dictionary and directory / Thomas M. Pankratz.
    p. cm.
    ISBN 1-56670-543-6 (alk. paper)
      1. Environmental engineering--Dictionaries. 2. Brand name products--Dictionaries. 3.
Trademarks--Dictionaries. 4. Environmental engineering--Directories. I. Title.

TD9 .P36 2000
628--dc21   00-044356

This book contains information obtained from authentic and highly regarded sources. Reprinted material is quoted with permission, and sources are indicated. A wide variety of references are listed. Reasonable efforts have been made to publish reliable data and information, but the author and the publisher cannot assume responsibility for the validity of all materials or for the consequences of their use.

Neither this book nor any part may be reproduced or transmitted in any form or by any means, electronic or mechanical, including photocopying, microfilming, and recording, or by any information storage or retrieval system, without prior permission in writing from the publisher.

The consent of CRC Press LLC does not extend to copying for general distribution, for promotion, for creating new works, or for resale. Specific permission must be obtained in writing from CRC Press LLC for such copying.

Direct all inquiries to CRC Press LLC, 2000 N.W. Corporate Blvd., Boca Raton, Florida 33431.

Trademark Notice: Product or corporate names may be trademarks or registered trademarks, and are used only for identification and explanation, without intent to infringe.

Preface

This book has been written to help professionals, students, and lay people identify the increasing number of terms in the fields of environmental engineering and science.

More than 8000 terms, acronyms, and abbreviations applying to wastewater, potable water, industrial water treatment, seawater desalination, air pollution, incineration, and hazardous waste remediation have been defined.

The most unique feature of this book is the inclusion of more than 3000 trademarks and brand names. Many of these commercial terms for proprietary products or processes are so common or descriptive that they have fallen into general use. This confusion is compounded by the fact that many terms contain similar prefixes (e.g., bio-, enviro-, hydra-, hydro-, etc.) and it is often difficult to tell them apart.

This book originates from *Screening Equipment Handbook*, first published in 1988, whose glossary contains a list of screening-related trademarks and brand names along with their company affiliation. Even though that list was relatively short, a surprisingly large number of companies had come and gone or changed their names through mergers or acquisitions. This led to an expanded directory entitled, *The Dictionary of Water and Wastewater Treatment Trademarks and Brand Names*, published in 1991, and which contained 1200 commercial terms.

The Concise Dictionary of Environmental Engineering followed in 1996. In addition to the 2200 commercial terms, it was further expanded to include 3000 generic environmental engineering terms. Shortly after it was published, the environmental equipment manufacturing industry began a consolidation led by USFilter, Waterlink, Baker Hughes, ITT, F.B. Leopold, and others that has resulted in changes to 43% of the terms included in the 1996 edition.

During the research for this book, many other books, magazines, dictionaries, glossaries, buyer's guides, catalogs, brochures, and technical papers were reviewed to locate new terms and their definitions. Although there are too many references to list, I would like to acknowledge the help of these publications and their authors.

In addition to technically reviewing this book, John B. Tonner was especially helpful with his suggestions, advice, research assistance, and computer wizardry. Regardless of when I would call, John was always available to help. His www.world-wide-water.com Web site also proved to be a valuable research tool.

I would like to acknowledge the libraries that were used in my research. They include the M.D. Anderson Library at the University of Houston, the Helen Hall Library in League City, Texas, the Houston Public Library Central Branch, and the library at King Fahd University of Petroleum and Mining in Dhahran, Saudi Arabia.

I also recognize USFilter and Alfa-Laval for their support.

I'm grateful for the assistance of the many friends and colleagues who suggested new terms and challenged old ones, helped with definitions, provided encouragement, or assisted in the book's production. Some of these people include Robert W. Brown, Gordon Carter, Bill Copa, Chad Dannemann, Jim Force, Jack Gardiner, Duane Germenis, Stacie Jones, John Meidl, Mack Moore, Chad Pankratz, Bill Perpich, Barb Petroff, Jim Symons, Mark Wilson, and Joe Zuback.

Like the first edition, published in 1996, much of my work on this book took place while traveling; the rest was done in the evenings and weekends. I would never have been able to finish without the continued patience and support of my wife, Julie, and our children, Chad, Sarah, Mike, and Katie.

This book is dedicated to my wife, Julie Lynn Pankratz, and our grandson, Gabriel R. Suarez, who was born the same day this book was completed.

Tom Pankratz

Introduction

This dictionary contains terms used in the fields of environmental engineering and environmental science, and the definitions provided relate to their use in an environmental context only.

The commercial terms represent company brand names or trademarks, and have been italicized to differentiate them from the technical terms in general usage. Whenever appropriate, the use of ™ or ® has been included following the name of the entry, although terms may be registered trademarks even though they do not include either symbol. It is also possible that some of the entries listed as trademarks may not be registered or properly used by the manufacturers listed in connection with them.

Brand names and trademarks often evolve and take different forms. Variations in the use of capitalization, hyphens, or symbols often occur over time. The representation of the words included in this book reflects the latest version seen in use and are assumed to be the preferred form.

Commercial acronyms are included if they are registered trademarks or commonly used abbreviations of company names. Nonregistered product model numbers and trademarks that are the same as the name of a company are not always included. Many definitions were extrapolated from stories, advertisements, or product brochures and were not directly corroborated by the company listed as being responsible for the term.

The company name included in the definition of a commercial term usually represents the company that manufactures that particular product or process. In some cases, the listed company may only market, distribute, or license the product.

In several instances, the same brand name has been listed more than once to describe different products or processes from different companies. The author is unaware of any dispute involving these cases and is simply reporting that the companies identified have used the term for the product described. In some cases, the term may be dormant, obsolete, or no longer available from the company listed.

Company addresses, phone numbers, and e-mail addresses listed in the Manufacturer's Directory were confirmed over a period of several years. Some contact information may have changed, especially with the recent telephone area code changes in many parts of the U.S. Readers are cautioned that an incorrect phone number, address, or e-mail address does not mean that a company is no longer in business.

There are a few cases where a company whose name is listed in a definition is not included in the Manufacturer's Directory. If current contact information for a company could not be located, the out-of-date information was not included.

Terms have been arranged alphabetically using current word processing software.

In general, terms related to plumbing, household products, computer programs, or software have not been included.

All of the terms have been listed in good faith. A reasonable attempt has been made to confirm all definitions and, in the case of commercial terms, verify the companies responsible for the listings. The author apologizes for any omissions or errors.

If you are aware of any changes or additions that should be included in subsequent editions, please send them to Tom Pankratz, P.O. Box 75064, Houston, Texas USA, 77234-5064.

Foreword

The areas of environmental engineering and sciences and their related business activities have grown to the point that they overlap the professional and private lives of almost everyone. As environmental issues become more complicated, so does the vocabulary required to understand and discuss them. This *Environmental Engineering Dictionary and Directory* defines many terms that did not even exist a decade ago.

My own field of water reclamation and reuse is an example of a relatively new area of environmental engineering that has fostered the introduction of many new terms and technologies.

When considering advanced treatment of municipal and industrial wastewaters, a repeated thesis has been that such a high quality effluent should be put to beneficial use rather than simply wasted. Today, technically proven treatment and purification processes exist to provide treated water of almost any quality desired. This offers a realistic framework for considering water reclamation and reuse in many parts of the world that are experiencing water shortages. Nonpotable water reuse applications, such as agricultural and landscape irrigation, toilet flushing in large office buildings, and water for aesthetic and environmental purposes have become major options for planned water reuse.

Water reuse provides innovative and alternative options for agriculture, municipalities, and industries. However, water reuse is only one alternative in planning to meet future water resource needs. Conservation, efficient management and use of existing water supplies, and the development of new water resources based on watershed management or seawater desalination are examples of other alternatives.

As the field of environmental engineering continues to develop, so will the vocabulary required for its discussion and study. Our need to understand the environment and to better appreciate our relationship with nature is greater now than at any time in our history. Thus Tom's book is particularly timely and relevant.

Takashi Asano, Ph.D., P.E.
Adjunct Professor
Department of Civil and Environmental Engineering
University of California at Davis

A

Å   See "Angstrom (Å)."

A&I   Alternative and Innovative.

A/O®   Wastewater treatment process for biological removal of nitrogen by USFilter/Krüger.

A²/O®   Biological treatment process for phosphorus and nitrogen removal by USFilter/Krüger.

A²C™   Biological wastewater treatment system by Baker Process — Municipal Division.

A·I·R   Photocatalytic process to destroy VOCs by Trojan Technologies, Inc.

AA   See "atomic absorption spectrophotometry (AA)."

AAEE   American Academy of Environmental Engineers.

AAP   Asbestos Action Program.

AAPCO   American Association of Pesticide Control Officials.

AAQS   Ambient air quality standards.

AARC   Alliance for Acid Rain Control.

AAS   American Association for the Advancement of Science.

ABA1000®   Alumina oxide for phosphate reduction by Selecto, Inc.

ABA2000®   Alumina oxide for lead and heavy metals removal by Selecto, Inc.

ABA8000®   Alumina oxide for fluoride removal by Selecto, Inc.

abandoned well   A well whose use has been permanently discontinued or which is in a state of such disrepair that it cannot be used for its intended purpose.

abatement   Reducing the degree or intensity of, or eliminating, pollution.

abattoir   A place where animals are slaughtered for their meat and meat byproducts.

ABC Filter™   Automatic backwashable cartridge filter by USFilter/Rockford.

Abcor®   Ultrafiltration membrane product by Koch Membrane Systems, Inc.

ABF   Activated bio-filtration wastewater treatment system by Infilco Degremont, Inc.

ABF   Traveling bridge type automatic backwashing gravity sand filter by Aqua-Aerobic Systems, Inc.

abiocoen   All of the geologic, climatic, and other nonliving elements of an ecosystem.

abiotic   Nonliving elements in the environment.

ABJ™   ABJ product group of Sanitaire Corp.

ablation   The combined processes of glacial melting and evaporation which results in a net loss of ice.

ablation zone   The lower part of a glacier where the net loss of ice exceeds the net gain.

ABS   (1) Acrylonitrile-butadiene-styrene. A black plastic material, used in the manufacture of pipes and other components. (2) Alkyl-benzene-sulfonate. A surfactant formerly used in synthetic detergents that resisted biological breakdown.

absolute filter rating   A filter rating which indicates that 99.9% of the particles larger than a specified size will be removed by the filter.

absolute humidity   The total amount of water vapor present in the air, measured in grams per cubic meter.

absolute pressure   The total pressure in a system, equal to the sum of the gage pressure and atmospheric pressure.

absolute purity water   Water with a specific resistance of 18.3 megohm-cm at 25°C.

absolute zero   The lowest temperature possible; 0° on the Kelvin scale or approximately -273°C (-459.7°F).

absorbate   A substance used to soak up another substance.

absorbed dose   The amount of a chemical that enters the body of an exposed organism.

absorbent   Any substance that exhibits the properties of absorption.

absorption   The process of transferring molecules of gas, liquid, or a dissolved substance to the surface of a solid where it is bound by chemical or physical forces.

absorption field   A trench or pit filled with gravel or loose rock designed to absorb septic tank effluent.

ABW®   Traveling bridge type gravity sand filter by Infilco Degremont, Inc.

abyssal zone   A zone of deep oceanic waters, generally deeper than 2000 meters and between the hadal and bathyal zones where light does not penetrate.

AC   See "activated carbon."

AC®   Industrial wastewater treatment unit by Colloid Environmental Technologies Co.

ACA   American Conservation Association.

acaricide   A pesticide used to kill spiders, ticks, or mites.

ACBM   Asbestos-containing building material.

Accelapak®   Modular water treatment plant by Infilco Degremont, Inc.

Accelator®   Solids contact clarifier with primary and secondary mixing zones by Infilco Degremont, Inc.

Accelo Hi-Cap   Filter underdrain block formerly offered by Infilco Degremont, Inc.

Accelo-Biox®   Modular wastewater treatment plant by Infilco Degremont, Inc.

Accel-o-Fac™   Sewage treatment plant design by Lake Aid Systems.

acceptable risk   The level of risk associated with minimal adverse effects, usually determined by a risk analysis.

Access Analytical   Former name of IDEXX Laboratories, Inc.

accessory species   Species found in less than half but more than one quarter of the area covered by a plant community.

accident site   The location of an unexpected occurrence, failure or loss, either at a plant or along a transportation route, resulting in a release of hazardous materials.

acclimatization   The physiological and behavioral adjustments of an organism to changes in its environment.

Accofloc®   Ion exchange media by Colloid Environmental Technologies Co.

accretion   The increase in size of an inorganic body by the addition or accumulation of particles.

ACCU®   Air sampler by Rupprecht & Patashnick.

Accuguard™   Automated pH electrode cleaning and calibration module by BIF.

Accu-Mag   Electromagnetic flow meter by USFilter/Wallace & Tiernan.

accumulation zone   The upper part of a glacier where net gain in ice exceeds the net loss.

accumulator   A tank installed in a circulating water system to allow for fluctuations in flow, temperature, pressure, or other variations in operation.

AccuPac®   Cross-corrugated surface media for biological wastewater treatment by Brentwood Industries, Inc.

Accura-flo®   Flumes for measuring flows by Composite Structures.

Accu-Tab™   Tablet chlorination system by Hammonds and PPG Industries, Inc..

Accuvac   Chemical reagents in vacuum vials for chemical analysis of fluids by Hach Co.

ACE rule   See "Any Credible Evidence rule (ACE rule)."

acetaldehyde   An organic chemical formed during the disinfection of water, most commonly associated with the use of ozone as disinfectant. Chemical formula is CH_3CHO.

acetic acid   A weak, organic acid contained in vinegar and used in the manufacture of organic chemicals and plastics. Also called "ethanoic acid." Chemical formula is CH_3COOH.

acetone   A colorless, volatile liquid used in organic synthesis and as a commercial solvent. Chemical formula is CH_3COCH_3.

ACFM   Actual cubic feet per minute.

ACFTD   Air cleaner fine test dust. Dust used to calibrate particle counters.

ACGIH   American Conference of Governmental Industrial Hygienists.

ACH   (1) See "air changes per hour (ACH)." (2) See "aluminum chlorohydrate (ACH)."

ACI   American Concrete Institute.

acid   (1) A substance that can react with a base to form a salt. (2) A substance that can donate a hydrogen ion or proton.

acid deposition   See "acid rain."

acid mine drainage   Drainage of water from areas that have been mined for coal or other mineral ores, usually having a low pH due to contact with sulfur-bearing material.

acid neutralizing capacity (ANC)   Measure of the ability of water or soil to resist changes in pH.

acid rain   Precipitation having an unusually low pH, generally attributed to the absorption of sulfur dioxide pollution in air. Also known as "acid deposition."

acid shock   The biological disruption of an aquatic system that results from rapid acidification.

acid-forming bacteria   Microbes that can metabolize complex organic compounds under anaerobic conditions, leading to the production of methane.

acidic   The condition of water or soil that contains a sufficient amount of acid substances to lower the pH below 7.0.

acidity   The capacity of an aqueous solution to neutralize a base.

acidophil   (1) A cell or substance easily stained by acid dyes. (2) An organism that has an affinity for, and grows in, an acidic environment. Also spelled "acidophile."

ACL   Alternate concentration limit.

ACM   Asbestos-containing material.

ACM®   Thin film composite reverse osmosis membrane by TriSep Corp.

Acme   Former screening equipment manufacturer.

ACM-LP™   Low pressure thin film composite reverse osmosis membrane by TriSep Corp.

ACMS™   Seawater reverse osmosis membrane by TriSep Corp.

ACoE   U.S. Army Corps of Engineers.

Acousticair®   Low noise blowers by Tuthill Pneumatics Group.

ACP   Air Carcinogen Policy.

acre-foot   The volume of water that would cover a one acre area one foot deep. Equivalent to approximately 1233.6 cubic meters or 325,900 gallons.

acrolein   An aldehyde compound used as a microbiocide in the manufacture of organic chemicals.

Acro-Pac®   Packaged seawater reverse osmosis system by Aqua-Chem, Inc.

ACS-Plus   High purity chemicals for laboratory use by Hach Co.

ACT-100®   Double wall fiberglass laminated steel underground tank by Steel Tank Institute.

ACT™   Combined aeration technologies by Aeration Industries, Inc.

Acticarbone®   Activated carbon by Elf Atochem North America, Inc.

Actifil®   Packing media for biological reactors by Sanitaire Corp.

Actiflo®   High rate sedimentation process for water and wastewater treatment by USFilter/Krüger.

Actifloc™   Modular, high-rate water treatment plant by USFilter/General Filter.

Actinomycetes   A group of bacteria that share some features with fungi, and recognized as a source of musty or earthy odors in drinking water.

activated alumina   A partially dehydrated form of aluminum oxide frequently used as an adsorbent. Chemical formula is Al_2O_3.

Activated Biofilm Method   Fixed film wastewater treatment system by JDV Equipment Corp.

activated biofilter   A fixed film biological wastewater treatment process where a portion of the secondary sludge is returned to the reactor influent.

activated carbon (AC)   A highly adsorbent form of carbon used to remove dissolved organic matter from water and wastewater or odors and toxic substances from gaseous emissions.

activated charcoal   See "activated carbon."

activated sludge   The biologically active solids in an activated sludge process wastewater treatment plant.

activated sludge process   A biological wastewater treatment process where a mixture of wastewater and biologically enriched sludge is mixed and aerated to facilitate aerobic decomposition by microbes.

activation energy   The energy required to initiate a process or reaction.

Activator   Package wastewater treatment plant by Pollution Control, Inc.

activator   A chemical added to a pesticide to increase its activity.

Activator III   Oil recovery product by Sybron Chemicals, Inc.

active ingredient   In any pesticide product, the component that kills, or otherwise controls, target pests.

active life   The period of operation of a facility that begins with initial receipt of a solid waste and ends at completion of closure activities.

active portion   Any area of a facility where treatment, storage, or disposal operations continue to be conducted.

active solar heating   A heating system that derives heat from the sun's rays and incorporates active devices such as pumps or blowers to move heat from the point of collection to the point of storage or use.

Activol™   Wastewater grease emulsifier by Probiotic Solutions.

Acumem   Reverse osmosis product formerly offered by USFilter.

Acumer®   Water treatment polymers by Rohm and Haas, Co.

acute exposure   A single exposure to a toxic substance, usually lasting no longer than a day, which results in severe biological harm or death.

acute toxicity   A poisonous effect produced by a single short-term exposure which results in severe biological harm or death.

Acutec   Gas detection system by USFilter/Wallace & Tiernan.

ACWA   American Clean Water Association.

ACWM   (1) Asbestos-containing waste material. (2) See "asbestos-containing waste materials (ACUM)."

AD   Dry blending and dilution system by Komax Systems, Inc.

ADA   American Desalting Association. Formerly "NWSIA."

ADAM   Acryloyl ethyl dimethyl amine.

adaptation   Changes in an organism's structure or habits which help it adjust to its surroundings.

Adcat™   Oxidation catalyst systems for air pollution control by Goal Line Environmental Technologies.

Addigest®   Package extended aeration wastewater treatment plant by Smith & Loveless, Inc.

additive   A chemical substance incorporated into another substance to improve or preserve its quality.

additive mortality   The total mortality caused by different factors affecting a population for a given period of time including predation, fire, or catastrophes.

add-on control device   An air pollution control device such as carbon absorber or incinerator that reduces the pollution in an exhaust gas.

adenosine diphosphate (ADP)   A compound involved in the mobilization of energy in cellular metabolism. Energy is stored by adding a phosphate group to ADP to produce ATP.

adenosine triphosphate (ATP)   A high-energy phosphate compound that serves as the prime energy carrier in living organisms. Energy is released when ATP is converted to ADP and phosphate.

adenovirus   A waterborne pathogen that causes upper respiratory infections and gastroenteritis.

ADF   See "average daily flow (ADF)."

adhesion   The force of molecular attraction between unlike molecules.

ADI   Acceptable daily intake.

adiabatic lapse rate   The constant rate at which temperatures decrease as altitude increases. In a dry atmosphere the dry adiabatic lapse rate (DALR) is approximately $-1.00°C/100$ m rise.

adit   A horizontal passageway into a mine to provide access or drainage.

AdjustAir®   Adjustable coarse bubble air diffuser by USFilter/Diffused Air Products Group.

Adjust-O-Pitch   Mixing propellers with adjustable pitch blades by Walker Process Equipment.

administrative order   An EPA-issued order directing an individual, business, or other entity to take corrective action or refrain from an activity.

admixture   (1) A material or substance added in mixing. (2) A substance other than cement, aggregate, or water that is mixed with concrete.

Ad-Ox   Polishing scrubber for odor abatement by Purafil, Inc.

ADP   See "adenosine diphosphate (ADP)."

Adpec   Horizontal vacuum filter by Komline-Sanderson Engineering Corp.

ADR   (1) Accidental rectal discharge. (2) Alternative dispute resolution.

ADROWPU®   Water purification unit by Zenon Environmental, Inc.

Adsep™   Chromatographic process for separating organic and inorganic compounds by USFilter/Rockford.

Adsolv   Activated carbon VOC control system by CSM Worldwide/RaySolv.

adsorbable organic halides (AOX)   The gross measurement of all chlorinated organic compounds in an effluent.

adsorbate   A material adsorbed on the surface of another.

adsorbent   A material used to adsorb substances to its surface.

adsorption   The process of transferring a substance from a liquid to the surface of a solid where it is bound by chemical or physical forces.

Adsorption Clarifier™   Upflow buoyant media flocculator/clarifier by USFilter/Microfloc.

adulterants   Chemical impurities or substances that by law do not belong in a food or pesticide.

adulterated   (1) Any pesticide whose strength or purity falls below the quality stated on its label. (2) A food, feed, or product that contains illegal pesticide residues.

Advance®   Chlorine gas feeder systems by Capital Controls Co.

Advanced Fluidized Composting™   A combined biological and chemical sludge treatment process by USFilter/Industrial Wastewater Systems.

advanced oxidation process (AOP)   A process using a combination of disinfectants such as ozone and hydrogen peroxide to oxidize toxic organic compounds to nontoxic form.

advanced secondary treatment   Secondary wastewater treatment with enhanced solids separation.

advanced treatment plant (ATP)   A treatment facility using processes that provide treatment to levels greater than that of a conventional plant.

advanced wastewater treatment (AWT)   Treatment processes designed to remove pollutants such as phosphorus, nitrogen, and a high percentage of suspended solids which are not adequately removed by conventional secondary treatment processes.

advection   The transfer of heat by horizontal currents of air.

Advent®   Package water treatment plant by Infilco Degremont, Inc.

advisory   A nonregulatory document that communicates risk information to those who may have to make risk management decisions.

AEM   Acoustic emission monitoring.

aeolian deposit   Soil deposited by the wind.

Aeralater®   Packaged iron and manganese removal system by USFilter/General Filter.

aerated pile composting   A composting method where municipal wastewater solids are mixed with a bulking material and the mixture placed over a forced air ventilation system.

aeration   The addition of air or oxygen to water or wastewater, usually by mechanical means, to increase dissolved oxygen levels and maintain aerobic conditions.

Aeration Engineering Resources Corp.   Former name of Aercor product group of Sanitaire Corp.

Aeration Panel™   Fine bubble membrane diffuser panel by Parkson Corp.

aerator   A device used to introduce air or oxygen into water or wastewater.

Aerators, Inc.   Former name of USFilter/Aerator Products.

Aercor   Aeration and packing products by Sanitaire Corp.

Aer-Degritter   Aerated grit removal system by USFilter/Headworks Products.

AerFlare   Air diffuser by Walker Process Equipment.

Aergrid™   Floor grid aeration system by Aeration Technologies, Inc.

Aermax®   Fine pore aeration diffuser by Aeration Technologies, Inc.

Aero-Accelator®   Circular, packaged activated sludge treatment plant by Infilco Degremont, Inc.

aeroallergen   An allergen transported by air.

aerobe   An organism that requires free oxygen for respiration.

aerobic   Condition characterized by the presence of free oxygen.

aerobic digestion   Sludge stabilization process involving direct oxidation of biodegradable matter and oxidation of microbial cellular material.

aerobic treatment   Process by which microbes decompose complex organic compounds in the presence of oxygen and use the liberated energy for reproduction and growth.

Aeroburn   Wastewater treatment plant by Walker Process Equipment.

Aerocleve   Former manufacturer whose product line is now offered by Chemineer, Inc.

Aeroductor   Aerated grit removal system by Lakeside Equipment Corp.

Aero-Filter   Rotary distributor by Lakeside Equipment Corp.

Aer-O-Flo   Wastewater treatment equipment product line by Purestream, Inc.

aerogel   A substance formed by the suspension of small bubbles of gas in a liquid or solid.

Aero-Max   Tubular membrane diffuser by Aeration Research Company.

Aero-Mod®   Wastewater treatment product line by Waterlink/Aero-Mod Systems.

Aeropure   Activated carbon vapor filtration system by American Norit Company, Inc.

AeroScrub   Flue gas scrubber by Aeropulse, Inc.

Aerosep®   Multi-stage aerosol separation system by Kimre, Inc.

aerosols   A suspension of collodial particles in air or another gas.

Aero-Surf   Air driven rotating biological contactor by USFilter/Envirex.

Aero-Terra-Aqua   ATA Technologies, Inc.

Aerotherm   In-vessel composting system by Fairfield Service Co.

AerResearch   Aeration diffuser product line by Aeration Research Company.

Aershear™   Coarse bubble diffuser by Aeration Technologies, Inc.

Aertec™   Air diffuser product line by Aeration Technologies, Inc.

Aertube™   Static tube aerators by Aeration Technologies, Inc.

Aethalometer™   An aerosol black carbon monitor for measuring agricultural burning, diesel emissions and speciation of fine particles by Andersen Instruments, Inc.

AF Series™   Backwashable filter and carbon treatment units by Asahi America.

AFC™   Advanced fluidized composting process by USFilter/Industrial Wastewater Systems.

AFD   Adjustable frequency drive.

affected public   The people who live and/or work near a hazardous waste site.

affluent stream   A stream or river flowing into a larger river or lake.

afforestation   The process of establishing a forest where one did not previously exist.

AFO   Air fail open.

AFOs   Animal feeding operations.

AFPA   American Forest and Paper Association. Also called "AF&PA."

AfterBlend   Output booster for chemical feed system by USFilter/Stranco.

afterburner   A device used to reduce air emissions by incinerating organic matter in a gas stream.

aftercondenser   A condenser installed as the last stage of an evaporator venting system to minimize atmospheric steam discharge.

aftergrowth   The increase of bacterial density in treated distribution water caused by growth of bacteria released from pipewall biofilm and sediments.

aftershock   Any seismic tremor following a main earthquake event and originating at or near the same place.

AFX™   Ozone instrumentation products by IN USA, Inc.

agar   A gelatinous substance extracted from a red algae, commonly used as a medium for laboratory cultivation of bacteria.

agar plate   A circular glass plate, containing agar or another nutrient-medium, used to culture microorganisms.

AGC   Association of General Contractors of America.

age tank   A tank used to store a chemical solution of known concentration for feed to a chemical feeder. Also called a day tank.

agent   Any substance, organism, or active force that produces an effect or change.

Agent Orange   A dioxin-containing toxic herbicide used as a defoliant during the Vietnam War.

aggessive water   Water having corrosive qualities.

agglomerate   To gather fine particles into a larger mass.

aggrade   To build up the level or slope of a river bed or valley by the deposit of sediment.

AGI   Acute gastrointestinal illnesses.

Agidisc®   Disc filter with integrated agitation system by Baker Process.

Agisac   Sock-type screening sack by Copa Group.

AGMA   American Gear Manufacturers Association.

agrichemical   Any inorganic, artificial, or manufactured chemical substances used in agricultural processes, usually in the form of fertilizers, herbicides, and pesticides.

agricultural pollution   Farming wastes, including runoff and leaching of pesticides and fertilizers; erosion and dust from plowing; improper disposal of animal manure and carcasses; crop residues; and debris.

agrochemical   A substance such as a fertilizer or insecticide used in agriculture.

agro-ecosystem   Land used for crops, pasture, and livestock; the adjacent uncultivated land that supports other vegetation and wildlife; the associated atmosphere; and the underlying soils, groundwater, and drainage networks.

agronomy   Branch of agriculture that deals with the raising of crops and the care of the soil.

AHERA   Asbestos Hazard Emergency Response Act.

AhlFloat™   Dissolved air flotation by USFilter/Industrial Wastewater Systems.

Ahlstrom Aquaflow   Former name of wastewater equipment product group acquired by USFilter/Industrial Wastewater Systems.

AhlSurf™   Surface aerator by USFilter/Industrial Wastewater Systems.

A-horizon   Topsoil, or the uppermost layer of soil, containing the highest accumulation of mineral and organic matter.

AHS   See "aquatic humic substances (AHS)."

AIChE   American Institute of Chemical Engineers.

AIDS (Acquired Immune Deficiency Syndrome)   A usually fatal disease in which the human body can no longer defend itself against infections. AIDS is caused by the HIV retrovirus which is transmitted by the exchange of body fluids.

AIHA   American Industrial Hygiene Association.

air   The mixture of gases, primarily oxygen and nitrogen, that surrounds the earth and forms its atmosphere.

air binding   (1) A condition where air enters filter media and harms both the filtration and backwash processes. (2) The obstruction of water flow in a pipeline or pump due to the entrapment of air.

air bound   See "air binding."

air changes per hour (ACH)   The movement of a volume of air in a given period of time; if a building has one air change per hour, it means that all of the air in the building will be replaced in a one-hour period.

Air Comb®   Coarse bubble diffuser by Amwell, Inc.

air contaminant   Smoke, dust, fume, gas, odor, mist, radioactive substance vapor, pollen, or any combination thereof.

air curtain   A vertical barrier of air bubbled upward through water to contain oil spills or discourage fish from entering polluted water.

air diffuser   A device designed to transfer atmospheric oxygen into a liquid.

air gap   An open vertical gap or empty space that separates drinking water supply from another water system in a treatment plant or other location to protect it from contamination by backflow or backsiphonage.

Air Grid™   Sand filter air scour system by Roberts Filter Group.

air lock   The condition that occurs when air accumlates in a high point of a pipeline, reducing or blocking the flow of water.

air mass   A large body of air that has uniform properties such as temperature and humidity.

air padding   Pumping dry air into a container to assist with the withdrawal of liquid or to force a liquified gas such as chlorine out of the container.

air pollutant   Airborne gases, liquids, or solids that may be hazardous to animal or plant life.

air pollution   The presence in the atmosphere of any airborne gases, liquids, or solids that may be hazardous to animal or plant life.

air pollution episode   A period of abnormally high concentration of air pollutants, often due to low winds and temperature inversion, that can cause illness and death.

air purifying respirator (APR)   A respirator that uses physical and/or chemical means to filter air breathed by the user.

Air Quality Control Region   Federally designated area that is required to meet and maintain federal ambient air quality standards.

air quality criteria   The levels of pollution and lengths of exposure above which adverse health and welfare effects may occur.

air quality related value (AQRV)   A value referring to the reduction in visibility that may be caused by a new air emission.

Air Quality Standards   The level of pollutants prescribed by regulations that may not be exceeded during a given time in a defined area.

air scour   The agitation of granular filter media with air during the filter backwash cycle.

Air Seal   Coarse bubble diffuser by Jet, Inc.

Air Shuttle   Wastewater aeration system by Meurer Industries, Inc.

air stripping   The process of removing volatile and semivolatile contaminants from liquid by passing air and liquid countercurrently through a packed tower.

air toxics   Any air pollutant for which a NAAQS does not exist that may reasonably be anticipated to cause cancer, developmental effects, reproductive dysfunctions, neurological disorders, heritable gene mutations, or other serious or irreversible chronic or acute health effects in humans.

Airamic®   Air/gas diffuser by Ferro Corp.

AiRanger   Tank level measurement system by Milltronics, Inc.

Airbeam™   Aluminum aeration basin cover by Enviroquip, Inc.

Airbrush™   Rotor aerator by United Industries, Inc.

AirCirc™   Assembly used to maintain the vortex within a wastewater degritter by Fluidyne Corp.

Airco   Former name of BOC Gases.

Aircushion   Flotation clarifier by Wilfley Weber, Inc.

Aire-O₂® Propeller aspirator aerator by Aeration Industries, Inc.

Air-Grit Aerated grit removal system by Walker Process Equipment.

AirJection™ Aeration system by Mazzei Injector Corp.

AirLance™ In-vessel composting technology by CBI Walker, Inc. (licensee) and American Bio Tech, Inc. (licensor).

airlift A device for pumping liquid by injecting air near the bottom of a riser pipe submerged in the liquid to be pumped, lowering the specific gravity of the fluid mixture, and allowing it to rise up the riser pump.

Airmaster Floating aerator by Airmaster Aerator.

AirMASTER Wet particulate scubber by Fen-Tech Environmental, Inc.

Air-Mix® Pulsed bed filter surface cleaning process by USFilter/Zimpro.

Airmizer Air diffuser by USFilter/Diffused Air Products Group.

Airoflo™ Rotor aerator by S&N Airoflo, Inc.

Air-O-Lator® Floating aerator by Air-O-Lator Corp.

AiroPump Air lift pump by Walker Process Equipment.

AirOXAL® Pure oxygen process by Air Liquide America.

AirRaider™ MTBE pump and treat system by Product Level Control, Inc.

AirRide Density contolling system for compost in-feed by the former Waste Solutions.

Airsep Aerated grit collector system by USFilter/Aerator Products.

AirSep® Pressure swing adsorption oxygen systems by AirSep Corp.

AirTainer™ Tank cover by NuTech Environmental Corp.

air-to-cloth ratio Bag house application criteria which is the ratio of the air flow line to net cloth area.

air-to-water ratio Air stripping application criteria indicating the volume of air required per volume of water to remove volatile contaminants.

Airvac® Vacuum sewage collection system by Airvac, Inc.

AISC American Institute of Steel Construction.

AISI American Iron and Steel Institute.

AIWPS Advanced Integrated Wastewater Pond System.

Akta Klor Sodium chlorite solution by Vulcan Performance Chemicals.

Aktivox Selective oxidant with specificity for sulfides by Vulcan Performance Chemicals.

AL Acceptable level.

alabaster A compact, fine-grained gypsum material.

alachlor A herbicide used mainly to control weeds in corn and soybean fields and marketed under the trade name Lasso.

ALAPCO Association of Local Air Pollution Control Officers.

Alar Trade name for a pesticide used primarily to make apples redder, firmer, and less likely to fall off trees prematurely.

ALARA As low as reasonably achievable.

Albrivap High temperature additive for evaporators by Albright & Wilson Americas.

Alcofix® Inorganic coagulant by Ciba Specialty Chemicals.

alcohol A class of compounds containing the hydroxyl group OH.

aldehyde A class of organic compounds produced by the oxidation of alcohols containing a CHO group including formaldehyde and acetylaldehyde.

aldicarb   An insecticide made from ethyl isocyanate and sold under the trade name Temik.

aldrin   An insecticide and suspected carcinogen which has been banned for agricultural use by the U.S. Environmental Protection Agency (EPA).

alga   The singular form of "algae."

algae   Primitive, free-floating, plant-like aquatic organisms whose biological activity affect the pH and dissolved oxygen content in a water body.

Algae Sweep Automation   Automated clarifier algae sweep system by Ford Hall Co., Inc.

AlgaeMonitor   On-line fluorometer to monitor relative algae levels in a water system by Turner Designs.

algal bloom   A rapidly reproducing floating colony of algae that may cover a stream, lake, or reservoir, creating a nuisance condition.

AlgaSORB®   Ion exchange medium for heavy metal removal by Bio-Recovery Systems, Inc.

algicide   Any substance used to kill algae. Also "algaecide."

aliphatic compounds   Organic compounds with carbon atoms arranged in a straight or branched chain, rather than a ring.

aliquot   The amount of a sample used for analysis.

Alizair®   Fixed growth biotreat system for removal of odorous compounds from wastewater by USFilter/Krüger and OTV.

alkali   A substance with highly basic properties.

alkali metals   The elements lithium, sodium, potassium, rubidium, and cesium.

alkaline   Water containing sufficient amounts of alkalinity to raise the pH above 7.0.

alkaline soil   Soil with a pH greater than 7.0.

alkalinity   The ability of a water to neutralize an acid due to the presence of carbonate, bicarbonate, and hydroxide ions.

Alkalinity First™   Sodium bicarbonate by Church & Dwight Co. Inc.

alkaloid   Any of a group of organic compounds containing nitrogen and having alkaline properties including caffeine, cocaine, morphine, nicotine, and strychnine.

alkalosis   Abnormal condition in the body resulting from an excess alkalinity and an increased blood and tissue pH, often caused by exposure to low-oxygen air found at high altitudes.

Alka-Pro®   Process control system for biological wastewater treatment systems by USFilter/Davco.

All Climate™   Field-erected package wastewater plant by USFilter/Davco.

allergen   A substance inducing an allergic state or reaction.

allergy   A hypersensitivity to a particular substance such as food, pollen, or dust that causes the immune system to overreact to the substance.

Allied Colloids   Former name of Ciba Specialty Chemicals.

Allis-Chalmers   Former name of A-C Compressor Corp.

Allison   Internally fed rotating drum screen by KRC (Hewitt) Inc.

allotropy   The ability of an element to exist naturally in different forms in the same solid, liquid, or gaseous state.

alluvial soil   Soil formed of material that was carried by flowing water before being deposited.

alpha factor   The ratio of oxygen transfer coefficients for water and wastewater at the same temperature and pressure; used in the sizing of aeration equipment.

alpha particle   A positively charged atomic particle composed of two neutrons and two protons released by some atoms undergoing radioactive decay.

alpha radiation   A stream of positively charged particles released from radioactive isotopes.

alpha ray   A stream of particles emitting from the nucleus of a helium atom undergoing disintegration.

alpine tundra   An ecosystem found at high altitudes, above the timberline, characterized by mosses, lichens, and low-growing herbaceous plants.

ALR   Action leakage rate.

alternative energy   Energy obtained from sources other than traditional fossil fuels or nuclear energy, and which are usually renewable and nonpolluting. Alternative energy sources include solar energy, wave power, wind power, geothermal power, and biomass fuels.

alum   Common name for aluminum sulfate, frequently used as a coagulant in water and wastewater treatment. Chemical formula is $Al_2(SO_4)_3$.

alum sludge   Sludge resulting from treatment process where alum is used as a coagulant.

Alumadome   Self supporting aluminum covers for circular tanks by Conservatek Industries, Inc.

Alumavault   Self supporting aluminum covers for rectangular tanks by Conservatek Industries, Inc.

alumina   A form of aluminum oxide, also called "corundum" or "emery." Chemical formula is Al_2O_3.

aluminum   A lightweight, nonferrous metal with good corrosion resistance and electrical and thermal conductivity. Chemical symbol is Al.

aluminum chlorohydrate (ACH)   Water treatment coagulant.

aluminum sulfate   See "alum."

AlumStor   Modular liquid storage tank and feed system by ModuTank, Inc.

Alundum   Porous diffuser dome by Sanitaire Corp.

Alusil 70™   Custom-activated zeolites by Selecto, Inc.

amalgam   An alloy of mercury with one or more other metals. Silver amalgam is used as a dental filling.

Amazon™   Pulse jet baghouse filter bags by W.L. Gore & Associates, Inc.

Amberjet®   Ion exchange resins by Rohm & Haas, Co.

Amberlite®   Ion exchange resins by Rohm & Haas, Co.

Amberpack™   Ion exchange resins by Rohm & Haas, Co.

Ambersomb®   Carbonaceous adsorbent for VOC removal by Rohm & Haas Co.

ambient air quality   A general term used to describe the condition of the outdoor air.

amendment   Organic material or bulking agent such as wood chips or sawdust added to municipal solids in a composting operation to promote uniform air flow.

Amerfloc® Polyelectrolyte used to enhance liquid/solid separation by Ashland Chemical, Drew Industrial.

America Norit Former name of Norit Americas Inc.

American Well Works Former name of Amwell Inc.

Ameroid Polyelectrolyte used to enhance liquid/solid separation by Ashland Chemical, Drew Division.

Amersep® Coagulants and metal precipitants used in water and wastewater treatment by Ashland Chemical, Drew Industrial.

amines A group of compounds that are derivatives of ammonia with a hydrocarbon group replacing one or more of the hydrogen atoms.

amino acid A group of organic acids containing an amino group (NH_2) and a carboxyl group (COOH) that link together to form the proteins necessary for all life.

Aminodan™ Process Process water treatment and recovery process by Biothane Corp.

AmmonAsorb™ Activated carbon product by Waterlink/Barnebey Sutcliffe.

ammonia A compound of hydrogen and nitrogen that occurs extensively in nature. Chemical formula is NH_3.

ammonia nitrogen The quantity of elemental nitrogen present in the form of ammonia and the ammonium ion.

ammoniator A device used to feed gaseous ammonia.

ammonification Bacterial decomposition of organic nitrogen to ammonia.

ammonium ion A form of ammonia found in solution, the ion NH_4+.

ammonium nitrate A cystalline salt used as a source of nitrogen in the manufacture of some fertilizers and explosives. Chemical formula is NH_4NO_3.

amoeba A single-celled protozoan microbe. Also spelled "ameba."

amoebiasis See "amoebic dysentery." Also spelled "amebiases."

amoebic dysentery A form of dysentery caused by a protozoan parasite, usually resulting from poor sanitary conditions and transmitted by contaminated food or water. Also spelled "amebic dysentery."

amoebicide Any substance used to kill amoebas, either medicinally for treatment of amoebic dysentery or to sterilize water. Also spelled "amebicide."

amorphous Non-crystalline, having no definite shape or form.

AMOS Air Management Oversight System.

amperometric titrator Titration device containing an internal indicator or electrometric device to show when the reactions are complete.

amphibian (1) A member of the Amphibia class of cold blooded, scaleless vertebrates which includes frogs and salamanders. (2) A plant or animal that can live both on land or in water.

Amphidrome™ Fixed-film sequencing batch biological filter by Tetra Process Technologies.

Ampho-Mag™ Magnesia compound to absorb and buffer chemical spills by Premier Chemicals.

amphoteric Capable of reacting in water either as a weak acid or weak base.

AMSA Association of Metropolitan Sewer(age) Agencies.

anabatic wind   A localized wind that flows up valley or mountainous slopes, usually in the afternoon, caused by the replacement of cool valley air with the warmer, heated air above it.

anabiosis   A temporary state of suspended animation undergone by some aquatic organisms to survive periods of drought.

anabolism   The process in a plant or animal in which food is metabolized into complex compounds such as protein and living tissue.

anaerobe   An organism that can thrive in the absence of free oxygen.

anaerobic   Condition characterized by the absence of free oxygen.

anaerobic digestion   Sludge stabilization process where the organic material in biological sludges are converted to methane and carbon dioxide in an airtight reactor.

Anaerobic Selector Process   Biological system for phosphorus and BOD removal by USFilter/Davco.

Analite   Portable turbidimeters by Advanced Polymer Systems.

analyte   The chemical for which a sample is analyzed.

ANC   See "acid neutralizing capacity (ANC)."

An-CAT®   Polymer processing control unit by Norchem Industries.

ancillary equipment   Devices such as piping, flanges, valves, and pumps which are used in conjunction with treatment process or system.

Anco   Batch mixers by Enviropax, Inc.

anemometer   An instrument used to measure the force or velocity of wind.

aneroid barometer   An instrument used to measure atmospheric pressure which operates on the movement of a thin metal plate rather than the rise and fall of mercury.

angle of repose   The maximum angle the inclined surface of a loosely divided material can make with the horizontal.

Angstrom (Å)   A unit of measure equivalent to one ten-thousandth of a micron.

anhydride   A chemical compound derived by the elimination of water.

anhydrite   See "calcium sulfate."

anhydrous   A compound or substance that does not contain water.

animal unit (AU)   A unit for establishing grazing pressures from different kinds of grazing animals which consume similar kinds of forage, where 1AU is equivalent to the weight of 1 cow and 1 calf set at 454 kg.

anion   A negatively charged ion that migrates to the anode when an electrical potential is applied to a solution.

anion exchange   An ion exchange process in which anions in solution are exchanged for other anions bound to an ion exchange resin or medium.

anionic polymer   A polyelectrolyte with a net negative electrical charge.

Anitron   Biological fluidized bed wastewater treatment system by USFilter/Krüger.

ANM™   Nanofiltration softening membrane by TriSep Corp.

annelid   A wormlike organism, including earthworms and leeches, which have long, segmented bodies.

Annubar®   Mass flow monitoring system by Dieterich Standard.

anode   The positive electrode where current leaves an electrolytic solution.

anodic protection   Electrochemical corrosion protection achieved through the use of an anode having a higher electrode potential than the metal to be protected.

anoxia   A condition in which the blood oxygen level is less than normal.

anoxic   Condition characterized by the absence of free oxygen.

ANOX-R   Advanced treatment industrial wastewater treatment system by USFilter/Davco.

ANPRM   Advance Notice for Public Rulemaking. An EPA program for holding public forums for potential changes to the federal Water Quality Standard regulations.

ANSI   American National Standards Institute.

antagonism   Interference or inhibition of the effect of one chemical by the action of another.

antarctic   Characteristic of the region, climate, or vegetation south of the Antarctic Circle.

Antarctic Circle   An imaginary circle parallel to the equator at latitude 66°34′ south.

Anthozoa   A class of saltwater polyps including corals and sea anemones.

anthracite   A hard, black coal containing a high percentage of fixed carbon, a low percentage of volatile matter, and burns with little or no smoke.

anthracosis   A lung disease more commonly known as "black lung disease."

Anthrafilt®   Filter anthracite by Unifilt Corp.

anthropodust   Fugitive dust generated from human activities.

anthropogenic climate change   Human-induced global warming caused primarily by carbon dioxide, methane, and nitrous oxide emissions from engine exhausts, as well as respiration of humans and animals.

anthropogenic compounds   Compounds created by human beings, often relatively resistant to biodegradation.

antichlors   Reagents such as sulfur dioxide, sodium bisulfite, and sodium thiosulfate which can be used to remove excess chlorine residuals from water by conversion to an inert salt.

anti-degradation clause   See "prevention of significant deterioration (PSD)."

antifoam agent   A surface active agent used to reduce or prevent foaming.

antifoulant   An additive or dispersant that prevents fouling and/or the formation of scale.

antigen   A substance capable of stimulating an immune response.

antiknock additive   A compound, usually tetraethyl lead, added to gasoline to minimize engine pre-ignition and its' accompanying knocking or pinging. Pollution from the release of these compounds in auto emissions led to the introduction of nonleaded gasoline.

antioxidant   A substance that slows down or prevents oxidation of another substance.

antiscalant   An additive that prevents the formation of inorganic scale.

Any Credible Evidence rule (ACE rule)   Another name for EPA Clean Air Act "Credible Evidence rule."

AOB   Ammonia-oxidizing bacteria.

AOC   See "assimilable organic carbon (AOC)."

AOPs   See "advanced oxidation processes (AOP)."

AOT   Advanced oxidation technologies.

AOX   See "adsorbable organic halides (AOX)."

APCA   Air Pollution Control Association.

APER   Air pollution emissions report.

APF   See "assigned protective factor (APF)."

APHA   American Public Health Association.

API   American Petroleum Institute.

API gravity   An index inversely related to specific gravity used to identify liquid hydrocarbons.

API separator   Rectangular basin in which wastewater flows horizontally while free oil rises and is skimmed from the surface.

APOVAC®   Anti-pollution vacuum system for solvent recovery by Rosenmund, Inc.

apparent color   The color in water caused by the presence of suspended solids.

approach velocity   The average water velocity of fluid in a channel upstream of a screen or other obstruction.

APR   See "air purifying respirator (APR)."

APTI   Air Pollution Training Institute.

APWA   American Public Works Association.

AQCCT   Air-quality criteria and control techniques.

AQCP   Air Quality Control Program.

AQCR   Air Quality Control Region.

AQRV   See "air quality related value (AQRV)."

AQTX   Aquatic toxicity.

Aqua Bear   Medium density foam pipeline cleaners by Girard Industries, Inc.

Aqua Criss Cross   Medium density coated foam pipeline cleaners by Girard Industries, Inc.

Aqua Guard®   Self cleaning bar screen by Parkson Corp.

Aqua Pigs™   Polyurethane foam pipeline cleaners by Girard Industries, Inc.

Aqua Swab   Soft polyurethane pipeline cleaners by Girard Industries, Inc.

Aqua UV™   Ultraviolet water disinfection systems by Trojan Technologies, Inc.

Aqua-4™   Surface water treatment plant by Smith & Loveless, Inc.

AquaABF™   Traveling bridge type automatic backwash package filter by Aqua-Aerobic Systems, Inc.

Aquabelt®   Gravity belt thickener by Ashbrook Corp. (U.S.) Simon-Hartley, Ltd. (U.K.).

AquaCalc   Open channel flow computer by JBS Instruments.

AquaCAM-D   Aerator/mixer/decanter for use in sequencing batch reactor by Aqua-Aerobic Systems, Inc.

Aqua-Carb   Activated carbon by USFilter/Westates.

Aqua-Cat®   Sulfide conversion process to control odors by USFilter/Gas Technologies.

Aqua-Cell™   Package water treatment plant by WesTech Engineering Inc.

Aquaclaire™   Wastewater treatment systems by DAS International, Inc.

Aquacode   Membrane vessels by Spaulding Composites Co.

aquaculture The managed production of fish or shellfish in a pond or lagoon.
AquaDDM® Direct drive mixers by Aqua-Aerobics Systems, Inc.
AquaDecant® Floating decanter by AquaTurbo Systems.
Aquadene™ Corrosion and scale control products by Stiles-Kem Division, Met Pro Corp.
AquaDisk™ Woven cloth tertiary filter by Aqua-Aerobic Systems, Inc.
Aquafeed® Reverse osmosis antiscalants by B.F. Goodrich Co.
Aqua-Fer™ Well water treatment plant designed for iron removal by Smith & Loveless, Inc.
Aqua-Fix™ Dissolved ionic contaminant removal system by ATA Technologies Corp.
Aquaflow Former name of product group acquired by USFilter/Industrial Wastewater Systems.
Aqua-Jet® Direct drive aerators by Aqua-Aerobics Systems, Inc.
Aqua-Lator® High speed floating aerator by USFilter/Aerator Products.
AquaLift® Screw pump by Parkson Corp.
Aquamag® Magnesium hydroxide by Premier Chemicals.
AquaMax™ Desalination evaporator antiscalant by BetzDearborn, Inc.
Aquamite® Electrodialysis water treatment systems by Ionics, Inc.
AquaMJA™ Manifold jet aerator by Aqua-Aerobic Systems, Inc.
Aquaport® Seawater desalination system by Ambient Technologies, Inc.
Aquaray® Ultraviolet disinfection system by Infilco Degremont, Inc.
Aquaritrol® Coagulant control system using a programmable controller by USFilter/Microfloc.
Aquarius® Modular water treatment plant by USFilter/Microfloc.
AquaSBR® Sequencing batch reactor by Aqua-Aerobics Systems, Inc.
Aquascan On-line VOC monitor by Sentex Systems, Inc.
Aqua-Scrub™ Powdered activated carbon adsorption system by USFilter/Westates.
AquaSEAL™ Sequencing extended aeration lagoon system by Aqua-Aerobics Systems, Inc.
Aqua-Sensor Control system for water softener regeneration by Culligan International Corp.
Aquashade® Aquatic plant growth control by Applied Biochemists, Inc.
Aqua-Shear Mixer by Flow Process Technologies, Inc.
Aquasorb® Carbon adsorption treatment systems by Hadley Industries.
Aquasource® Membrane system by Infilco Degremont, Inc.
Aquaspir® Shaftless dewatering screw by Andritz-Ruthner, Inc.
Aquastore® Storage tanks by A.O. Smith Engineered Storage Products.
Aquatair Packaged biological wastewater treatment plant by BCA Industrial Controls, Ltd.
Aquatec® Solid media contact filter vessels by Colloid Environmental Technologies Co.
Aquatech Systems Former name of Aqualytics, Inc.
aquatic humic substances (AHS) Humic substances in true solution that exhibit colloidal properties.
Aquatreat™ Sequencing batch reactor by EnviroSystems Supply.

***Aqua-Trim*™**   Tray-type air stripper by Delta Cooling Towers, Inc.

***AquaTurbo*®**   Surface aerator and mixer products by AquaTurbo Systems.

Aquavap   Vapor compression type evaporator by Licon, Inc.

Aqua-View   Particle measuring system by Particle Measuring Systems, Inc.

***Aquaward*®**   Tablet feeder disinfection system by Exceltec International Corp.

***Aquazur V*®**   Rapid sand gravity filter by Infilco Degremont, Inc.

aqueduct   A conduit for carrying running water.

aqueous   Something made up of, similar to, or containing water.

aqueous chlorine   Term used to describe chlorine or chlorine compounds dissolved in water, often mistakenly called "liquid chlorine."

aqueous solution   A solution in which water is the solvent.

aquiclude   A low-permeability underground rock formation that absorbs water slowly but will not allow its free passage.

aquifer   A subsurface geological formation containing a large quantity of water.

aquifuge   An underground layer of impermeable rock that will not allow the free passage of groundwater.

***Aquilair*®**   Oxidation process for removal of VOC compounds from wastewater by USFilter/Krüger (North America) and OTV.

***AQuit*™**   Sulfide control product by USFilter/Davis Process.

***Aquox*™**   Potassium permanganate by Nalon Chemical, distributed by American International Chemical, Inc.

arable   Land capable of being farmed.

ARAR   Applicable or relevant and appropriate requirements. Cleanup standards, control standards and other substantive environmental protection requirements, criteria, and limitations promulgated under federal, state, and local laws.

***Arc Screen*™**   Self cleaning curved bar screen by Infilco Degremont, Inc.

ARCC   American Rivers Conservation Council.

Archimedes' principle   The principle of buoyancy which states that the force on a submerged body acts vertically upward through the center of gravity of the displaced fluid and is equal to the weight of the fluid displaced.

Archimedes' screw pump   See "screw pump."

archipelago   A group or chain of many islands.

arctic   Characteristic of the region, climate, or vegetation north of the Arctic Circle.

Arctic Circle   An imaginary circle parallel to the equator at latitude 66°34' north.

Arcticaer   High speed surface aerator with submersible motor by USFilter/Aerator Products.

area source   Any small source of non-natural air pollution that is released over a relatively small area but which cannot be classified as a point source, including vehicles and other small engines, small businesses and household activities.

Argo Scientific   Former name of BetzDearborn-Argo District.

ARI   Sulfur recovery and odor control product line by USFilter/Gas Technologies.

ARI   See "average rainfall intensity (ARI)."

***Aria*™**   Water treatment system by Pall Corp.

***Aries*™**   Filtration air scour technology by Roberts Filter Group.

arithmetic mean   The sum of a set of observations divided by the number of observations.

Arm & Hammer®   Sodium bicarbonate by Church & Dwight Co. Inc.

Armco   Former sluice gate manufacturer acquired by Hydro Gate Corp.

Arna®   Ultraviolet disinfection system by Arlat, Inc.

aromatics   A group of hydrocarbon compounds, including benzene, containing a closed ring structure.

array   (1) A staged arrangement of membrane elements in a system. (2) A group of solar collection devices arranged in a suitable pattern to efficiently collect solar energy.

Arro-Care®   Reverse osmosis maintenance and support services by USFilter/Rockford.

Arro-Cleaning®   Membrane care services by USFilter/Rockford.

Arrowhead®   Water treatment product line by USFilter/Rockford.

arroyo   A stream or watercourse that is often dry.

ARS™   Air regulated siphon by John Meunier, Inc.

arsenic   A naturally occurring element that is toxic to humans at very low levels. Chemical formula is As.

arsenicals   Pesticides containing arsenic.

Artco   Applied Regenerative Technologies Co.

artesian water   Bottled water from a well that taps a confined aquifer which is located above the normal water table.

artesian well   A well with sufficient pressure to produce water without pumping.

arthropod   Any of the largest phylum of invertebrate animals including insects and crustaceans having segmented bodies, exoskeletons, and jointed legs occurring in pairs.

Arus Andritz   Former name of Andritz-Ruthner, Inc.

ASA   Algae sweep automation system by Ford Hall Co., Inc.

ASB   Aerated stabilization basin.

asbestos   A mineral fiber, positively identified as a carcinogen, that does not conduct heat or electricity and formerly in wide use in the building industry for thermal insulation, soundproofing, roofing, and electrical insulation.

asbestos cement pipe   Pipe manufactured of a mixture of asbestos fiber and Portland cement.

asbestos-containing material   Construction materials that contain more than one percent asbestos.

asbestos-containing waste materials (ACWM)   Mill tailings or any waste that contains commercial asbestos and is generated by a source covered by the Clean Air Act Asbestos NESHAPS.

asbestosis   A chronic lung disease caused by exposure or inhalation of asbestos fibers.

as-built drawings   The original plans and specifications prepared for construction and corrected to reflect how a facility was actually built or installed.

A-scale sound level (dBA)   A modification of the decibel scale that approximates the sensitivity of the human ear used to note the instensity or annoyance of noise pollution.

ASCE   American Society of Civil Engineers.

Asdor   Former name of USFilter/Asdor.

ASDWA   Association of State Drinking Water Administrators.

aseptic   The state of being free of pathogenic organisms.

ash   The nonvolatile inorganic solids that remain after incineration.

Ashaire   Aerator product line by USFilter/Aerator Products.

Ashbrook-Simon-Hartley   Former name of Ashbrook Corp.

Ashfix™   A process to stabilize heavy metals in sludge and ash by Ashland Chemical, Drew Industrial.

ASHRAE   American Society of Heating, Refrigeration, and Air Conditioning Engineers.

Asiatic clam   A fresh water clam originally found in Southeast Asia and introduced in the U.S. in 1938.

ASME   American Society of Mechanical Engineers.

Aspergillus fumigatus   Airborne fungi that may result from composting operations which may cause human ear, lung, and sinus infections.

asphalt-rubber   A mixture of ground rubber and bituminous concrete used as a pavement interlayer to reduce stress and prevent cracking.

asphyxiant   A vapor or gas such as carbon monoxide that can cause unconsciousness or death by suffocation.

Aspi-Jet™   Aspirating aerator by Aqua-Aerobic Systems, Inc.

aspirating aerator   Aeration device using a motor driven propeller to draw atmospheric air into the turbulence caused by the propeller to cause the formation of small bubbles.

aspirator   A hydraulic device that creates a negative pressure by forcing liquid through a restriction and increasing the velocity head.

ASQC   American Society for Quality Control.

ASR   See "atmosphere supplying respirator (ASR)."

assay   A test for a particular chemical or effect.

assigned protective factor (APF)   Respirator designation indicating how much contaminant will be reduced with proper respiratory fit and wear. An APF of 5 indicates that if a cubic meter of air contains 10 mg of contaminant, the effective exposure would be 10/5 or 2 mg/cu m.

assimilable organic carbon (AOC)   The portion of dissolved organic carbon that is easily used by microbes as a carbon source.

assimilation   The ability of a body of water to purify itself of pollutants.

assimilative capacity   The ability of a water body to receive wastewater and toxic materials without deleterious effects on aquatic life or the humans who consume the water.

AST   Aboveground storage tanks.

ASTM   American Society for Testing and Materials.

Astrasand®   Continuously cleaned sand filter by Astraco Water Engineering.

Astraseparator®   Inclined plate settler by Astraco Water Engineering.

asymmetric membrane   Membranes that are not reversible and can only desalinate efficiently in one direction.

Asymmetrix™   Membrane cartridge filter by USFilter/Filtration & Separation.

AT®   Ozone generator by Ozonia North America.

ATAD   See "autothermal thermophilic aerobic digestion (ATAD)."

Atara Former manufacturer of digester gas mixing equipment whose product line was acquired by Infilco Degremont, Inc.

***ATD*™** Autotherm aerobic sludge digestion system by CBI Walker, Inc.

Athos Wet air oxidation process by USFilter/Krüger (North America) and OTV.

ATI Analytical Technology, Inc.

***ATL*®** Aero Tec Laboratories, Inc.

Atlantes Chemical Systems Former name of Hoffland Environmental, Inc.

***ATLAS*™** Advanced technology lagoon aeration system by Environmental Dynamics Inc.

atm See "atmosphere."

atmometer An instrument used to measure the evaporative capacity of the air.

atmosphere (1) The gaseous region that surrounds the earth. (2) A unit of pressure equal to 1.0333 kg/sq cm or 14.7 psi. Abbreviated "atm."

atmosphere supplying respirator (ASR) A respirator which supplies air or oxygen to the user.

atmospheric corrosion Corrosion resulting from exposure to the atmosphere.

atmospheric pressure The force exerted by the weight of the atmosphere above the point of measurement. At sea level, this pressure equals 101.3 kPa (14.7 psi), 760 mm (29.9 inches) of mercury column, or 10.3 m (33.9 feet) of water column.

Atochem Former name of Elf Atochem North America, Inc.

atoll A circular reef enclosing a shallow lagoon.

atom The smallest unit of an element that retains the characteristics of that element.

***Atomerator*®** Pressure aerator for iron oxidation by USFilter/General Filter.

atomic absorption spectrophotometry (AA) A highly sensitive instrumental technique for measuring trace quantities of elements in water.

atomize To divide a liquid into extremely fine particles.

atomizer An instrument through which a liquid is sprayed to produce a fine mist.

ATP (1) Average transmembrane pressure. (2) See "adenosine triphosphate (ATP)." (3) See "advanced treatment plant (ATP)."

***ATP*™** Aerobic thermophilic process sludge treatment system by CBI Walker, Inc.

***ATS*™** Sorbent media product line by Engelhard Corp.

ATSDR Agency for Toxic Substances and Disease Registry.

attached growth process See "fixed film process."

attainment area A geographic area in which the levels of a criteria air pollutant meet the health-based National Ambient Air Quality Standard for that pollutant.

attenuation The process by which a compound is reduced in concentration over time, through absorption, adsorption, degradation, dilution, and/or transformation.

attractant A chemical or agent that lures insects or other pests by stimulating their sense of smell.

AU See "animal unit (AU)."

***Auger Monster*®** Modular screen/grinder/solids removal unit for wastewater headworks system by JWC Environmental.

***Aurora UV*™** Ultraviolet disinfection system by Calgon Carbon Corp.

Austgen-Biojet Former name of ABJ product group of Sanitaire Corp.

***Auto5*™** Stack sampling system by Andersen Instruments, Inc.

AutoBelt   Rotary vacuum filter formerly offered by Walker Process Equipment.

autochthonous   A rock formation, organic matter, or substance that is produced or originates in the place where it is found.

autoclave   A device that sterilizes materials by exposure to pressurized steam.

Auto-Clean   Dissolved oxygen sensor by Analytical Technology, Inc.

Auto-Cleanse   Self-cleaning pump station with differential pressure activated flush valve by ITT Flygt Corp.

Autocon   Product group integrated within USFilter/Control Systems.

Auto-Dox   Controlled aeration weir by Purestream, Inc.

autogenous combustion   Burning that occurs when the heat of combustion of a wet organic material or sludge is sufficient to vaporize the water and maintain combustion without auxiliary fuel.

autogenous temperature   Equilibrium temperature in sludge combustion where the heat input from the fuel equals the heat losses, and combustion is self-supporting.

autoignition temperature   The minimum temperature at which a substance ignites without application of a flame or spark.

Auto-Jet®   Pressure leaf filter by USFilter/Whittier.

Autojust   Automatic feedwell gate controller for circular sludge collectors by USFilter/Aerator Products.

automatic sampling   Collecting samples of prescribed volume over a defined time period by an apparatus designed to operate remotely without direct manual control.

auto-oxidation   A self-induced oxidation process.

Autopress   High throughput automatic filter press by USFilter/Dewatering Systems.

***Auto-Pulse*™**   Tubular backpulse filter by USFilter/Whittier.

Auto-Rake®   Reciprocating rake bar screen by Franklin Miller, Inc.

autoreactive   A compound that is reactive under normal conditions and does not require heat, additional compounds, or a change in conditions.

Auto-Retreat   Automatic bar screen control system by Infilco Degremont, Inc.

***AutoSDI*™**   Portable, computer-based silt density index instrument by King Lee Technologies.

***Auto-Shell*™**   Granular media filter by USFilter/Whittier.

***Auto-Shok*™**   Tube-type vertical pressure leaf filter by USFilter/Whittier.

***Auto-Skimmer*™**   Skimmer used to remove floating hydrocarbons from water wells by Science Application International Corp.

***AutoTherm*™**   Aerobic thermophilic digestion system by CBI Walker, Inc.

autothermal thermophilic aerobic digestion (ATAD)   A biological digestion system that converts soluble organics to lower-energy forms through anaerobic, fermentative, and aerobic processes at thermophilic temperatures.

autothermic combustion   See "autogenous combustion."

Autotravel   Traveling bridge sludge collectors by Simon-Hartley, Ltd.

Autotrol   Rotating biological contactors product line by USFilter/Envirex.

autotroph   Organism that derives its cell carbon from carbon dioxide.

Auto-Vac®   Rotary drum filter by Alar Engineering Corp.

autoxidation   A spontaneous process in which a compound is oxidized, usually in the presence of oxygen, but other compounds can act as electron acceptors.

available chlorine   A measure of the amount of chlorine available in chlorinated lime, hypochlorite compounds, and other materials used as a source of chlorine when compared with that of liquid or gaseous chlorine.

average daily flow (ADF)   The total flow past a point over a period of time divided by the number of days in that period.

average flow   The arithmatic average of flows measured at a given point.

average rainfall intensity (ARI)   A measurement of rainfall per unit of time.

Avery Filter   Recessed chamber filter press by Komline-Sanderson, Engineering Corp.

AVGF®   Automatic valveless gravity filter by USFilter/Warren.

AWI   Anthratech Western, Inc.

AWS   American Welding Society.

AWT   See "advanced wastewater treatment (AWT)."

AWWA   American Water Works Association.

AWWARF   American Water Works Association Research Foundation.

axial flow   The flow of fluid in the same direction as the axis of symmetry of a tank or basin.

axial flow pump   A type of centrifugal pump in which fluid flow remains parallel to the flow path and develops most of its head by the lifting action of the vanes.

Azenit   Biological wastewater treatment process technology by USFilter/Krüger (North America) and OTV.

azeotropic   A liquid of two or more substances that behaves like a single substance in that the vapor produced by partial evaporation has the same composition as the liquid.

Aztec®   Chlorine/oxidant residual analyzer by Capital Controls Co.

B

B-10 life   The rated life defining the number of revolutions that 90% of a group of identical bearings will complete before first evidence of failure develops. Also known as "L-10 life."

B2A™   Multi-media biological filter by USFilter/Krüger (North America) and OTV.

Babcock Water Engineering   Former name of CASS Water Engineering, Inc.

BAC   Biologically active carbon.

bacilli   Rod-shaped bacteria.

back pressure   Pressure due to a force operating in a direction opposite to that required.

backfill   The material used to refill a ditch or excavation, or the process of refilling.

backflow   Flow reversal in a water distribution system that may result in contamination due to a cross connection.

backflow prevention device   Device used to prevent cross connection, or backflow of nonpotable water into potable water system.

background concentration   The general level of air pollutants in a region with all local sources of pollution ignored.

background contamination   Contamination introduced accidentally into dilution waters, reagents, rinse water, or solvents which can be confused with constituents in the sample being analyzed.

background level   The concentration of pollutants in a definite area during a fixed period of time prior to starting up a source of emission undercontrol.

background organic matter (BOM)   Natural organic matter in a mixture with one or more specific organic chemicals.

background radiation   Nuclear radiation arising from within the human body and normal surroundings.

background soil pH   The pH of the soil prior to the addition of substances that alter the hydrogen ion concentration.

backsiphonage   A backflow of water of questionable quality that results from a negative pressure within water distribution system.

backwash   A high-rate reversal of flow for the purpose of cleaning or removing solids from a filter bed or screening medium.

backwash rate   The flow rate used during filter backwash when the direction of flow through the filter is reversed for cleaning.

BACM   Best available control measures.

BACT   Best available control technology.

bacteria   Microbes that decompose and stabilize organic matter in wastewater.

BADT   Best available demonstrated technology.

BAF   (1) See "biologically active filter." (2) Biologically activated foam.

baffle   An obstructing device or plate used to provide even distribution or to prevent short-circuiting or vortexing of flow entering a tank or vessel.

baffle chamber   In incinerator design, a chamber designed to promote the settling of fly ash and coarse particulate matter by changing the direction and/or reducing the velocity of the gases produced by the combustion of the refuse or sludge.

Baffleflow   Oil removal tank with permeable baffles to prevent short circuiting by Walker Process Equipment.

bagasse   The fibrous residue from crushed sugar cane or sugar beet after the extraction of sugar juice.

baghouse   An air emissions control device that uses a fabric or glass fiber filter to remove airborne particulates from a gas stream.

bailer   A long pipe with a check valve at the lower end used to remove a slurry or oil from the bottom or side of a well.

Bakflo   Barge mounted oil skimmer by Vikoma International Ltd.

balefill   A land disposal site where solid waste material is compacted and baled prior to disposal.

baler   A machine used to compress and bind solid recyclable materials such as cardboard or paper.

ball valve   A valve utilizing a rotating ball with a hole through it that allows straight-through flow in the open position.

ballast water   Water used in a ship's hold for stabilization often requiring treatment as an oily wastewater.

Ballasted Floc Reactor™ **(BFR)**   Reactor-clarifier system by USFilter/General Filter.

ballistic separator   A machine that sorts organic from inorganic matter for composting.

Bamag   Former name of Lurgi Bamag GmbH.

banana-blade mixer   High flow, low-shear mixer with rotating back-swept propeller.

band application   The spreading of pesticides, fertilizer, or other chemicals over, or next to, each row of plants in a field.

Bandit™   Raking machine by Brackett Geiger.

bandscreen   See "traveling water screen (TWS)."

bank sand   Sand excavated from a natural deposit, usually not suitable for use in a filter processing or grading.

bar   A unit of pressure equal to 0.9869 atmospheres, 10^6 dynes per square centimeter, 100,000 pascals, and 14.5 pounds per square inch.

bar screen   A screening device using a parallel set of stationary bars typically spaced at 25 mm (1 inch) to 50 mm (2 inches).

Bardenpho™   Biological wastewater treatment process for removal of nitrogen and phosphorus by Baker Process.

Barminutor®   Combination bar screen and comminuting device by Yeomans Chicago Corp.

barnacles   A marine crustacean with a calcareous shell that attaches itself to submerged objects.

Barnebey & Sutcliffe   Former name of Waterlink/Barnebey Sutcliffe.

barometer   An instrument used to measure atmospheric pressure.

barometric condenser   A condenser in which vapor is condensed by direct contact with water.

barometric damper   A pivoting plate used to regulate the amount of air entering a duct or flue to maintain a constant draft within an incinerator.

barometric leg   (1) A condensate discharge line submerged below the liquid level of an atmospheric tank. (2) A gravity tailpipe from a vacuum barometric condenser.

barometric pressure   Ambient or local pressure surrounding a gauge, evaporator shell, vent pipe, etc.

barrel (bbl)   42 U.S. gallons.

barriered landscape water renovation system (BLWRS)   A wastewater treatment and denitrification system where wastewater is applied to the top of a mound of soil overlaying a water barrier and microbes oxidize soluble organics as the water percolates through the soil.

barrier reef   A long ridge of coral built up from the sea floor that runs parallel to the mainland but is separated from it by a deep lagoon.

Barry Rake   Trash rake by Cross Machine, Inc.

BART   Best available retrofit technology.

Bartlett-Snow™   Rotary calciner for soil reclamation by ABB Raymond.

basal application   The application of pesticides on plant stems or tree trunks just above the soil line.

base   (1) A substance that can accept a proton. (2) A substance that can react with an acid to form a salt. (3) An alkaline substance.

baseline   A sample used as a comparative reference point when conducting further tests or calculations.

basic water requirement (BWR)   The amount of water required by humans for drinking, sanitation, bathing, and cooking needs to meet a minimum quality of life, and generally considered to be 50 liters per day per person.

basicity factor   Factor used to determine neutralization capabilities of alkaline reagents used to treat acidic wastes.

basket centrifuge   Batch-type centrifuge where sludge is introduced into a vertically mounted spinning basket and separation occurs as centrifugal force drives the solids to the wall of the basket.

basophil   A cell or substance easily stained by basic dyes. Also spelled "basophile."

Basys Biolfilter   Modular air pollution control filter by Basys Technologies.

BAT   See "best available technology (BAT)."

Batch Master   Package wastewater treatment system by USFilter/Industrial Wastewater Systems.

batch process   A noncontinuous treatment process in which a discrete quantity or batch of liquid is treated or produced at one time.

batch reactor   A reactor where the contents are completely mixed and flow is neither entering nor leaving the reactor vessel.

Batch-Master   Bottom discharge basket centrifuge by Baker Process/Ketema.

Batch-Miser   Horizontal plate filter by Baker Process/Ketema.

Batch-O-Matic   Bottom discharge basket centrifuge by Baker Process/Ketema.

BATEA   Best available technology economically achievable and available.

bathyal zone   The ecological zone of the ocean above the abyssal zone, generally between 200 and 3500 meters.

BATNEEC   Best available technology not entailing excessive cost.

battery limit   The boundary limits of equipment or a process unit that defines interconnecting points for electrical piping or wiring.

Bauer®   Screening equipment product line acquired by Andritz-Ruthner, Inc.

Baumé   Designation of hydrometer scale used to indicate specific gravity.

bauxite   Ore containing alumina monohydrate or alumina trihydrate which is the principal raw material for alumina production.

bbl   See "barrel (bbl)."

BCC   See "bioaccumulative chemicals of concern (BCC)."

BCF   (1) See "bioconcentration factor (BCF)." (2) Bead and crevice free.

BCL Screen   Back cleaned bar screen by Waste-Tech, Inc.

BCPCT   Best conventional pollutant control technology.

BCT   Best control technology.

BDAT   See "best demonstrated available technology (BDAT)."

BDCT   Best demonstrated control technology.

BDOC   See "biodegradable dissolved organic carbon (BDOC)."

BDT   Best demonstrated technology.

beachwell   A shallow intake well making use of beach sand and structure as a filter medium.

Bead Mover™   Ion exchange resin and filter media loading pump by IX Services Co.

Bead Thief™   Ion exchange resin core sampler by IX Services Co.

Beaufort scale   A numerical scale of wind force where a Beaufort force 0 wind is calm and a force 12 wind indicates hurricane force with winds in excess of 120 km/hr (75 mph).

bed depth   The depth of filter media or ion exchange resin contained in a vessel.

bed load   Sediment particles resting on or near the channel bottom that are pushed or rolled along by the flow of water.

bed volume (BV)   The volume occupied by filter media in a filter, or resin in an ion exchange device.

bedrock   Solid rock encountered below the mantle of loose rock and soil which occurs on the earth's surface.

Beggiatoa   Filamentous microbe, commonly associated with sludge bulking, that results from low dissolved oxygen levels and/or high sulfide levels.

BEI   See "biological exposure indexes (BEI)."

BEJ   Best engineering judgement.

Bekomat ®   Micro-processor-driven condensate trap by BEKO Condensate Systems Corp.

Bekosplit   Separation process for condensate emulsions by BEKO Condensate Systems Corp.

Bekox   Former name of USFilter/Bekox.

Belclene®   Scale control additive by BioLab, Inc.

Belcor   Organic corrosion control by BioLab, Inc.

Belgard®   Antiscalant for seawater evaporators by BioLab, Inc.

Belite®   Antifoaming agent by BioLab, Inc.

Bellacide   Algaecide by BioLab, Inc.

BelloZon   Chlorine dioxide generator by ProMinent Fluid Controls, Inc.

Beloit-Passavant   Company acquired by USFilter/Zimpro.

Belspere®   Chemical dispersant by BioLab, Inc.

belt conveyor   A device used to transport material, consisting of an endless belt that revolves around head and tail pulleys.

belt filter press   See "belt press."

belt press   A dewatering device utilizing two fabric belts revolving over a series of rollers to squeeze water from the sludge.

belt thickener   Mechanical sludge processing device that uses a revolving horizontal filter belt to pre-thicken sludge prior to dewatering and/or disposal.

bench test   A small scale test or study used to determine whether a technology is suitable for a particular application.

beneficial organism   A pollinating insect, pest predator, parasite, pathogen, or other biological control agent which functions naturally or as part of an integrated pest management program to control another pest.

benthal oxygen demand The oxygen demand exerted by microbes and other organisms living on, or in close association with, the organic mud and sludge deposits on the bottom of a river or stream.

benthic Relating to the bottom environment of a water body.

benthos Microbes and other organisms living on or in close association with the bottom of a water body.

BentoLiner™ Clay composite liner by SLT North America, Inc.

Bentomat™ Geotextile-bentonite liner by Colloid Environmental Technologies Co.

bentonite Colloidal clay-like mineral that can be used as a coagulant aid in water treatment systems. Sometimes used as the earth component or soil amendment for construction of a pond or landfill liner because of its low permeability.

benzene A colorless, flammable liquid with carcinogenic properties produced from coal tar and used as a solvent. Benzene is an aromatic hydrocarbon characterized by its six-sided ring structure with the chemical formula C_6H_6.

Bepex Former name of Hosokawa Bepex Corp.

berm A horizontal, earthen ridge or bank.

Bernoulli's Equation Energy equation commonly used to calculate head pressure, and considers velocity head, static head, and elevation.

beryllium An airborne metal hazardous to human health when inhaled that may be discharged by machine shops, ceramic and propellant plants, and foundries.

best available technology (BAT) The best technology, treatment techniques, or other means available after considering field, rather than solely laboratory conditions.

best demonstrated available technology (BDAT) A technology demonstrated in full-scale commercial operation and shown to have statistically better performance than other technologies.

best management practice (BMP) The schedules of activities, methods, measures, and other accepted industry management practices to prevent pollution of waters and facilitate compliance with applicable regulations.

Beta NOx 2000™ NOx oxidation/reduction unit by Duall Division, Met-Pro Corp.

beta particles Electrons emitted by a radioactive nucleus.

beta radiation A stream of beta particles released during radioactive decay.

Betz Laboratories Former name of BetzDearborn, Inc.

BevMAX™ Reverse osmosis system for beverage industry applications by USFilter.

BFI® Trademark of Browning-Ferris Industries, Inc.

BFR See "Ballasted Floc Reactor™ (BFR)."

B-Gon™ Mist eliminator by Kimre Inc.

BGS2 Sludge dryer/pelletizer formerly offered by Wheelabrator Water Technologies, Inc.

BHN Probiotic Lagoon sludge oxidation product line by Bio Huma Netics, Inc.

B-horizon The intermediate soil layer, usually having a high clay content, where minerals and other particles washed down from the A-horizon accumulate.

bhp Brake horsepower. The power developed by an engine as measured by a dynamometer applied to the shaft or flywheel.

Bibbigard   Torque limitor by Brunel Corp.

Bibo   Dewatering and drainage pumps by ITT Flygt Corp.

bicarbonate   A chemical compound containing an HCO_3 group.

bicarbonate alkalinity   Water alkalinity caused by bicarbonate ions.

bicarbonate hardness   Water hardness caused by calcium bicarbonate and magnesium bicarbonate.

BI-CHEM   Surfactant degradation product by Sybron Chemicals, Biochemical Division.

***Bi-Chem*®**   Selectively adapted bacterial cultures for wastewater treatment by Sybron Chemicals, Inc.

BIF   See "boilers and industrial furnaces (BIF)."

***BIF*®**   Product group of BIF.

biflow filter   Granular media filter characterized by water flow from both top and bottom to a collector located in the center of the filter bed.

***Big Dipper*®**   Grease removal unit by Thermaco, Inc.

bilharzia   Waterborne disease also known as "schistosomiasis."

binary fission   Asexual reproduction in some microbes where the parent organism splits into two independent organisms.

Bio Disk-10   Biological deodorizer by Neutraman, Inc.

Bio Dredge   A lagoon sludge oxidation process by Unisol.

***Bio Genesis*™**   Microbial formulation to reduce wastewater odors by Bio Huma Netics, Inc.

Bio Gro   Bio Gro Division of Wheelabrator Water Technologies, Inc.

Bio Jet-7   Organic solution of seven strains of live, non-toxic bacteria by Jet, Inc.

***Bio Max*™**   Floating piping system by Environmental Dynamics Inc.

***Bio*Fix*®**   Alkaline stabilization process for biosolids by Wheelabrator Water Technologies, Inc.

***Bio*Lime*®**   Agricultural liming agent by Wheelabrator Water Technologies, Inc.

***Bio/Scent*™**   Liquid odor neutralizer by Hinsilblon Laboratories.

***BioAccelerator*™**   Groundwater treatment unit by Biotrol.

bioaccumulants   Substances that increase in concentration in living organisms as they take in contaminated air, water, or food because the substances are very slowly metabolized or excreted. See also "biological magnification."

bioaccumulation   The process by which living organisms absorb and retain chemicals or elements from their environment, especially from their food.

bioaccumulative   A characteristic of a chemical whose rate of intake into a living organism is greater than its rate of excretion or metabolism.

bioaccumulative chemicals of concern (BCC)   Chemicals including mercury, dioxin, chlordane, DDT, and PCBs which accumulate in the environment and represent potential risks to human health, aquatic life, and wildlife.

Bio-Activation   Combination activated sludge and trickling filter wastewater treatment system by Amwell, Inc.

bioassay   An analytical method that uses living organisms to measure the effect of a substance, factor, or condition on the environment.

bioaugmentation   A method of cleaning up pollution by inoculating a site with specific contaminant-targeted microbes in high density.

Biobed   Wastewater treatment plant by Biothane Corp.

biobrick   A kiln-dried building brick with an organic component provided by municipal wastewater solids.

BioCam   Wastewater treatment system utilizing conditioned anaerobic methanogens by Parkson Corp.

***Biocarb*™**   Activated carbon by USFilter/Westates.

***Biocarbone*®**   Biological wastewater treatment process using immersed fixed bed filter by USFilter/Krüger.

Biocel   Modular biofiltration system for VOC reduction by Envirogen.

biochemical oxidation   Oxidative reactions brought about by biological activity which result in chemical combination of oxygen with organic matter.

biochemical oxygen demand (BOD)   A standard measure of wastewater strength that quantifies the oxygen consumed in a stated period of time; usually 5 days and at 20°C.

***Biocidal*™**   Sodium hypochlorite system by Scienco/FAST Systems.

biocide   A chemical used to inhibit or control the population of troublesome microbes.

Bio-Clarifier   Secondary clarifier with rotating sludge scoop for use with package rotating biological contactor by USFilter/Envirex.

Bioclean   Reverse osmosis membrane cleaner by BetzDearborn-Argo District.

***Bioclere*™**   Packaged wastewater treatment plants by Ekofinn Bioclere.

bioconcentration   The net increase in concentration of a substance in plants and animals above what is found in the natural surroundings.

bioconcentration factor (BCF)   The accumulation of chemicals that live in contaminated environments equal to the quotient of the concentration of a substance in aquatic organisms divided by the concentration in the water during the same time period.

***Biocontact*®**   Biological aerated filter by Ekokan.

biocontactor   A unit process such as an aeration basin, trickling filter, RBC, or digester where microbes degrade/transform organic matter.

bioconversion   The conversion of organic waste products into an energy resource through the action of microbes.

biocriteria   Quantitative and narrative goals for the aquatic community used within water programs.

***Biocube*™**   Aerobic biofilter for airborne odors and VOCs by AMETEK Rotron Biofiltration.

***Bio-D*®**   Bioremediation nutrients by Medina Products Bioremediation Division.

biodegradable   Term used to describe organic matter which can undergo biological decomposition.

biodegradable dissolved organic carbon (BDOC)   The portion of TOC that is easily degraded by microbes.

biodegradable organic matter (BOM)   The portion of organic matter in water that can be degraded by microbes.

***BIOdek*®**   Synthetic media for fixed film wastewater treatment reactors by Brentwood Industries, Inc.

***BioDen*™**   Anaerobic biological nitrate removal process by Nitrate Removal Technologies, LLC.

BioDenipho® Biological phosphorus and nitrogen removal process by USFilter/Krüger.

Biodenit® Biological denitrification process using immersed fixed bed filter by USFilter/Krüger.

Bio-Denitro® Biological nitrogen removal process by USFilter/Krüger.

biodiversity An environment where multiple organisms coexist.

BioDoc® Rotary distributor for trickling filters by WesTech Engineering Inc.

Bio-Drum Rotating drum containing biological filter media for wastewater treatment by JDV Equipment Corp.

Bio-Energizer Lagoon sludge oxidation system by Probiotic Solutions.

biofeasibility A bioremediation feasibility study done to determine the applicability and potential success of a bioremediation technique or procedure for a given site.

biofilm An accumulation of microbial growth.

BiofiltAIR™ Biological air filter by Biorem Technologies, Inc.

biofilter See "biological filter."

BioFlex® Moving aeration chains for wastewater treatment systems by Parkson Corp.

Bioflush Trash rake by E. Beaudrey & Co.

Biofor™ Biological fixed film wastewater treatment system by Infilco Degremont, Inc.

biofoul Presence and growth of organic matter in a water system.

BioFuser® Oxygen transfer and mixing units for wastewater treatment systems by Parkson Corp.

biogas The gases produced by the anaerobic decomposition of organic matter.

biogenesis The theory that living organisms arise only from other living organisms.

Bioglas Rigid, open cell foam silica biological oxidation media by the former Bioglas Corp.

Bioglas Alpha Package fixed film wastewater treatment plant by the former Bioglas Corp.

Biograte Floor grating by Brentwood Industries, Inc.

BioGuard ACS™ Biofouling inhibitor for reverse osmosis systems by Professional Water Technologies, Inc.

BioGuard® Influent cleaning system by Parkson Corp.

BioGuide® Process monitoring and control technology by BioChem Technology, Inc.

Bioken Former name of Filter Products.

biokinetics The branch of science that pertains to the study of living organisms.

Biolac® Extended aeration waste treatment process by Parkson Corp.

Biolift™ Waste activated sludge thickening system by Baker Process.

Biologic™ Nutrient supplement for wastewater treatment facilities by SciCorp Systems, Inc.

biological exposure indexes (BEI) Guidelines used for assessing the hazard posed to healthy workers by chemical substances present in the body.

biological filter A bed of sand, stone, or other media through which wastewater flows that depends on biological action for its effectiveness.

biological magnification   Refers to the process whereby certain substances such as pesticides or heavy metals move up the food chain, work their way into rivers or lakes, and are eaten by aquatic organisms such as fish, which in turn are eaten by large birds, animals, or humans.

biological process   The process by which the metabolic activities of bacteria and other microorganisms break down complex organic materials into simple, more stable substances.

biological treatment   A treatment technology that uses bacteria to consume organic waste.

biologically active filter (BAF)   A granular media filter, usually employing activated carbon or anthracite which relies on the growth of a biofilm to aid in the degradation of organic matter and/or ammonia.

Bio-Lysis®   Sludge reduction process by Kady International.

biomass   The mass of biological material contained in a system.

biome   A biological community or ecosystem characterized by a specific habitat and climate such as a tropical rain forest or a desert.

biomedical waste   Waste derived from the operation of hospital, laboratory, and health care facilities.

Biomizer™   Continuously sequencing reactor process by Environmental Dynamics Inc.

Bio-Module   Package wastewater treatment plant using rotating biological contactors by USFilter/Envirex.

BioMonitor™   Automated on-line BOD analyzer by Anatel Corp.

biomonitoring   The use of living organisms to test water quality at a discharge site or further downstream.

Bio-Net®   Rotating biological contactor by NSW Corp.

BIONOx™   Submersible aerator/mixer by ABS Pumps, Inc.

Bionutre   Nutrient removal process by USFilter/Envirex.

Bio-Nutri™   Nitrogen and phosphorus removal process by Smith & Loveless, Inc.

Bio-Ox™   Bioreactor for biological treatment of wastewater by SRE, Inc.

bio-oxidation   See "biochemical oxidation."

Bio-Pac   Package trickling filter formerly offered by USFilter/Envirex.

Bio-Pac SF#30   Trickling filter media by NSW Corp.

Biopaq ®   Upflow anaerobic sludge blanket process used for treatment of high-strength wastes by CBI Walker, Inc., (U.S. licensee), Biwater (U.K. licensee), and Paques B.V. (licensor).

BioPasteur®   Biosolids pasteurization system by USFilter/Krüger.

biopile   A remediation technique that involves the mounding of contaminated soil in a lined and covered pile with air and amendments circulating via pumping.

Bio-Pure   Water reclamation treatment plant by AquaClear Technologies Corp.

biopure water   Water that is sterile, pyrogen free and has a total solids content of less than 1 mg/L.

Biopuric   Technology to remove hydrogen sulfide from anaerobic digester biogas by Biothane Corp.

Bio-Reel™   Fixed film wastewater treatment system using coiled, corrugated tubing by Schreiber Corp.

bioremediation   The use of the natural ability of microbes to use waste materials in their metabolic processes and convert them into harmless endproducts.

Bio-S®   Bioremediation surfactant by Medina Products Bioremediation Division.

Bioscan 2   Monitoring technology for microbial activity by BetzDearborn, Inc.

Bioscrub™   Rotating biological contactor odor and VOC treatment system by CMS Group, Inc.

Bioscrubbers™   Biological-based system for removal of odors by WRc Process Engineering.

Biosep®   Membrane water/wastewater treatment process by USFilter/Krüger (North America) and OTV.

Bio-Separator   Floating flow diversion baffle for lagoons by ThermaFab, Inc.

bioslurping   An in situ remediation technique that involves extraction of both vapor and liquid from the subsurface, also known as "dual-phase extraction."

Bio-Sock   Fabric sock used to introduce bacterial cultures into a flow by Sybron Chemicals, Inc.

Biosock™   Biological culture application system by Sybron Chemicals, Inc.

biosolids   Solid organic matter recovered from municipal wastewater treatment that can be beneficially used, especially as a fertilizer. "Biosolids" are solids that have been stabilized within the treatment process, whereas "sludge" has not.

biosorption process   See "contact stabilization process."

Bio-Source™   Biocide by Avista Technologies.

biosparging   A groundwater remediation technology where compressed air is injected into a contaminated aquifer.

biosphere   The mass of living organisms found in a thin belt at the earth's surface.

BioSpiral   Rotating biological contactor formerly offered by Walker Process Equipment.

Biostart™   Liquid microbial concentrate by Advanced Microbial Systems, Inc.

biostat   A substance that inhibits biological growth without destroying the biomass.

biostimulation   A method of cleaning up pollution using indigenous microbes at a contaminated site and providing only fertilizers, nutrients, or special chemical compounds that speed the growth of the indigenous microbial population.

Biostyr®   Upflow mixed media reactor process for removal of nitrogen and suspended solids by USFilter/Krüger (North America) and OTV.

Biostyrene®   Floating filter media by USFilter/Krüger (North America) and OTV.

Bio-Surf   Rotating biological contactor process by USFilter/Envirex.

biota   All living organisms within a system.

Biotac™   Bioremediation system that delivers bacteria to wet wells by USFilter/Davis Process.

biotechnology   Techniques that use living organisms or parts of organisms to produce a variety of products to improve plants or animals or to develop microorganisms to remove toxics from bodies of water, or act as pesticides.

Biothane®   Anaerobic wastewater treatment process by Biothane Corp.

Bio-Tite®   Biofiltration odor control system by Thermacon Enviro Systems, Inc.

Bioton®   Biological VOC and odor control system by Monsanto Enviro-Chem Systems, Inc.

biotower   See "biological filter."

Biotox®   Regenerative thermal oxidation process by Biothermica International, Inc.

biotransformation   Conversion of a substance into other compounds by organisms; includes biodegradation.

biotrickling filter   Odor treatment system where air is scrubbed with recirculating liquid flowing over high-porosity packing materials covered with a thin film of sulfur-oxidizing microbes.

bioturbation   The net effect of the activity of benthic organisms at wastewater treatment plant discharges which may aid in the dispersion of contaminants and increase the exchange of oxygen and nutrients between the sediment and water.

bioventing   An in situ groundwater remediation technology where air is introduced into unsaturated soil to facilitate biodegradation of organic contaminants.

BioWeb™   Synthetic media providing a growth site for wastewater biomass by USFilter/Davco.

Biox™   Package water treatment plant by Bioscience, Inc.

Bioxide®   Sewage odor control product by USFilter/Davis Process.

Bioxide-AQ™   Odor and corrosion control product by USFilter/Davis Process.

BIPM   International Bureau of Weights and Measures.

bipolar membrane   A membrane composed of two distinct layers of oppositely charged materials.

bipolar membrane electrodialysis   See "water splitting."

Bird Machine   Sludge dewatering equipment product line of Baker Process.

Birm®   Granular filter media for removal of iron and manganese by Clack Corp.

bittern   The bitter liquid remaining after the crystallization of salt from a brine. See also "mother liquor."

Bitumastic®   Coal tar coating products by Carboline Co.

bituminous coal   A coal high in carbonaceous matter that yields a considerable amount of volatile waste matter when burned.

Blace Filter   A precoat filter system by Blace Filtronics, Inc.

Black Clawson   Former name of Thermal Black Clawson.

black liquor   Strong organic waste generated during kraft pulping process.

black lung disease   Common name for the lung disease "anthracosis" caused by prolonged inhalation of coal dust, which results in fibrosis, or scarring of lung tissue.

black sand   Discoloration of filter sand resulting from manganese deposits.

black water   (1) A condition in drinking water that results from the presence of excess oxidized manganese. (2) Water that contains animal, human, or food waste.

blank   A quality control sample representing a matrix and containing all the constituents except the analyte.

blanketing gas   Nitrogen or other inert gas used in a "gas blanket."

blast furnace   Furnace used in iron-making process in which hot blast air flows upward through the raw materials and exits at the furnace top.

Blastocystis   An intestinal protozoan parasite transmitted by contaminate food or water.

Blaw-Knox   Former name of Buffalo Technologies, Inc.

bleach   An oxidizing compound usually containing chlorine combined with calcium or sodium.

bleed   To draw accumulated liquid or gas from a line or container.

Blendmaster   Sludge mixer by McLanahan Corp.

***Blendrex*™**   Motionless mixer by LCI Corp.

blind flange   A flanged plate or blind used to close the end of a pipeline.

blinding   The reduction or cessation of flow through a filter resulting from solids restricting the filter openings.

Blizzard Adsorption System   Polymeric adsorbent technology for VOC abatement by On-Demand Environmental Systems, Inc.

BLM   U.S. Department of Interior's Bureau of Land Management.

Block & Hong process   Biological phosphorus removal process by USFilter/Krüger.

blood worm   The larval stage of the midge fly.

bloodborne pathogen   Pathogenic microbes present in human blood and other potentially infectious materials which can cause disease in humans.

bloom   See "algal bloom."

blowback   Filtrate blown out through the filter medium on a rotary vacuum filter by the air introduced to move the filter cake away from the cloth.

blowdown   A controlled discharge from a recirculating system designed to prevent a buildup of some material.

blower   Air conveying equipment that generates pressures up to 103 kPa (15 psi) commonly used for wastewater aeration systems.

***BLS*™**   Sludge reduction process by Kady International.

blue baby syndrome   See "methemoglobinemia."

blue vitriol   Common name for copper sulfate used to control algae.

blue-green algae   A group of aquatic organisms having a blue pigment in addition to a green-colored colorophyll, and often the cause of nuisance conditions in water bodies.

BLWRS   See "barriered landscape water renovation system (BLWRS)."

BMP   See "best management practice (BMP)."

BMR   Baseline monitoring report.

BNR   Biological nutrient removal.

Boa-Boom   Oil spill containment booms by Environetics, Inc.

***Boat*®**   Boat-shaped intrachannel clarifier by United Industries, Inc.

BOD   See "biochemical oxygen demand (BOD)."

BOD$_5$ Five-day carbonaceous or nitrification-inhibited BOD. See also "biochemical oxygen demand (BOD)."

***BOD-Seed*™**   Seed microbes for BOD testing by Sybron Chemicals, Inc.

BODu   See "ultimate BOD (BODU)."

body burden   The total radiation or toxic material present in the body at a point in time.

body feed   Coating material added to the influent of precoat filters during filtration cycle.

bog   Poorly drained land filled with decayed organic matter that is wet and spongy and unable to support an appreciable weight.

***BogenFilter*™**   Belt filter press by Klein America, Inc.

boil out   An evaporator cleaning process where wash water is boiled in an evaporator to remove scale deposits.

boiler   A vessel in which water is continually vaporized into steam by the application of heat.

boiler feedwater   Water which, in the best practice, is softened and/or demineralized, heated to nearly boiler temperature, and deaerated before being pumped into a steam boiler.

***Boilermate*®**   Packed column deaerator by Cleaver-Brooks.

boilers and industrial furnaces (BIF)   A category of thermal treatment operations, also including cement and aggregate kilns, asphalt, and smelting furnaces, whose combustion processes and air emissions are regulated.

boiling point   The temperature at which a liquid's vapor pressure equals the pressure acting on the liquid.

boiling point elevation (BPE)   The difference between the boiling point of a solution and the boiling point of pure water at the same pressure.

BOM   See "biodegradable organic matter (BOM)" and "background organic matter (BOM)."

bomb calorimeter   An instrument used to determine the heat content of sludge or other material.

bone char   A carbon-based adsorbent made by carbonizing animal bones.

BonoZon   Ozone generator by ProMinent Fluid Controls, Inc.

BOO   Build, own, operate.

BOOM   Build-own-operate-maintain.

boom   A floating barrier used to contain oil on a body of water.

booster pump   A pump used to raise the pressure of the fluid on its discharge side.

BOOT   Build-own-operate-transfer.

***Boothwall*™**   Dust collector cartridge filters by Dustex Corp.

BoR   U.S. Bureau of Reclamation. Also called "BuRec."

bore hole   A man-made hole in a geological formation.

***Bosker*™**   Trash rack cleaner by Brackett Green.

BOT   Build, operate, transfer.

botanical pesticide   A pesticide whose active ingredient is a plant-produced chemical such as nicotine or strychnine.

bottom ash   The noncombustible particles that fall to the bottom of a boiler furnace.

bottoming cycle   Cogeneration system where thermal heat is produced by the process and by-product electricity is then generated.

botulism   A severe form of food poisoning usually associated with development of a toxin produced by bacteria such as bacillus in improperly preserved or prepared food.

bound water   Water held on the surface or interior of colloidal particles.

Bouyoucos   A laboratory test procedure employing hydrometers to determine the fine particle size distribution in a slurry.

Bowser-Briggs   Former manufacturer of oil/water separation equipment.

Boyle's Law   The volume of a gas varies inversely with its pressure at constant temperature.

Boythorp Glass coated steel tanks by Klargestor.

BPE See "boiling point elevation (BPE)."

BPEO Best practical environmental option.

BPR (1) Biological phosphorous removal. (2) Boiling point rise.

Brackett Green Former name of Brackett Geiger.

brackish water Water containing low concentration of soluble salts, usually between 1000 and 10,000 mg/L.

branch sewer A sewer that receives wastewater from a small area and discharges into a main sewer serving more than one area.

Brandol® Cylindrical fine bubble diffusers by USFilter/Schumacher Filters.

brass A copper alloy containing up to 40% zinc.

braze To thermally bond metallic parts with a cuprous alloy.

break tank A storage tank at atmospheric pressure from which feed water is drawn prior to further treatment or use.

breakpoint chlorination Addition of chlorine to a water or wastewater until the chlorine demand has been satisfied. Further addition will result in a chlorine residual so that disinfection can be assured.

breakthrough That point in the granular media filter cycle when the filtrate turbidity begins to increase because the filter bed is full and no longer able to retain solids.

breakwater An offshore barrier, often connected to shore, which breaks the force of waves and provides shelter from wave action.

Breeze™ Compact air stripping unit by Aeromix Systems, Inc.

BRI See "building-related illness (BRI)."

brine Water saturated with, or containing a high concentration of, salts, usually in excess of 36,000 mg/L.

brine concentrator Term used to describe a vertical tube falling film evaporator employing special scale control techniques to maximize concentration of dissolved solids.

brine heater The heat input section of a multistage flash evaporator where feedwater is heated to the process' top temperature.

brine mud Waste material, often associated with well-drilling or mining, composed of mineral salts or other inorganic compounds.

brine staging See "reject staging."

British Thermal Unit (Btu) The quantity of heat required to raise the temperature of one pound of water by 1°F.

Brix scale A scale used in a hydrometer for measuring the concentration or density of sugar in solution.

broadcast application The spreading of pesticides over an entire area.

broad-crested weir A weir having a substantial crest width in the direction parallel to the direction of water flowing over it.

broke Paper waste generated prior to completion of the papermaking process.

bromate The highest oxidation state of the bromide ion which can be formed during the ozonation of waters containing bromide.

BromiCide® Microbiocide by BioLab Water Additives.

bromide   An inorganic ion found in surface water and groundwater that, when oxidized by chlorine or ozone, can result in the formation of bromide-substituted disinfection byproducts.

bromine   A halogen element used as a water disinfectant in combination with chlorine as a chlorine-bromide mixture. Chemical symbol is Br.

bronze   A copper-tin alloy, or any other copper alloy, that does not contain zinc or nickel as the principal alloying element.

brown coal   A common term for lignite.

brownfield   An inactive site or property being put back into productive economic use after the relevant environmental agencies agree contaminants present at the property no longer pose an unacceptable risk to human health or the environment.

Brownian motion   Erratic movement of colloidal particles that results from the impact of molecules and ions dissolved in the solution.

Brownie Buster   Organic solids agitator/separator by Enviro-Care Co.

Bruner   Product line by Culligan International.

Bruner-Matic®   Water treatment control center by Culligan International Corp.

brush aerator   Mechanical aeration device most frequently used in oxidation ditch wastewater treatment plants, consisting of a horizontal shaft with protruding paddles, that is rapidly rotated at the water surface. Also called a "rotor."

BS&W   Bottom sediments and water.

BTEX   Benzene, toluene, ethylbenzene, and xylene.

Btu   See "British Thermal Unit (Btu)."

BTU-Plus®   Filter media that incinerates to inert ash by Alar Engineering Corp.

BTX   Benzene, toluene, and xylene.

bubble point   The pressure at which air first passes through a wet membrane; the path being the channel of greatest pore size.

bubbler system   Common terminology for pneumatic-type differential level controller.

bubonic plague   An acute infectious disease usually transmitted from infected animals to humans by the bite of a rat flea.

buchner funnel   A laboratory funnel with a perforated bottom that utilizes a disposable filter paper to evaluate wastewater and sludge dewaterability.

bucket elevator   A conveying device consisting of a head and foot assembly that supports and drives an endless chain or belt to which buckets are attached.

Budd   Nonmetallic chain product line by Polychem Corp.

buffer   A substance that stabilizes the pH value of solutions.

buffer strips   Strips of grass or other erosion-resisting vegetation between or below cultivated strips or fields.

buffered   The ability to resist changes in pH.

buffering capacity   The capacity of a solution to resist a change in composition, especially changes in pH.

Buflovak®   Evaporator and crystallizer product line by Buffalo Technologies, Inc.

building-related illness (BRI)   Condition in which at least 20% of a building's occupants display symptoms of illness for more than 2 weeks and the cause of the illness can be traced to a specific building source.

bulk density   The density/volume ratio for a solid including the voids contained in the bulk material.

bulkhead   A partition of wood, rock, concrete, or steel used for protection from water, or to segregate sections of tanks or vessels.

bulking sludge   A poorly settling activated sludge that results from the predominance of filamentous organisms.

bulky waste   Large items of waste materials, such as appliances, furniture, large auto parts, trees, stumps.

***Bullseye*™**   Wastewater nutrient removal process by United Industries, Inc.

buoyancy   The tendency of a body to rise or float in a liquid.

BuRec   U.S. Bureau of Reclamation. Also called "BoR."

burette   A glass tube with fine gradations and bottom stopcock used to accurately measure and dispense fluids.

burning agents   Additives that improve the combustibility of the materials to which they are applied.

burning rate   The rate at which solid waste is incinerated or heat is released during incineration.

burnishing   A surface finishing process in which surface irregularities are displaced rather than removed.

bushing   (1) A short threaded tube which screws into a pipe fitting to reduce its size. (2) The bearing surface for pin rotation when a chain revolves around a sprocket.

butterfly valve   A valve equipped with a stem-operated disk that is rotated parallel to the liquid flow when opened and perpendicular to the flow when closed.

BV   See "bed volume (BV)."

***BVF*®**   Anaerobic wastewater treatment system by ADI Systems, Inc.

BWI   British Drinking Water Inspectorate.

BWR   See "basic water requirement (BWR)."

BWRO   Brackish water reverse osmosis.

bypass   A channel or pipe arranged to divert flow around a tank, treatment process, or control device.

byproduct   A material or substance that is not a primary product of a process and is not separately produced.

C

C   See "Celsius (C)."

C. parvum   See "*Cryptosporidium*."

C × T   The product of the "residual disinfection concentration" (C) in mg/L determined before or at the first customer, and the corresponding "disinfectant contact time" (T) in minutes. Also called "CT value."

C × T$_{99.9}$ The CT value required for 99.9% inactivation of *Giardia lamblia* cysts.

c/c   Center-to-center.

CA   See "cellulose acetate (CA)."

CA membrane   Cellulose acetate membrane.

CA•RE™   Spent cartridge filter recovery program by USFilter Corp.

CAA   See "Clean Air Act (CAA)."

CAAA   See "Clean Air Act Amendments (CAAA)."

CableTorq   Circular thickener with automatic torque load response system by GL&V/Dorr-Oliver, Inc.

CaCO₃   See "calcium carbonate."

CAD   Computer-aided design.

cadmium (Cd)   A heavy metal element that accumulates in the environment.

CADRE®   VOC destruction process by Vara International.

CAF®   Cavitation air flotation units by HydroCal, Inc.

CAFE   See "Corporate Average Fuel Economy Standard (CAFE)."

CAFOs   Concentrated animal feeding operations.

Cairox®   Potassium permanganate by Carus Chemical Co., Inc.

Cairox ZM®   Zebra mussel control technology by Carus Chemical Co., Inc.

caisson   Watertight structure used for underwater work.

cake   Dewatered sludge with a solids concentration sufficient to allow handling as a solid material.

cake filtration   Filtration classification for filters where solids are removed on the entering face of the granular media.

CakePress   Modular high pressure section of dewatering press by Parkson Corp.

Cal   Large calorie. See "calorie."

cal   Small calorie. See "calorie."

calandria   The heating element in an evaporator consisting of vertical tubes which act as the heating surface.

calcareous   Composed of, or containing, calcium compounds, particularly calcium carbonate.

calcify   To become stone-like or chalky due to deposition of calcium salts.

calcine   Ore, carbonate, mineral, or concentrate that has been roasted in an oxidizing atmosphere to remove sulfur or carbon dioxide.

calcined lime   See "quicklime."

calciner   A device in which the moisture and organic matter in phosphate rock is reduced in a combustion chamber.

calcining   Exposure of an inorganic compound to a high temperature to alter its form and drive off a substance that was originally part of the compound.

Calciquest   Liquid polyphosphate by Calciquest, Inc.

calcium carbonate   A white, chalky substance which is the principal hardness and scale-causing compound in water. Chemical formula is $CaCO_3$.

calcium carbonate equivalent (mg/L as CaCO₃)   A convenient unit of exchange for expressing all ions in water by comparing them to calcium carbonate which has a molecular weight of 100 and an equivalent weight of 50; signifies that the concentration of a dissolved mineral is chemically equivalent to the stated concentration of calcium carbonate.

calcium hardness   The portion of the total hardness attributed to calcium compounds.

calcium hydroxide   See "hydrated lime."

calcium hypochlorite   A chlorine compound frequently used as a water or wastewater disinfectant. Chemical formula is $Ca(OCl)$.

calcium sulfate   A white solid known as the mineral "anhydrite" with the chemical formula $CaSO_4$, and gypsum with the formula $CaSO_4 \cdot 2H_2O$.

calibration   The determination, checking, or rectifying of the gradation of any instrument giving quantitative measurements.

Callaway Chemical   Company acquired by Vulcan Performance Chemicals.

calorie   The amount of heat required to raise the temperature of one gram of water 1°C, also known as a "small calorie." A "large calorie" or "kilocalorie" is the amount of heat required to raise the temperature of a kilogram of water by 1°C.

Calver   Chemicals for use in the analysis of calcium in water by Hach Company.

Calvert   Manufacturer acquired by Monsanto Enviro-Chem Systems, Inc.

CAM   Carbon adsorption method.

CAM rule   See "Compliance Assurance Monitoring rule (CAM rule)."

CAMP   Continuous air monitoring program.

Camp Nozzle   Plastic strainer-type nozzle for filter underdrain by Walker Process Equipment.

Campylobacter **enteritis**   A waterborne gastrointestinal disorder.

cancer   A class of diseases characterized by the uncontrolled growth of cells.

Cannon™   Positive displacement digester mixer by Infilco Degremont, Inc.

Cannonball²   Portable multiple gas detector by Biosystems, Inc.

Cansorb   Activated carbon adsorber by TIGG Corp.

cap   A layer of clay or other impermeable material installed over the top of a closed landfill to prevent entry of rainwater and minimize leachate.

CAPA®   Caprolactone product by Solvay America.

capacitive deionization   An electrically regenerated electrosorption process capable of desalinating saline water.

capillarity   The ability of a soil to retain a film of water around soil particles and in pores through the action of surface tension.

capillary   (1) A slender hair-like structure or a very fine, small bore tube. (2) A blood vessel with very fine openings that joins the smallest arteries with the smallest veins.

capillary action   The movement or action of a liquid through interstices, capillary tubes, or other very fine openings due to the molecular attraction between molecules of the liquid for each and a solid surface.

capillary fringe   The zone of porous material above the zone of saturation containing water held by capillary action.

Capitox   Modular wastewater treatment plant by Simon-Hartley, Ltd.

Capozone®   Ozone generation system by Capital Controls Co.

Capsular®   Wet well mounted pump station by Smith & Loveless, Inc.

Captivated Sludge Process   Fixed film biological waste treatment system by the former Waste Solutions.

Captor®   Fixed film biological waste treatment system by the former Waste Solutions.

capture efficiency   The fraction of organic vapors generated by a process that are directed to an abatement or recovery device.

CAR™   Aerobic wastewater treatment system with covered reactor by ADI Systems, Inc.

Carball   Carbon dioxide generator formerly offered by Walker Process Equipment.

carbamates   A class of pesticides, herbacides, and fungicides developed as less-hazardous replacements for chlorinated hydrocarbons.

Carbo Dur™   Granular activated carbon by USFilter/Warren.

Carbo-Cor™   Crossflow membrane used for purification of water-based surface cleaners by Koch Membrane Systems Inc.

Carbofilt   Anthracite filter media by International Filter Media.

Carbolux®   Decontaminate system to remove crystallized contaminants from electroplating solutions by USFilter/Dewatering Systems.

carbon   An element present in many inorganic and all organic compounds.

carbon adsorption   The use of powdered or granular-activated carbon to remove refractory and other organic matter from water.

carbon black   An additive that prevents degradation of thermoplastics by ultraviolet light.

carbon chloroform extract (CCE)   The residue from a carbon chloroform extraction test.

carbon chloroform extraction   Test to determine organic matter in water where organics adsorbed on an activated carbon cartridge are extracted from the carbon by chloroform and weighed or analyzed.

carbon cycle   A graphical presentation of the movement of carbon among living and nonliving matter.

carbon dioxide   A noncombustible gas formed in animal respiration and the combustion and decomposition of organic matter. Chemical formula is CO_2.

carbon fixation   A process occurring in photosynthesis where atmospheric carbon dioxide gas is combined with hydrogen obtained from water molecules.

carbon monoxide   Colorless, odorless gas produced by incomplete combustion of organic fuels lethal to humans at concentrations exceeding 5000 mg/L. Chemical formula is CO.

carbon steel   A general purpose steel whose major properties depend on its 0.1 to 2% carbon content without substantial amounts of other alloying elements.

carbon-14   A naturally occurring radioactive isotope of carbon that emits beta particles when it undergoes radioactive decay.

carbonaceous   Of, pertaining to, or yielding carbon.

carbonaceous biochemical oxygen demand (CBOD)   The portion of biochemical oxygen demand where oxygen consumption is due to oxidation of carbon, usually measured after a sample has been incubated for 5 days. Also called "first-stage BOD."

carbonate   A compound containing the anion radical of carbonic acid CO_3.

carbonate alkalinity   Alkalinity resulting from the presence of carbonate ions.

carbonate hardness   The hardness in water caused by bicarbonates and the carbonates of calcium and magnesium.

carbonation   The diffusion of carbon dioxide gas through a liquid.

carbonator   A device used to carbonate or recarbonate water.

Carbonite   Anthracite filter media by Carbonite Filter Corp.

Carborundum   Former manufacturer of wastewater treatment equipment.

carboxylic   The functional group COOH found in all carboxylic acids.

carboxylic acid   Organic acids such as acetic, lactic, and citric acids which contain one or more COOH groups.

carboy   A large container used to store or transport liquid chemicals or water samples.

carcinogen   A cancer or tumor-causing agent.

cardinal points   The four principal points of a compass: north, south, east, and west.

Carrobic   Aerobic digester/thickener used with oxidation ditch wastewater treatment system by Baker Process.

Carrousel®   Biological oxidation/wastewater treatment system by DHV Water BV, licensed to Baker Process.

Carter   Product line of JDV Equipment Corp.

Cartermix   Anaerobic sludge digester mixing system by JDV Equipment Corp.

cartridge filter   A filter unit with cylindrical replaceable elements or cartridges.

Carulite®   Catalysts for VOC destruction by Carus Chemical Co.

Carver-Greenfield process   Multiple effect evaporation process to extract water from sludge.

CASAC   EPA's Clean Air Scientific Advisory Committee.

Cascade   The combined use of a gravity belt thickener followed by a belt filter press to dewater sludge by Gebr. Bellmer GmbH.

Cascade™   Biological filtering system using synthetic media by USFilter/General Filter.

cascade aeration   An aeration method using a series of steps to promote oxygen uptake in a flowing stream.

casing   A pipe or tube placed in a bore hole to support the sides of the hole and to prevent other fluids from entering or leaving the hole.

cask   A thick-walled container, usually lead, used to transport radioactive material.

CASS™   Cyclic activated sludge system for wastewater treatment system by CASS Water Engineering, Inc.

cast iron   A general description for a group of iron-carbon-silicon metallic products obtained by reducing iron ore with carbon at temperatures high enough to render the metal fluid and cast it in a mold.

CastKleen   Cast-in-place filter underdrain by Baker Process.

Cat Floc®   Cationic polymer to enhance solids/liquid separation by Calgon Corp.

catalyst   A substance that modifies or increases the rate of a chemical reaction without being consumed in the process.

catalytic converter   A device installed in the exhaust system of an internal combustion engine which utilizes catalytic action to oxidize hydrocarbon and carbon monoxide emissions to carbon dioxide.

catalytic cracking   The use of a catalyst during a cracking process.

catch basin   An open basin that serves as a collection point for stormwater runoff.

catchment   A barrel, cistern, or other container used to catch water.

catchment area   The area of land bounded by watersheds draining into a river, lake, or reservoir.

categorical exclusion A class of actions which either individually or cumulatively would not have a significant effect on the human environment and therefore would not require preparation of an environmental assessment or environmental impact statement.

categorical pretreatment standard A technology-based effluent limitation for an industrial facility discharging into a municipal sewer system.

category I contaminant U.S. EPA contaminant category indicating sufficient evidence of carcinogenicity via ingestion in humans or animals exists to warrant classification as "known or probable human carcinogens via ingestion."

category II contaminant U.S. EPA contaminant category for which limited evidence of carcinogenicity via ingestion exists to warrant classification as "possible human carcinogens via ingestion."

category III contaminant U.S. EPA contaminant category of substances for which insufficient or no evidence of carcinogenicity via ingestion exists.

catenary bar screen Mechanical screening device using revolving chain-mounted rakes to clean a stationary bar rack.

cathode The negative electrode where the current leaves an electrolytic solution.

cathodic protection Electrochemical corrosion protection achieved by imposing an electrical potential to counteract the galvanic potential between dissimilar metals which would lead to corrosion.

cation A positively charged ion that migrates to the cathode when an electrical potential is applied to a solution.

cation exchange The ion exchange process in which cations in solution are exchanged for other cations bound to an ion exchange resin or medium.

cation load factor y The sum of calcium, magnesium, sodium, and potassium expressed as calcium carbonate equivalents.

cationic polymer A polyelectrolyte with a net positive electrical charge.

Cat-Ox™ Catalytic oxidation system by Catalytic Combustion Corp.

caustic Alkaline or basic.

caustic scrubbing An air pollution control process using a solution of sodium hydroxide to remove sulfur dioxide from flue gases.

caustic soda Common term for sodium hydroxide. Chemical formula is $NaOH$.

cavitation (1) A selective corrosion that results from the collapse of air or vapor bubbles with sufficient force to cause metal loss or pitting. (2) The action of a pump attempting to discharge more water than suction can provide.

CBG Clean burning gasoline.

CBOD See "carbonaceous biochemical oxygen demand (CBOD)."

cc See "cubic centimeter (cc)."

CCB Coal combustion by-products.

CCC Streaming current coagulation control center by Milton Roy Co.

CCC Compromised container caps.

CCE See "carbon chloroform extract (CCE)."

CCOHS Canadian Centre for Occupational Health and Safety.

CCPP Calcium carbonate precipitation potential.

CCS2000™ Clarifier control system by Drexelbrook Engineering Co.

CD   See "corona discharge method (CD)."

CD ozone generation   Ozone discharge technology most frequently used in the potable water industry.

CDC   See "Centers for Disease Control (CDC)."

CDI®   Continuous deionization process which regenerates resins with electricity by USFilter/Lowell.

CDT   Capacitive deionization technology.

CE rule   See "Credible Evidence rule (CE rule)."

CE-Bauer   Former screening equipment supplier acquired by Andritz-Ruthner, Inc.

Cecarbon®   Granular activated carbon by Elf Atochem North America, Inc.

Cecasorb®   Adsorbent canisters containing activated carbon by Elf Atochem North America, Inc.

CEDI   Continuous electrodeionization.

Celatom®   Diatomite filter aid by Eagle-Picher Minerals, Inc.

CELdek   Synthetic media for evaporative cooling systems by Munters.

Celgard®   Microporous, flat sheet, and hollow fiber membranes by Celgard LLC.

cells   (1) In solid waste disposal, holes where waste is dumped, compacted, and covered with layers of dirt on a daily basis. (2) The smallest structural part of living matter capable of functioning as an independent unit.

cellulose acetate (CA)   A plastic material used to make the cellulosic-type semipermeable reverse osmosis membranes.

Celsius (C)   The SI temperature scale on which 0° is the freezing point and 100° is the boiling point of water. Often referred to as the "centigrade" scale.

CEM   See "continuous emissions monitoring (CEM)."

CEMA   Conveyor Equipment Manufacturers Association.

CEMcat™   Continuous emissions monitor by Advanced Sensor Devices, Inc.

cement   A powder that when mixed with water binds a stone and sand mixture into strong concrete when dry.

cement kiln dust   Alkaline material produced during the manufacture of cement that may be used to stabilize sludge.

cementing   The process of pumping a cement slurry into a drilled hole and/or forced behind the casing.

CEMS   Continuous emissions monitoring systems.

Censys™   Water and wastewater treatment products by USFilter/Lowell.

Centaur®   Activated carbon by Calgon Carbon Corp.

Centers for Disease Control (CDC)   A U.S. Department of Health agency responsible for surveillance of disease patterns, developing disease control and prevention procedures, and public health education.

Center-Slung   Basket centrifuge by Baker Process/Ketema.

centigrade   Colloquial term for the Celsius scale of temperature measurement.

centipose   A unit of the dynamic viscosity of a liquid. The dynamic viscosity of water at 20°C is 1 centipose.

Centrac   Metering pump by Milton Roy Co.

Centra-flo™   Continuously backwashed gravity sand filter by Applied Process Technology, Inc.

centrate   The liquid remaining after solids have been removed in a centrifuge.

Centri-Cleaner®   Liquid cyclone by Andritz-Ruthner, Inc.

Centrico   Sludge dewatering decanter centrifuge by Westfalia Separator, Inc.

Centridry®   Biosolids dewatering and drying process by Baker Process.

CentriField®   Wet scrubber by Entoleter, Inc.

centrifugal collector   A mechanical system using centrifugal force to remove aerosols from a gas stream or to dewater sludge.

centrifugal pump   A pump with a high speed impeller that relies on centrifugal force to throw incoming liquid to the periphery of the impeller housing where velocity is converted to head pressure.

centrifugation   The use of centrifugal force to separate solids from liquids based on density differences.

centrifuge   A dewatering device relying on centrifugal force to separate particles of varying density such as water and solids.

Centripress®   Solid bowl centrifuge by Baker Process.

CenTROL®   Gravity cluster sand filter by USFilter/General Filter.

Centrox®   Aspirating aerators by Hazleton Environmental, Inc.

CEQ   Council on Environmental Quality.

Cerabar   Pressure transmitter by Endress+Hauser.

Ceraflo®   Ceramic membrane filters by USFilter/General Filter.

CeramicÅ   Ceramic filter tube by Coors Ceramics Co.

CERCLA   Comprehensive Environmental Response, Compensation, and Liability Act. Also known as Superfund.

CERMS   Continuous emissions rate monitoring system.

cesspool   A covered tank with open joints constructed in permeable soil to receive raw domestic wastewater and allow partially treated effluent to seep into the surrounding soil while solids are contained and undergo digestion.

CETCO   Colloid Environmental Technologies Co.

CETCOfloc™   Specialty water treatment chemicals by Colloid Environmental Technologies Co.

CF200™   Fine band screen by Brackett Geiger.

CFB   Circulating fluidized bed.

CFC   See "chlorofluorocarbon (CFC)."

CFR   Code of Federal Regulations.

cfs   Cubic feet per second.

CFSTR   Continuous flow, stirred-tank reactor.

CFU   See "colony forming units (CFU)."

CGMP   Current good manufacturing practice.

Chabelco   Chain products marketed by USFilter/Envirex.

chain and flight collector   A sludge collector mechanism utilized in rectangular sedimentation basins or clarifiers.

chain of custody   The documentation maintained regarding all personnel involved in the handling, storage, and analysis of hazardous waste samples from the point where samples are prepared though the point of final disposal.

Chainbelt   Former name of USFilter/Envirex parent company.

Chainsaver Rim   Sludge collector sprockets with wear rim by Jeffrey Chain Corp.

chamber   A compartment or space enclosed by walls; often prefixed by a descriptive word indicating its function, such as grit chamber, screen chamber, discharge chamber, or flushing chamber.

channel   (1) A perceptible natural or artificial waterway that contains moving water or forms a connecting link between two bodies of water. (2) The deep portion of a river or waterway where the main current flows. (3) The part of a body of water deep enough to be used for navigation through an area otherwise too shallow for navigation.

Channel Flow   Sewage disintegrator by C&H Waste Processing.

Channel Master®   Fine screen by Hans Huber GmbH.

Channel Mitt™   Shaftless spiral screening and dewatering device by WesTech Engineering Inc.

Channel Monster®   In-channel solids reduction unit by JWC Environmental.

Channel Piranha   Sewage shredder by ZMI/Portec Chemical Processing.

ChannelAire™   Submersible aerator/mixer by ABS Pumps, Inc.

channeling   A condition that occurs in a filter or other packed bed when water finds furrows or channels through which it can flow without effective contact with the bed.

char   To reduce to charcoal by burning.

characteristic hazardous waste   A waste material declared hazardous because it exhibits ignitable, corrosive, reactive, or toxic characteristics.

charge density   In a polyelectrolyte, the charge density is the mole ratio of the charged monomers to noncharged monomers.

Chargepac®   Coagulant used in water and wastewater treatment by Ashland Chemical, Drew Industrial.

Charles' Law   The volume of gas at constant pressure varies in direct proportion to the absolute temperature.

check valve   A valve that opens in the direction of normal flow and closes with flow reversal.

Check Well   Well level measuring device by Drexelbrook Engineering Company.

chelating agent   A compound that is soluble in water and combines with metal ions to keep them in solution. See also "sequestering agent."

chelation   A chemical complexing of metallic cations with organic compounds to prevent precipitation of the metals. See also "sequestration."

Chem-Clean®   Grease and oil removal system for tertiary filtration by USFilter/Zimpro.

Chem-Feed®   Chemical metering injection pump by Blue-White Industries.

Chem-Fine™   Pleated filter cartridge by USFilter/Filtration & Separation.

Chem-Flex®   Portable holding tank by Aero Tec Laboratories, Inc.

Chem-Gard®   Direct and magnetically driven centrifugal pumps by Vanton Pump & Equipment Corp.

chemical feeder   A device used to dispense chemicals at a predetermined rate.

chemical fixation   The transformation of a chemical compound to a new, nontoxic form.

chemical oxidation   The oxidation of compounds in water or wastewater by chemical means. Typical oxidants include ozone, chlorine, or potassium permanganate.

chemical oxygen demand (COD)   A measurement of biodegradable and nonbiodegradable (refractory) organic matter, widely used as a means of measuring the strength of domestic and industrial wastewaters.

chemical sludge   Sludge resulting from chemical treatment processes of inorganic wastes which are not biologically active.

chemical treatment   Any water or wastewater treatment process involving the addition of chemicals to obtain a desired result such as precipitation, coagulation, flocculation, sludge conditioning, disinfection, or odor control.

chemically emulsified oil particles   These are usually less than 1 micron in size and will not separate or rise to the surface no matter how much time is allowed.

Chemidisk™   Rotating biological contactor by CMS Group, Inc.

Cheminjector-D®   Diaphragm-metering pump by PennProcess Technologies, Inc.

chemisorption   The formation of an irreversible chemical bond between the sorbate molecule and the surface of the adsorbent.

Chemix   Dry polymer mixing and feeding unit by Semblex, Inc.

chemnet   Mutual aid network of chemical shippers and contractors who assign a contracted emergency response company to provide technical support if a representative of the firm whose chemicals are involved in an incident is not readily available.

chemocline   A zone of a lake or reservoir in which the concentration of dissolved substances changes rapidly with depth.

Chemomat   Electrochemical membrane cell separation system by Ionics, Inc.

chemostat   An apparatus designed to grow bacteria cultures at controlled rates.

chemosterilant   A chemical that controls pests by preventing reproduction.

chemotrophs   Organisms that extract energy from organic and inorganic oxidation/reduction reactions.

Chem-Scale™   Weighing scale for vertical chemical tanks by Force Flow Equipment.

ChemScan   Process analyzers by Applied Spectometry.

ChemSensor®   VOC monitor by Osmonics, Inc.

ChemSpec®   Air sampler by Rupprecht & Patashnick Inc.

Chemterc   The industry-sponsored Chemical Transportation Emergency Center which provides information and/or emergency assistance to emergency responders.

Chem-Tower®   Bulk chemical feeding unit by Smith & Loveless, Inc.

Chemtrac®   Streaming current monitoring systems by Chemtrac Systems, Inc.

Chemtube®   Diaphragm-metering pump by USFilter/Wallace & Tiernan.

Chernobyl   Ukrainian town that was the site of a 1986 nuclear power plant accident where radiation to the environment was released.

Chevron™   Clarifier tube settlers by USFilter/Warren.

Chicago Pump   Product group of Yeomans Chicago Corp.

chicane   A plow or other obstacle used on a belt thickener or belt press to mix or turn sludge to facilitate sludge dewatering.

chimney effect   The tendency of air or gas in a vertical passage to rise when it is heated because its density is lower than the surrounding air or gas.

Chi-X®   Odor control product by NuTech Environmental Corp.

chloramines   Disinfecting compounds containing nitrogen, chlorine, and hydrogen formed by the reaction between hypochlorous acid, ammonia, and/or organic amines in water. Also called "combined available chlorine."

Chlor-A-Vac™   Gas induction systems by Capital Controls Co.

Chlorgen+™   Chlorine gas generator by Inchen USA, Inc.

chloride   (1) The ionic form of the element chlorine where the atom has gained one electron, whose chemical symbol is Cl^-. (2) Any salt containing the Cl^- anion.

chlorinated   (1) The condition of water or wastewater that has been treated with chlorine. (2) A description of an organic compound to which chlorine atoms have been added.

chlorination   The addition of chlorine to a water or wastewater, usually for the purpose of disinfection.

chlorinator   A metering device used to add chlorine gas or solutions to water or wastewater.

chlorine   An oxidant commonly used as a disinfectant in water and wastewater treatment. Chemical formula is Cl_2.

chlorine contact chamber   A detention chamber to diffuse chlorine through water or wastewater while providing adequate contact time for disinfection.

chlorine demand   The difference in the amount of chlorine added to a water or wastewater and the amount of residual chlorine remaining after a specific contact duration, usually 15 minutes.

chlorine dose   The amount of chlorine applied to a liquid, usually expressed in milligrams per liter (mg/L) or pounds per million gallons (lb/mil gal).

chlorine residual   The amount of chlorine remaining in water after application at some prior time; the difference between the total chlorine added and that consumed by oxidizable matter. See "free chlorine residual."

chlorine tablets   Common term for pellets of solidified chlorine compounds such as calcium hypochlorite used for water disinfection.

chlorine toxicity   The detrimental effects on biota caused by the inherent properties of chlorine.

chlorite   Any salt of chlorous acid containing the monovalent radical ClO_2.

ChlorMaster™   Sodium hypochlorite generation system by Pepcon Systems, Inc.

Chloro-Cat™   Catalytic oxidizer by Global Technologies.

chlorofluorocarbon (CFC)   Ozone-depleting compounds containing carbon and one or more halogens, usually fluorine, chlorine, or bromine which have been used as commercial refrigerants and propellants in aerosol sprays.

chloroform   A trihalomethane formed by the reaction of chlorine and organic material in water. Chemical formula is $CHCl_3$.

Chloromatic   Electrolytic chlorine generator by Brinecell, Inc.

Chloropac®   Hypochlorite generation system by USFilter/Electrocatalytic.

chlorophenoxy   A class of herbicides that may be found in domestic water supplies and may cause adverse health effects.

chlorosis   Discoloration of normally green plant parts caused by disease, lack of nutrients, or various air pollutants.

Chlor-Scale™   Weighing device for chlorine ton containers by Force Flow Equipment.

Chlortrol   Residual chlorine analyzer by Bailey-Fischer & Porter.

cholera   An acute, highly infectious disease of the gastrointestinal tract caused by the waterborne bacterium *Vibrio cholerae*.

cholinesterase   An enzyme necessary to control the proper transmission of nerve impulses within the body, and whose inhibition is characteristic of the toxicity of some classes of pesticides, e.g., organophosphates.

chopper pump   Pump that chops solids between the impeller and fixed cutter bar.

C-horizon   The unaltered soil layer underlying the B-horizon containing a minimum of soil fauna and flora.

chromatography   The separation of a mixture into its component compounds according to their relative affinity for a solvent system or column media.

Chromaver   Chemical reagents used to determine presence of chromates in solutions by Hach Company.

chromium   See "heavy metals."

chronic effect   An adverse effect on a human or animal in which symptoms recur frequently or develop slowly over a long period of time.

chronic toxicity test   Test method used to determine the concentration of a substance that produces an adverse effect on a test organism over an extended period of time.

Ci   See "curie (Ci)."

CICA   Confederation of International Contractors Association.

Cide-Trak™   Biocide monitoring system by Azur Environmental.

CIP   See "clean-in-place (CIP)."

Cipolletti weir   A weir having a trapezoidal-shaped notch.

circle of influence   The circular outer edge of a depression produced in the water table by the pumping of water from a well. See also "cone of influence" and "cone of depression."

Circox®   High-rate aerobic reactor by CBI Walker, Inc.

CirculAire™   Aspirating aerator by Aeration Industries, Inc.

Circuline   Circular sludge collector product line by USFilter/Envirex.

Circumfed   Dissolved air flotation unit by Tenco Hydro, Inc.

cistern   A small covered tank for storing water, usually placed underground.

CITES   Convention on International Trade in Endangered Species.

citric acid   A crystalline acid present in citrus fruits. Chemical formula is $C_6H_8O_7 \cdot H_2O$.

CIWEM   The Chartered Institution of Water and Environmental Management. Also known as "IWEM."

CIX™   Ion exchange wastewater treatment/metals recovery system by Kinetico Engineered Systems, Inc.

Cl₂   See "chlorine."

CLAM®   Cleansimatic liquid analysis meter by Monitek Technologies, Inc.

Clar+Ion®   Cationic coagulants and flocculants by General Chemical Corp.

Claraetor   Circular clarifier with aeration compartment formerly offered by GL&V/Dorr-Oliver, Inc.

ClarAtor®   Clarifier technology by Waterlink/Aero-Mod Systems.

Claribloc®   Compact physical/chemical water treatment plant by USFilter/Krüger (North America) and OTV.

ClariCone™   Solids contact clarifier by CBI Walker, Inc.

clarification   Any process or combination of processes whose primary purpose is to reduce the concentration of suspended matter in a liquid.

clarifier   A quiescent tank used to remove suspended solids by gravity settling. Also called sedimentation or settling basins, they are usually equipped with a motor driven chain and flight or rake mechanism to collect settled sludge and move it to a final removal point.

Clariflo®   Water and wastewater treatment process technology by USFilter/Krüger (North America) and OTV.

Clari-Float®   Package wastewater treatment plant including dissolved air flotation by Tenco Hydro, Inc.

ClariFloc®   Polyelectrolyte used to enhance liquid/solid separation by Polydyne, Inc.

Clariflocculator   Combination clarifier and flocculator by GL&V/Dorr-Oliver, Inc.

ClariFlow   Upflow clarifier products by Walker Process Equipment.

Clarigester   Two-story tank combining clarification and digestion by GL&V/Dorr-Oliver, Inc.

Clarion®   Absorption media by Colloid Environmental Technologies Co.

Claripak   Upflow, inclined plate clarifier by USFilter/Aerator Products.

Clarisep™   Oily wastewater ultrafiltration system by Pall Corp.

ClariShear™   Floating sludge collector by Techniflo Systems.

ClariThickener™   Combination clarifier and thickener by Baker Process.

Clari-Trac®   Track-mounted siphon sludge removal system for rectangular clarifiers by F.B. Leopold Co., Inc.

Clari-Vac®   Floating bridge type siphon sludge removal unit for rectangular clarifiers by F.B. Leopold Co., Inc.

Clar-i-vator®   Solids contact clarifier by Smith & Loveless, Inc.

Clar-O-Floc™   Hopper-bottom clarifier by Alar Engineering Corp.

Clar-Vac   Induced and dissolved air flotation systems by Dontech, Inc.

classifier   A device used to separate constituents according to relative sizes or densities.

clathrate   A compound formed by the inclusion of molecules in cavities formed by crystal lattices.

clay   A fine-grained earthy material that is plastic when wet, rigid when dried, and vitrified when fired to high temperatures.

clay liner   A layer of low permeability soil added to the bottom and sides of an earthen basin for use as a disposal site or pond to limit infiltration to the underlying rock and soil strata.

ClaySorb   Granular, organically modified clay filtration medium for removing emulsified oil and grease by TurnKey Solutions, Inc.

Clean Air Act (CAA)   U.S. Federal law requiring the EPA to set air pollutant emission standards.

Clean Air Act Amendments (CAAA)   Amendments issued in 1990 to expand the EPA's enforcement powers and place restrictions on air emissions.

Clean Chemicals   High purity chemicals for laboratory use only by Hach Company.

clean fuels   Blends or substitutes for gasoline fuels, including compressed natural gas, methanol, ethanol, liquefied petroleum gas, and others.

Clean Shot   Pneumatic solids delivery system by USFilter/CPC.

Clean Squeeze™   Screenings washer and compactor by Schreiber Corp.

Clean Water Act (CWA)   1972 U.S. federal law regulating surface water discharges; updated in 1987.

Clean-A-Matic   Self-cleaning basket strainer by GA Industries, Inc.

clean-in-place (CIP)   A method of cleaning a filter medium or membrane to restore its performance without removing it from the system.

Cleantec   Grit recycling system by Brackett Geiger.

clear cutting   The practice of completely felling a stand of trees, usually followed by the replanting of a single species.

Clear View™   Continuous emissions monitoring system by Goal Line Environmental Technologies.

Clearcon   Circular clarifier product line by Vulcan Industries, Inc.

Clearflo   Cylindrical clarifier by Roberts Filter Group.

Clear-Flow™   Solids contact clarifier by Hi-Tech Environmental, Inc.

Clearigate   Algaecide by Applied Biochemists, Inc.

clearwell   A tank or reservoir of filtered water used to backwash a filter.

Climber®   Reciprocating rake bar screen by Infilco Degremont, Inc. (USA) and Brackett Geiger (Europe).

ClimbeRack™   Bar screen gear rack that eliminates need for lubrication by Infilco Degremont, Inc.

clinker   A fused byproduct of the combustion of coal or other solid fuels.

clino   See "clinoptilolite."

clinoptilolite   A naturally-occurring clay that can be used in an ion exchange process for ammonia removal.

Cloromat®   Sodium hypochlorite generator by Ionics, Inc.

ClorTec®   Hypochlorite-generating systems by ClorTec.

close-coupled pump   A pump coupled directly to a motor without gearing or belting.

closed cycle cooling system   A cooling water system in which heat is transferred by recirculating water contained within the system, producing a relatively small blowdown stream of concentrated solids.

Clostridium botulinum   Anaerobic microbe that causes botulism.

closure plan   Written plan to decommission and secure a hazardous waste management facility.

cloud   A mass of small water droplets in the atmosphere which is not of sufficient size to fall to the earth.

Cloud Chamber Scrubber™   Wet scrubber technology by Tri-Mer Corp.

cloud seeding   The artificial introduction of chemicals such as silver iodide or dry ice into clouds to induce rain.

CLP   Contract laboratory program.

CLR Process   Closed loop reactor oxidation ditch process by Lakeside Equipment Corp.

Cluster Rules   An industry-specific, integrated regulation that controls the release of both air and water pollutants.

CMA   Chemical Manufacturers Association.

CMF   Continuous microfiltration process.

CMF-S   Continuous microfiltration system by USFilter/Memcor.

CMP   Chemical mechanical polishing.

CO   Carbon monoxide.

CO_2   See "carbon dioxide."

Coagblender   Turbine type in-line and open channel mixers by USFilter/Aerator Products.

coagulant   A chemical added to initially destabilize, aggregate, and bind together colloids and emulsions to improve settleability, filterability, or drainability.

coagulant aid   A material that improves the effectiveness of a coagulant by forming larger or heavier particles, speeds the reactions, or permits reduced coagulant dosage. Often referred to as a "flocculant."

coagulation   The destabilization and initial aggregation of finely divided suspended solids by the addition of a polyelectrolyte or a biological process.

coal gasification   The conversion of solid coal to a gas mixture to be used as fuel.

coal pile runoff   Rainfall runoff from or through a coal storage pile.

coalesce   The merging of two droplets to form a single, larger droplet.

Coanda effect   The tendency of a liquid coming out of a nozzle or orifice to travel close to the wall contour even if the wall curves away from the jet's axis.

Coanda Tulip®   Clarifier inlet distributor by Hans Huber GmbH.

coarse bubble aeration   An aeration system that utilizes submerged diffusers which release relatively large bubbles.

coarse sand   Sand particles, usually larger than 0.5 mm.

coarse screen   A screening device, usually having openings greater than 6 mm (0.25″).

coastal reclamation   Reclaiming land from shallow coastal areas of the sea by dumping rubble and refuse or constructing breakwaters, sea walls, and drainage of the enclosed area.

Coastal Zone Management Act (CZMA)   Act requiring all federal agencies and permittees who conduct activities affecting a state's coastal zone to comply with an approved state coastal zone management program.

COC   See "cycles of concentration (COC)."

cocci   Sphere-shaped bacteria.

COD   See "chemical oxygen demand (COD)."

CODcr   Notation used for "chemical oxygen demand" as determined using potassium dichromate and sulfuric acid.

CodeLine™   Membrane pressure vessel housing by Advanced Structures, Inc.

codisposal   A method of sludge disposal where the sludge is mixed with sludges from different processes or with sorted refuse and incinerated, composted, or treated by pyrolysis prior to final disposal.

CODmn   Notation used for "chemical oxygen demand" as determined using permanganate.

coefficient   A numerical quantity, determined by experimental or analytical methods, interposed in a formula that expresses the relationship between two or more variables to include the effect of special conditions or to correct a theoretical relationship to one found by experiment or actual practice.

coefficient of haze (COH)   A measure of air visibility determined by the darkness of the stain remaining on white paper after it has been used to filter air.

Coex Seal™   Containment liner by National Seal Co.

cofferdam   A temporary dam, usually of sheet piling built to provide access to an area that is normally submerged.

Co-fire   Burning of two fuels in the same combustion unit, e.g., coal and natural gas or oil and coal.

Cog Rake   Reciprocating rake bar screen by USFilter/Headworks Products.

CogBridge   Traveling bridge sludge collector by Walker Process Equipment.

cogen   See "cogeneration."

cogeneration   A power system that simultaneously produces both electrical and thermal energy from the same source.

COH   See "coefficient of haze (COH)."

cohort   People assembled on the basis of a common characteristic and followed or traced over a period of time.

cohort study   An epidemiological study where population subgroups with a common exposure to a suspected disease-causing agent are studied over time to determine the risk of developing disease.

Coilfilter   Rotary vacuum belt filter by Komline-Sanderson Engineering Corp.

coke   The solid carbon residue resulting from the distillation of coal or petroleum.

coke oven   An industrial process which converts coal into coke, one of the basic materials used in blast furnaces for the conversion of iron ore into iron.

coke tray aerator   An aerator where water is sprayed or flows over coke-filled trays.

cold lime-soda softening   Lime-soda softening process of water treatment at ambient temperatures.

coliform bacteria   A group of rod-shaped bacteria living in the intestines of humans and other warm-blooded animals and shed in their fecal material, and whose presence in water indicates that the water has received contamination of an intestinal origin.

coliform index   A rating of the purity of water based on a count of fecal bacteria.

Colilert®   Reagent used to detect and identify coliforms and *E. coli* by IDEXX Laboratories, Inc.

coliphage   A bacterial virus that uses *E. coli* as its host cell.

ColiSure   Coliform presence/absence test medium by Millipore Corp.

Collectaire   Airlift activated sludge removal system formerly offered by USFilter/Envirex.

collection main   The public sewer to which a building service or individual system is connected.

collection system   In wastewater, a system of conduits, generally underground pipes, that receives and conveys sanitary wastewater and/or storm water. In water supply, a system of conduits or canals used to capture a water supply and convey it to a common point.

collector chain   Chain used to convey the scraper in a rectangular sludge collector.

Collision Scrubber™   Air pollution control scrubber by Monsanto Enviro-Chem Systems, Inc.

colloid   Suspended solid with a diameter less than one micron that can not be removed by sedimentation alone.

Colloidair Separator™   Open basin DAF by USFilterCorp.

colloidal   Resembling or made up of colloids.

colmatage   The reversible portion of flux decline in a membrane separation system.

colony forming units (CFU)   The number of bacteria present in a sample as determined in a laboratory plate count test where the number of visible bacteria colony units present are counted.

color   Water condition resulting from presence of colloidal material (see "apparent color") or organic matter (see "true color") measured by visual comparison with lab prepared standards.

color throw   The discharge of color to the effluent of a filter or ion exchange system.

color units (CU)   The unit used to report the color of water.

colorimeter   A photoelectric instrument used to measure the amount of light of a specific wavelength absorbed by a solution.

Color-Katch™   Flocculant/coagulant by Kem-Tron.

ColOX™   Fixed film aerobic bioreactor system by Tetra Process Technologies.

Combi®   Self-contained, prefabricated headworks system by Waterlink Separations, Inc.

Combi-Guard   Packaged screening unit by Andritz Sprout-Bauer S.A.

combination chain   Chain used in conveyor applications having cast block links with steel pins and connecting bars.

combined available chlorine   The concentration of chlorine combined with ammonia as chloramine, and still available to oxidize organic matter.

combined cycle generation   A gas turbine generator system where heat from turbine generator exhaust gases are recovered by a steam generating unit whose steam is used to drive a steam turbine generator.

combined sewer   A sewer used to receive sanitary wastewater, storm water, and surface water.

combined sewer overflow (CSO)   Flow from a combined sewer that exceeds the capacity of the sewer system and is discharged directly to a receiving water during certain rainfall conditions.

Combu-Changer®   Regenerative oxidizer to control VOC emissions by ABB Air Preheater, Inc.

combustible liquid   Any liquid having a flash point at or above 38°C (100°F) and below 93°C (200°F).

combustibles   Materials that can be ignited at a specific temperature in the presence of air to release heat.

combustion gases   The mixture of gases and vapors produced by burning.

combustion product   Substance produced during the burning or oxidation of a material.

Combustrol®   Fly ash conditioning treatment technology marketed by Wheelabrator Air Pollution Control, Inc.

Comet   Electrically driven rotary distributor for fixed film reactor by Simon-Hartley, Ltd.

commercial waste   Solid waste from non-manufacturing establishments such as office buildings, markets, restaurants, and stores.

commercial waste management facility   A treatment, storage, disposal, or transfer facility which accepts waste from a variety of sources, as compared to a private facility which normally manages a limited waste stream generated by its own operations.

commercial water use   Potable water use in commercial enterprise providing salable goods or services.

commingled recyclables   Mixed recyclables that are collected together.

comminute   To crush, grind, or pulverize something into minute particles.

comminutor   A circular screen with cutters that grinds large sewage solids into smaller particles.

Common Sense Initiative (CSI)   A U.S. EPA program that encourages industry-specific, rather than pollution-specific, environmental protection measures.

common wall construction   A construction technique where adjacent concrete basins share a common wall to reduce construction costs.

community water system (CWS)   A public water system serving at least 25 year-round residents or having 15 or more connections used by year-round residents.

Compact CDI®   Continuous deionization product by USFilter/Lowell.

Compact RO   Reverse osmosis product by USFilter/Lowell.

compaction   (1) The reduction of the bulk of solid waste by rolling and tamping. (2) The reduction in thickness of a filter medium or membrane as a result of pressure.

Compaplate®   Wide-gap welded heat exchanger for high fouling service by Alfa-Laval Separation, Inc.

compensated hardness   A calculated value based on total hardness, magnesium-to-calcium ratio, and sodium concentration used to correct for reductions in hardness removal capacity in zeolite exchange water softeners.

competitive inhibition   The situation that occurs when two compounds compete for the same enzyme, leading to interference in metabolism of one by the other.

Completaire   Package waste treatment plant with complete mix activated sludge formerly offered by USFilter/Envirex.

complete treatment   A method of treating water that consists of the addition of coagulant chemicals, flash mixing, coagulation-flocculation, sedimentation, and filtration. Also called conventional filtration.

completed test   The third and last step in the analysis of water and wastewater for the presence of fecal bacteria. Positive cultures from previous tests are inoculated and a gram stain is performed on isolated colonies.

CompleTreator   Package trickling filter waste treatment plant formerly offered by GL&V/Dorr-Oliver, Inc.

complexing   The formation of a complex compound.

Compliance Assurance Monitoring rule (CAM rule) The EPA Clear Air Act rule that lets a regulated industry select the means to demonstrate that its operation is within permit conditions.

compliance coal Any coal that emits less than 1.2 pounds of sulfur dioxide per million Btu when burned. Also known as "low sulfur coal."

compliance cycle The 9-year calendar cycle, beginning January 1, 1993, during which public water systems must monitor. Each cycle consists of three 3-year compliance periods.

Compliance Master™ Oil/water separator by Mercer International Inc.

compliance schedule A negotiated agreement between a pollution source and a government agency that specifies dates and procedures by which a source will reduce emissions and, thereby, comply with a regulation.

Compmaster™ Composting computer process control system by USFilter/CPC.

Component Clarifier Standard group of clarifier components and options that can be matched with an application's requirements by Baker Process.

composite sample A combination of individual samples of water or wastewater taken at preselected intervals to minimize the effect of variability of individual samples.

compost The end product of composting.

Compost-A-Matic Sludge composting system by Farmer Automatic of America, Inc.

composting Stabilization process relying on the aerobic decomposition of organic matter in sludge by bacteria and fungi.

compound A substance consisting of two or more independent elements that can only be separated by chemical reactions.

Compound 146 Polyurethane material used for sprocket tooth insert by USFilter/Rex & Link-Belt Products.

compression settling Phenomenon referring to sedimentation of particles in a concentrated suspension where further settling can occur only by compression of the existing structure of settled particles. Also called "type IV settling."

compressor A mechanical device used to increase the pressure of a gas or vapor.

concentration (1) The amount of a substance dissolved or suspended in a unit volume of solution. (2) The process of increasing the amount of a substance per unit volume of solution.

concentration factor A number indicating the number of times a solution may be concentrated.

concentration polarization A phenomenon in which solutes form a dense, polarized layer next to a membrane surface which eventually restricts flow through the membrane.

concentration ratio The ratio of the concentration of solids in a water system to those of the dilute makeup water added to the system.

Concord™ Catalytic incinerators by Catalytic Products International.

concrete A mixture of water, sand, stone, and a binder that hardens to a stone-like mass.

condensate Water obtained by evaporation and subsequent condensation.

condensate polishing Treatment of condensate water to achieve required purity.

condensation The change in state from vapor to liquid; the opposite of evaporation.

Condense-A-Hood   Air/odor collecting hoods by Bedminster Bioconversion Corp.

condenser   A heat transfer device used to cool steam and convert it from the vapor to liquid phase.

Conden-Sorb™   VOC abatement system by M&W Industries, Inc.

conditioning   Pretreatment of a wastewater or sludge, usually be means of chemicals, to facilitate removal of water in a subsequent thickening or dewatering process.

conductance   (1) A measure of a solution's electrical conductivity that is equal to the reciprocal of the electrical resistance. (2) A rapid method of estimating the dissolved-solids content of a water sample by determining the capacity of a water sample to carry an electrical current.

conduction   The transfer of heat from one body to another by direct contact.

conductivity   The ability of a substance to conduct electricity; directly related to the mineral content of water.

cone of depression   A depression in the water table that develops around a pumped well.

cone of influence   The depression, roughly conical in shape, produced in the water table by the pumping of water from a well.

Cone Screen   Internally fed rotary fine screen by Andritz-Ruthner, Inc. (Western Hemisphere) and USFilter/Contra-Shear.

Conesep™   External regeneration technology for ion exchange systems by Glegg Water Conditioning Co.

confidence interval   A computed interval with a specified probability that contains the estimated value of a parameter.

Configurator®   Computerized process design and equipment selection tool by USFilter/Rockford.

confined aquifer   An aquifer in which groundwater is confined under pressure which is significantly greater than atmospheric pressure.

confined space   A potentially hazardous space which has limited openings for entry and exit, unfavorable natural ventilation, and/or is not designed for continuous worker occupancy.

confirmed test   The second of three steps in the analysis of water and wastewater for the presence of fecal bacteria. Positive cultures from the first, "presumptive test," are inoculated and examined for fermentation after incubation. If fermentation is present, a third stage, the "completed test," is performed.

confluence   The point where the flow of streams or rivers meet.

confluent growth   A continuous bacterial growth covering the filtration area of a membrane filter, in which bacterial colonies are not discrete.

congeal   To thicken, jell, or solidify, usually by cooling or freezing.

congenital   Existing at or before birth, usually referring to an abnormal trait or disorder.

conjunctive use   The combined management of surface and groundwater supplies to provide a greater yield and/or more stable cost structure than operating two independent supplies.

connate water   Water trapped in sedimentary rocks during their formation, also known as "fossil water."

Conoscreen   Rotating disc microscreen by Nuove Energie.

consent decree   A binding agreement by two parties in a lawsuit which settles all questions raised in the case and does not require additional judicial action.

conservation   The careful and organized management and use of natural resources.

Conservation Reserve Enhancement Program (CREP)   The U.S. Department of Agriculture program which gives farmers financial incentives for instituting conservation practices on their land.

Consolidated Electric   Product group integrated within USFilter/Control Systems.

constant-rate filtration   Filter operation where flow through the filter is maintained at a constant rate by an adjustable effluent control valve.

constructed conveyance   Artificial waterways including ditches, culverts, flumes, canals, or natural waterbodies that are altered by humans.

consumptive waste   Water that returns to the atmosphere without beneficial use.

ContaClarifier™   Upflow buoyant media clarifier by Roberts Filter Group.

Contac-Pac   Circular steel contact aeration package waste treatment plant formerly offered by USFilter/Envirex.

contact condenser   A device in which steam is condensed through direct contact with a cooling liquid.

contact flocculation   A water treatment process in which coagulated water passes through a coarse media to enhance floc formation prior to filtration or sedimentation.

contact pesticide   A chemical that kills pests when it touches them instead of when it is ingested.

contact process   Wastewater treatment process where diffused air is bubbled over fixed media surfaces.

contact stabilization process   Modification of the activated sludge process where raw wastewater is aerated with activated sludge for a short time prior to solids removal and continued aeration in a stabilization tank. Also called "biosorption process."

contact time   The time in which a chemical is in contact with another reacting chemical or constituent.

Container Filter   Sedimentation basin by Flo Trend Systems, Inc.

contaminant   Any foreign component present in another substance.

contamination   The degradation of natural water, air, or soil quality resulting from human activity.

Cont-Flo™   Back cleaned reciprocating rake bar screen by John Meunier, Inc.

continental divide   A watershed boundary separating rivers flowing in one direction from those flowing in an opposite direction.

continental drift   A theory that continents shift their positions as a result of currents in the molten rocks of the earth's mantle.

Continental®   Water treatment products and systems by USFilter/Lowell.

contingency plan   A document setting out an organized, planned, and coordinated course of action to be followed in case of fire, explosion, or release of hazardous waste constituents which could threaten human health or the environment.

contingent valuation survey (CVS)   A survey technique for assigning value to injured natural resources based on respondents willingness to support various resources in monetary terms.

continuous discharge   A routine release to the environment that occurs without interruption, except for infrequent shutdowns for maintenance and process changes.

continuous emissions monitoring   The continuous measurement of pollutants emitted into the atmosphere from combustion or industrial processes.

continuous sample   A flow of water from a particular place in a plant to the location where samples are collected for testing; may be used to obtain grab or composite samples.

Continuous-Flo   Traveling bridge filter by USFilter/Zimpro.

contract labs   Laboratories under contract to the EPA that analyze samples taken from waste, soil, air, and water or carry out research projects.

contract operations   Private operation of municipal facilities such as water and wastewater treatment plants.

contracted weir   A rectangular notched weir having a crest width narrower than the channel within which it is installed.

Contraflo®   Solids contact reactor clarifier by USFilter/General Filter.

Contraflux®   Countercurrent-activated carbon adsorption unit by Graver Co.

contrail   A visible trail of cloud-like condensed water vapor often forming in the wake of an airplane flying in clear skies. Also called a "vapor trail."

Contra-Shear®   Screening equipment product line by Andritz-Ruthner, Inc. (Western Hemisphere) and USFilter/Contra-Shear.

Contreat®   Aerobic wastewater treatment package plant by EnviroSystems Supply.

control technique guidelines (CTG)   EPA documents issued to assist state and local pollution control authorities to achieve and maintain air quality standards for certain sources through reasonably available control technologies.

controlled reaction   A chemical reaction under temperature and pressure conditions maintained within safe limits to produce a desired product or process.

convection   The transfer of heat by a moving fluid such as air or water.

conventional systems   Systems that have been traditionally used to collect municipal wastewater in gravity sewers and convey it to a central primary or secondary treatment plant prior to discharge to surface waters.

conversion   See "recovery."

Convertofuser®   Wide band coarse bubble diffuser with fine sheath by USFilter/Diffused Air Products Group.

conveyance loss   Water loss in pipes, channels, conduits, or ditches by leakage or evaporation.

cooling pond   A pond where water is cooled by contact with air prior to reuse or discharge.

cooling tower   An open water recirculating device that uses fans or natural draft to draw or force ambient air through the device to cool warm water by direct contact.

cooling tower blowdown   A sidestream of water discharged from a cooling tower recirculation system to prevent scaling or precipitation of saturated salts or minerals.

cooling water   Water used, usually in a condenser, to reduce the temperature of liquids or gases.

COP™   Clarifier optimization program by WesTech Engineering, Inc.

Copa CSO Screen   CSO screen by Waste-Tech, Inc.

Copa Screen   Packaged screening and dewatering unit by Longwood Engineering Co, Ltd.

CopaClarifier   Secondary clarifiers with filter brushes by Copa Group.

Copa-NILL   Tipping bucket CSO tank flush system by Waste-Tech, Inc. (U.S.) and Copa Group (U.K.).

Copasacs   Fine screening sack by NSW Corp. (U.S.) and Copa Group (U.K.).

Copasocks   Sock-type screening sack by Copa Group.

Copatrawl   Sock-type screening sack by NSW Corp. (U.S.) and Copa Group (U.K.).

Copawash   Rotating boom wash system for stormwater tanks by Waste-Tech, Inc. (U.S.) and Copa Group (U.K.).

CopaWets   Chemical coagulation and flocculation process for wastewater treatment by Copa Group.

coping   The top or covering of an exterior masonry wall.

Coplastix®   Synthetic composite sluice gates and stop logs by Ashbrook Corp. (U.S.) and Simon-Hartley, Ltd. (U.K.).

copper sulfate   Chemical used for algae control, also called "blue vitriol." Chemical formula is $CuSO_4$.

copperas   Common name for ferrous sulfate heptahydrate, a common coagulant. Chemical formula is $FeSO_4 \cdot 7H_2O$.

copper-nickel   A copper alloy containing 10 to 30% nickel to increase resistance to corrosion and stress corrosion cracking.

coral   Calcium carbonate skeletal structures of the *Anthozoa* class of marine polyps often deposited in large masses forming reefs and atolls in tropical seas.

coral reef   A ridge composed primarily of coral lying at or near the surface of a tropical sea.

Core Separator   Particulate control device for gaseous emissions by LSR Technologies, Inc.

corner sweep   Scraper used to remove sludge from the corner of a square clarifier.

corona   (1) The layer of ionized gas surrounding the sun. (2) A sometimes visible electric discharge resulting from a partial electric breakdown in a gas.

corona discharge method (CD)   A method of producing ozone where air or oxygen is passed between two electodes and the subsequent application of a high voltage results in a corona discharge, producing ozone as a result of power dissipation.

Corosex®   Processed magnesia used in filters to neutralize acidity by Clack Corp.

Cor-Pak®   Catalytic oxidizer system by ABB Air Preheater, Inc.

corporate average fuel economy standard (CAFE)   A 1978 standard that enhanced the national fuel conservation effort, imposing a miles-per-gallon floor for motor vehicles.

corrosion   Attack on material through chemical or electrochemical reaction with surrounding medium.

corrosive   The characteristic of a chemical agent that reacts with the surface of a material causing it to deteriorate or wear away.

corrugated plate interceptor (CPI)   Oil separation device utilizing inclined corrugated plates to separate free nonemulsified oil and water based on their density difference.

Corten   High strength, low alloy steel with enhanced atmospheric corrosion resistance by US Steel Corp.

corundum   See "alumina."

Cosmos®   Suspended solids monitoring system by GLI International.

cost recovery   A legal process by which potentially responsible parties who contributed to contamination at a Superfund site can be required to reimburse the trust fund for money spent during any cleanup actions by the federal government.

Costar   Product line of by Corning, Inc.

Counter Current®   Aeration process using rotating diffusers suspended from a rotating bridge by Schreiber Corp.

coupon test   A method of determining the rate of corrosion or scale formation by placing metal strips, or coupons, of a known weight in a tank or pipe.

cover material   Soil or other suitable material used to cover solid wastes in a sanitary or secure landfill.

Covertite   Clear span wastewater treatment tank cover by Thermacon Enviro Systems, Inc.

CPC   Chemical protective clothing.

CPC Engineering   Former name of USFilter/CPC.

CPI   (1) See "corrugated plate interceptor (CPI)." (2) Chemical process industry.

CPM   Critical path method.

CPSC   Consumer Products Safety Commission.

CPVC   Chlorinated polyvinyl chloride. A chlorinated form of PVC that provides increased heat resistance.

cracking   A thermal process in which petroleum distillates or residues are broken down into products having lower boiling points and of altered chemical constitutions.

cradle-to-grave   Hazardous waste management concept that attempts to track hazardous waste from its generation point (cradle) to its ultimate disposal point (grave).

Crane®   Water treatment product line by Cochrane Inc.

Crawler   Bar screen by Vulcan Industries, Inc.

Credible Evidence rule (CE Rule)   The EPA Clear Air Act rule allowing industry, regulatory agencies, or the public to use any credible evidence or information to show compliance or non-compliance with emissions standards and limitations.

creek   (1) A stream, usually flowing into a river, or serving as a natural drainage course for a basin. (2) A narrow tidal inlet through a coastal marsh.

Crenothrix polyspora   A genus of filamentous bacteria that utilize iron in their metabolism and cause staining, plugging, and taste and odor problems in water systems. See also "iron bacteria."

CREP   See "Conservation Reserve Enhancement Program."

crest gate   A gate installed on the crest of a spillway or dam used to vary the discharge.

crevice corrosion   Localized corrosion in narrow crevices filled with liquid.

Crisafulli   Former name of SRS Crisafulli, Inc.

criteria pollutants   The major air pollutants, including carbon monoxide, hydrocarbons, lead, nitrogen dioxide, sulfur dioxide, ozone, and suspended particulates, for which the U.S. EPA has established ambient air quality standards.

critical flow   The rate of flow of a fluid equal to the speed of sound in that fluid.

critical pitting temperature   A value used to compare a material's resistance to pitting corrosion.

critical point   The combination of pressure and temperature at which point a gas and liquid become indistinguishable.

critical pressure   The minimum pressure necessary to liquefy a gas that is at critical temperature.

critical temperature   The temperature above which a gas cannot be liquefied solely by an increase in pressure.

Cromaglass   Batch treatment wastewater system by Cromaglass Corp.

cross collector   A mechanical sludge collector mechanism extending the width of one or more longitudinal sedimentation basins, and used to consolidate and convey accumulated sludge to a final removal point.

cross connection   A physical connection in a plumbing system through which a potable water supply could be contaminated.

Cross/Counteflo   Inclined plate clarifier by USFilter/Zimpro.

Cross-Flo®   Inclined static screen by Kason Corp.

crossflow filtration   Method of filtration where the feed stream flows parallel to the surface of the filter medium and only a portion of the feed passes through the filter.

Crossflow Fouling Index™   Membrane fouling test index by BetzDearborn-Argo District.

crosslinkage   The degree of bonding of a monomer or set of monomers to form an insoluble, three-dimensional resin matrix.

Crouzat™   Water treatment products and systems by USFilter Corp.

Crown®   Self-priming sewage pump product line by Crane Pumps & Systems.

Crown Press™   Sludge dewatering test device by Phipps & Bird.

CRP®   Continuous recirculation sludge mixing process for anaerobic digesters by USFilter/Envirex.

CRT   Cell residence time.

crude oil   Unrefined petroleum as produced from underground formations.

crumb rubber   Ground or shredded rubber produced by shredding used in automobile tires that can be recycled in asphalt-rubber or other products.

crypto   See "*Cryptosporidium.*"

cryptosporidiosis   Gastrointestinal disease caused by the ingestion of waterborne *Cryptosporidium parvum*, often resulting from drinking water contaminated by runoff from pastures or farmland.

Cryptosporidium   A protozoan parasite that can live in the intestines of humans and animals.

Cryptosporidium parvum   A species of *Cryptosporidium* known to infect humans.

crystal   A homogenous chemical substance that has a definite geometric shape with fixed angles between its faces, having distinct edges or faces.

Crystalactor   A device which softens water, removes phosphates, and recovers heavy metals from wastewater by DHV Water BV.

crystalline   Having a regular molecular structure evidenced by crystals.

crystallization   The process of forming crystals.

crystallizer   Common term for a forced circulation evaporator.

CSA   Canadian Standards Association.

CSF   Coagulation, sedimentation, filtration.

CSI   See "Common Sense Initiative (CSI)."

CSO   Combined sewer overflow.

CSP   Concrete surface profile.

CSPE   Chlorosulphonated polyethylene.

CSS™   Wet scrubber technology by Tri-Mer Corp.

CSTR   Completely stirred tank reactor.

CT value   See "C × T."

CU   See "color units (CU)."

cubic centimeter (cc)   A volume measurement equal in capacity to one milliliter (ml). One quart is approximately 946 cubic centimeters.

cubic meter (m³)   A volume measurement equal to 1000 liters or 264.2 gallons. One cubic meter of water weighs approximately 1 metric ton.

Cullar®   Activated carbon filter by Culligan International Corp.

cullet   Crushed glass.

Cullex   Softening resin by Culligan International Corp.

Cullsorb   Greensand filter by Culligan International Corp.

culm   Coal dust or anthracite tailings.

cultural eutrophication   Increasing rate at which water bodies "die" by pollution from human activities.

culture   A microbial growth developed by furnishing sufficient nutrients in a suitable environment.

culvert   An enclosed channel serving as a continuation of an open stream where a stream meets a roadway or other barrier.

cumulative exposure   The summation of exposures of an organism to a chemical over a period of time.

cup screen   A single entry, double exit drum screen.

cupric   Of, or containing copper.

cupric sulfate   Copper sulfate.

cupro-nickel   An alloy of copper containing up to 40% nickel.

curb stop   A water service shutoff valve located in a water service pipe near the curb and between the water main and the building.

curie (Ci)   A unit of radioactivity equal to 3.7×10^{10} disintegrations per second; 1 gram of radium has 1 Ci of radioactivity.

Currie Clarifier   Circular clarifier with aeration compartment formerly offered by GL&V/Dorr-Oliver, Inc.

curtain wall   An external wall that is not load bearing, usually refering to a wall that extends down below the surface of the water to prevent floating objects from entering a screen forebay.

cutaneous   Relating to the skin.

Cutrine®-Plus   Algaecide/herbicide by Applied Biochemists, Inc.

Cuver   Chemical used in detect waterborne copper by Hach Company.

CVOC   Chlorinated volatile organic compound.

CVS   See "contingent valuation survey (CVS)."

CWA   See "Clean Water Act (CWA)."

CWAs   Chemical warfare agents.

CWS   See "community water system (CWS)."

CW-SRF   Clean Water State Revolving Fund.

cwt   Hundredweight.

Cyanamer®   Scale inhibitors and dispersants by Cytec Industries, Inc.

cyanazine   A common and potentially carcinogenic herbicide sometimes found in drinking water.

cyanide   A compound containing a CN group, usually extremely poisonous, often used in electroplating and other chemical processes.

Cybreak™   Emulsion breaker by Cytec Industries, Inc.

Cycle-Let®   Wastewater treatment and recycling system by Zenon Environmental, Inc.

cycles of concentration (COC)   The ratio of the total dissolved solids concentration in a recirculating water system to the total dissolved solids concentration of the makeup water.

Cyclesorb®   Granular activated carbon adsorption system by Calgon Carbon Corp.

Cyclo®   Speed reducer and gearmotor product line by Sumitomo Machinery Corp.

Cyclo Blower   Air blower by Gardner Denver Blower Division.

Cyclo Grit Washer   Inclined screw type grit washer and dewatering unit by Baker Process.

Cyclo/Phram®   Rotary plunger metering pump by BIF.

CycloClean™   Hydrocyclone separator by Krebs Engineers.

Cyclofloc   Method for increasing clarifier rise rates by USFilter/Krüger.

Cy-Clo-Grit   Prefabricated cyclonic grit collector by Waste-Tech, Inc.

Cyclo-Hearth   Multiple hearth furnace by USFilter/Zimpro.

Cyclo-Jet™   Floating self-aspirating aerator by Waterlink Biological Systems.

Cyclone™   Coarse bubble diffuser by Aeromix Systems, Inc.

Cyclospora   A family of protozoan organisms that are thought to cause waterborne disease.

CycloSpray®   Fixed media filter by Lighthouse Separation Systems, Inc.

Cyclotherm   Sludge heat exchanger by USFilter/Envirex.

Cyclo-Treat™   Cyclone separator by USFilter/Envirex.

Cygnet   Rotary distributor for fixed film reactor by Simon-Hartley, Ltd.

cyst   A resting stage formed by some bacteria and protozoa in which the whole cell is surrounded by a protective layer.

cytotoxin   Any material toxic to cells.

CZMA   See "Coastal Zone Management Act (CZMA)."

D

D Tech™   Environmental field test kits by Strategic Diagnostics, Inc.

D/DBP   Disinfectant/disinfection by product.

D/DBP Rule (D/DBPR)   Proposed U.S. EPA rule to limit the maximum contaminant level of trihalomethanes.

D/T level   The dilution-to-threshold level of a water sample which indicates the number of dilutions of clean, odor-free air necessary to reduce an odor to a level that most people cannot detect.

D-20   Filter underdrain nozzle by Infilco Degremont, Inc.

Dac Floc   Polyelectrolyte used to enhance liquid/solid separation by Dacar Chemical Co.

DAF   See "dissolved air flotation (DAF)."

DAFT   Dissolved air floatation thickener.

daily cover   Cover material spread and compacted on the top and side slopes of compacted solid waste at the end of each day to control fire, moisture, and erosion and to assure an aesthetic appearance.

Dakota   Belt filter press by HydroCal, Inc.

DALR   Dry adiabatic lapse rate. See "adiabatic lapse rate."

dalton   A nominal unit of weight equal to that of a single hydrogen atom; 1×10^{-24} grams.

Dalton's Law of Partial Pressure   In a mixture of gases, each gas exerts pressure independently of the others and the pressure of each gas is proportional to the amount of that gas in the mixture.

Dangler   Fine bubble aeration system by MixAir Technologies, Inc.

darcy   A unit of measure used to indicate permeability, standardized by the American Petroleum Institute.

DataGator™   Sewer flow metering system by TN Technologies, Inc.

DataRAM™   Continuous measurement airborne particulates monitor by MIE, Inc.

Davco   Former name of USFilter/Davco.

Davis Process   Former name of USFilter/Davis Process.

Davis Water & Waste   Former name of USFilter/Davco.

Davis-D²™   Coarse bubble diffuser by USFilter/Davco.

Davis-OM™   Oval membrane air diffuser by USFilter/Davco.

Davy Bamag   Former name of Lurgi Bamag GmbH.

day tank   Tank used to store chemicals or diluted polymer solution for 24 hours or less.

dB   See "decibel (dB)."

dBA   See "A-scale sound level (dBA)."

DBC Plus®   Bacterial culture for wastewater treatment by Enviroflow Inc.

DBO   Design-build-operate. A method of project implementation where a contractor is responsible for a facility's design, construction, and long-term operation.

DBOO   Design-build-own-operate. A method of project implementation where a contractor is responsible for a facility's design, construction, ownership, and operation.

DBP   See "disinfection byproduct (DBP)."

DBP₀   Instantaneous disinfection byproduct concentration.

DBPFP   See "disinfection byproduct formation potential (DBPFP)."

DBPP   See "disinfection byproduct precursor (DBPP)."

DBS™   Clarifier and thickener drive units by DBS Manufacturing, Inc.

DC   Direct current.

D-Chlor™   Dechlorination system using sodium sulfite tablets by Exceltec International Corp.

DCI System Six™   Open channel UV system by Bailey-Fischer & Porter.

DCP   Dissolved concentration potential.

DCS   See "distributed control system (DCS)."

DDT   Dichlorodiphenyltrichloroethane. A chlorinated hydrocarbon insecticide banned in many countries because of its persistence in the environment and accumulation in the food chain.

DDW   Distilled deionized water.

DE   See "diatomaceous earth (DE)."

dead storage   The stored reservoir water remaining at a level below the reservoir's lowest discharge level, and therefore unavailable for use.

deadleg   The part of a pipeline in which fluid does not flow and may stagnate.

deaerator   Device used to remove dissolved gases from solution.

dealkalization   Any process that removes or reduces alkalinity of water.

dealkalizer   Ion exchange unit with strong anion bed used to reduce bicarbonate alkalinity.

DeAmine™   Odor control product by NuTech Environmental Corp.

deashing   See "demineralizing."

decant   Separation of a liquid from settled solids by pouring or drawing off the upper layer of liquid after the solids have settled.

decanter centrifuge   See "solid bowl centrifuge."

decarbonator   A device used to remove alkalinity from solution by conversion to CO_2 prior to air stripping.

decat water   Water that has had the hardness minerals removed by a water softener.

decay products   Degraded radioactive materials, often referred to as "daughters" or "progeny."

DeCelerating Flo   Gravity sand filter by CBI Walker, Inc.

dechlorination   The partial or complete reduction of residual chlorine by any chemical or physical process.

decibel (dB)   The unit for measuring sound pressure levels. It is the logarithm of the ratio of the sound's intensity to the intensity of the weakest audible sound.

declining-rate filtration   Filter operation where the rate of flow through the filter declines and the level of the liquid above the filter bed rises throughout the length of the filter run.

decontamination   The process of reducing or eliminating the presence of harmful substances, such as infectious agents, so as to reduce the likelihood of disease transmission from those substances.

Dee Fo®   Foam control additive by Ultra Additives, Inc.

deep bed filter   Granular media filter with a sand or anthracite filter bed that is deeper than a conventional filter, i.e., 0.9m (3 feet), and often up to 1.8m (6 feet) deep.

Deep Bubble™   Corrosion control aeration system by Lowry Aeration Systems, Inc.

Deep Draw   Airlift diffuser for lagoons by Wilfley Weber, Inc.

deep well injection   Technique where raw or treated wastes are discharged through a properly designed well into a geological stratum.

DeepAer   Aeration system with rectangular eductor tubes by Walker Process Equipment.

DeepBed   Granular media filter bed used as fixed film reactor by Tetra Process Technologies.

Defined Substrate Technology™   Reagent system designed to promote growth of target microbe by Environetics, Inc.

Deflectofuser®   Coarse bubble air diffuser by USFilter/Diffused Air Products Group.

deflocculating agent   A material added to a suspension to prevent settling.

defoliant   A chemical applied to plants which causes them to lose their leaves.

deforestation   The permanent clearing of forest land and its conversion to nonforest uses.

degasifier   Device used to remove dissolved gases from solution, usually by means of an air-stripping column.

degradation   The biological breakdown of organic substances.

dehydrate   The physical or chemical process where water in combination with other matter is removed.

DeHydro®   Vacuum assisted sludge drying bed by Infilco Degremont, Inc.

deinking   The process of removing ink from secondary fibers.

deionization (DI)   The process of removing ions from water, most commonly through an ion exchange process.

dekSPRAY   Cooling tower nozzles by Brentwood Industries, Inc.

Delaval Filter   Precoat condensate polishing filter by Idreco USA, Ltd.

deliquescent   The ability of a dry solid to absorb water from the air and soften or dissolve as a result.

delist   The use of a petition process to have a facility's toxic designation rescinded.

DelPAC   Polyaluminum coagulants by Delta Chemical Corp.

Delrin   High molecular weight acetal resin polymer material by E.I. Dupont De Nemours, Inc.

Delt™   Traveling water screen chain by USFilter/Rex & Link-Belt Products.

Delta™   Mixing technology by Calgon Carbon Corp.

delta   The flat alluvial area at the mouth of some rivers where an accumulation of river sediment is deposited in a sea or lake.

Delta G®   Parallel plate separator by Smith & Loveless, Inc.

delta P   Differential pressure.

delta T   Differential temperature.

DeltaFlow™   Water purification plant by Ecolochem, Inc.

Delta-Pak®   Mass transfer packing media by Delta Cooling Towers, Inc.

Deltapilot   Hydrostatic level measurement device by Endress+Hauser.

Delta-Stak®   Inclined plate clarifier by Baker Process.

Delumper®   Solids disintegrator and crusher products by Franklin Miller, Inc.

demin   Common abbreviation for "demineralizer."

demineralizing   The process of removing minerals from water, most commonly through an ion exchange process.

Demister®   Mist eliminator by Koch-Otto York.

dendrimer   Hyperbranch macromolecule to control inorganic water constituents to prevent crystal formation.

Denite®   Denitrification process using a granular media fixed film reactor bed by Tetra Process Technologies.

denitIR®   Biological denitrification system by Baker Process.

denitrification   Biological process in which nitrates are converted to nitrogen.

Denitri-Filt™   Biological denitrification filter by USFilter/Davco.

Densadeg®   Thickener-clarifier unit with lamella zone and sludge recirculation by Infilco Degremont, Inc.

Densator®   High density solids contact clarifier with primary and secondary mixing zones by Infilco Degremont, Inc.

dense, nonaqueous phase liquid (DNAPL)   A liquid that is immiscible in and denser than water.

density   The ratio of the mass of an object to its volume.

density current   A flow of water through a larger body of water which retains its unmixed identity due to a difference in density.

Densludge   Digestion system with primary thickening unit by GL&V/Dorr-Oliver, Inc.

Dentrol   Sludge density controller by Walker Process Equipment.

Deox/2000®   Dechlorination analyzer by USFilter/Wallace & Tiernan.

Deox®   Oxygen and dissolved gas removal technology by Ecolochem, Inc.

deoxyribonucleic acid (DNA)   The macromolecule that contains the hereditary material vital to reproduction.

De-Pac™   Disposal dewatering filter by PacTec, Inc.

Department of Energy (DOE)   U.S. federal agency responsible for research and development of energy technology.

Department of Transportation (DOT)   U.S. federal agency responsible for regulating transport of hazardous and nonhazardous materials.

depletion curve   In hydraulics, a graphical representation of water depletion from storage-stream channels, surface soil, and groundwater.

Deplution   Wastewater treatment products by JDV Equipment Corp.

Depolox®   Chlorine, pH, and fluoride analyzer USFilter/Wallace & Tiernan.

depth filtration   Filtration classification for filters where solids are removed within a granular media bed.

Depurator   Induced air flotation unit by Baker Hughes Process Systems.

dermal   Used or applied to the skin.

dermal exposure   Contact between a chemical and the skin.

dermal toxicity   The ability of a pesticide or toxic chemical to poison people or animals by contact with the skin. See "contact pesticide."

DES   A synthetic estrogen, diethylstilbestrol is used as a growth stimulant in food animals. Residues in meat are thought to be carcinogenic.

Desal™   Desalination equipment and systems product line by Osmonics Desal.

Desalator™   Reverse osmosis product line by USFilter/Bekox.

desalination   The process of removing dissolved salts from water.

Desalination Systems   Former name of Osmonics Desal.

desalinization   See "desalination."

desalting   See "desalination."

DeSander®   Hydrocyclone separator by Krebs Engineers.

desert   A region characterized by a climatic pattern where evaporation exceeds precipitation.

desertification   The process where the biological productivity of land is reduced, resulting in desert-like conditions.

desiccant   A substance capable of absorbing moisture, used as a drying agent.

design criteria   (1) Engineering guidelines specifying construction details and materials. (2) Objectives, results, or limits that must be met by a facility, structure, or process in performance of its intended functions.

design storm   The magnitude of a storm upon which the design of a storm water facility is based, usually expressed in terms of the probability of the occurrence over a period of years.

designated pollutant   Air pollutants including acid mist, total reduced sulfur, and fluorides which are neither criteria nor hazardous pollutants, as described in the Clean Air Act, but for which new source performance standards exist.

designated uses   Those water uses, including cold water fisheries, public water supply, and irrigation, identified in state water quality standards that must be achieved and maintained as required under the Clean Water Act.

designer bugs   Common term for microbes developed through biotechnology that can degrade specific toxic chemicals at their source in toxic waste dumps or in groundwater.

desilting basin   A settling basin used for sedimentation of silt, usually for storm-water flows.

desorption   The release of an adsorbed solute from an adsorbent.

destination facility   The facility to which regulated medical waste is shipped for treatment and destruction, incineration, and/or disposal.

destratification   Vertical mixing within a lake or reservoir to totally or partially eliminate separate layers of temperature, plant, or animal life.

destroyed medical waste   Regulated medical waste that has been ruined, torn apart, or mutilated through thermal treatment, melting, shredding, grinding, tearing, or breaking so that it is no longer generally recognized as medical waste, but has not yet been treated.

destruction and removal efficiency (DRE)   An expression of hazardous waste incinerator efficiency stated as the percentage of incoming principal organic hazardous components destroyed during incineration.

Destrux®   Heavy duty rotary knife cutter to granulate scrap material by Franklin Miller, Inc.

desulfurization   Removal of sulfur from fossil fuels to reduce pollution.

detectable leak rate   The smallest leak from a storage tank expressed in terms of gallons-or liters-per-hour that a test can reliably discern with a certain probability of detection or false alarm.

detention time   The theoretical time required to displace the contents of a tank or unit at a given rate of discharge.

detergent   Synthetic washing agent that helps to remove dirt and oil and may contain compounds which kill useful bacteria and encourage algae growth when present in wastewater that reaches receiving waters.

detoxification   Treatment to remove a toxic material.

Detritor   Grit removal unit with reciprocating raking mechanism by GL&V/Dorr-Oliver, Inc.

detritus   (1) Decaying organic matter such as root hairs, stems, and leaves usually found on the bottom of a water body. (2) Grit or fragments of rock or minerals.

detritus tank   Square tank grit chamber incorporating a revolving rake to scrape settled grit to a sump for removal.

Developure   Depth filter by Osmonics, Inc.

dew   Water droplets that form on cool surfaces following condensation of atmospheric water vapor.

dew point   The temperature to which air with a given concentration of water vapor must be cooled to result in condensation of the vapor.

Dewa   Water and wastewater treatment systems by DWT-Engineering Oy.

dewater   (1) To extract a portion of the water present in a sludge or slurry. (2) To drain or remove water from an enclosure.

dewatered sludge   The solids remaining after the free water has been removed from a wet sludge.

dewatering lagoon   A lagoon constructed with a sand and underdrain bottom.

dewatering table   See "belt thickener."

Dewplan   Former name of USFilter/Dewplan.

DF   Dilution factor.

DFR   Dynamic fixed film reactor by Schreiber Corp.

DFT   See "dry film thickness (DFT)."

DFT Classic®   Wound fiber filter cartridge by USFilter/Filtration & Separation.

DGF   Dissolved gas flotation. See "dissolved air flotation (DAF)."

DHC   See "dirt holding capacity (DHC)."

DHHS   Department of Health and Human Services.

DI   See "deionization."

dialysis   The separation of substances from solution on the basis of molecular size by diffusion through a semipermeable membrane.

Diamite Series™   Membrane-cleaning liquids for removal of organic and inorganic foulants by King Lee Technologies.

Diamond Gate   Screenings press by Andritz-Ruthner, Inc. (Western Hemisphere) and USFilter/Contra-Shear.

Diamond Seal™   Metering gate by Tetra Process Technologies.

diarrhea   Frequent and excessive discharging of the bowels, producing abnormally thin, watery feces, usually a symptom of gastrointestinal upset or infection.

diatom   A unicellular algae with a yellowish brown color and siliceous shell.

diatomaceous earth (DE)   Skeletal deposits of diatoms used as filter aids and a filter medium.

diatomaceous earth filter   Water treatment filter that uses a layer of diatomaceous earth as the filter medium.

diatomite   See "diatomaceous earth."

diazinon   A common organophosphate insecticide.

dibenzofurans   A group of highly toxic organic compounds.

DIC   Dissolved inorganic carbon.

dieldrin   A chlorinated hydrocarbon pesticide formerly used for moth-proofing and agricultural insect control, now banned in the U.S.

dielectric heater   A solid waste treatment heating system where an alternating electric field of high frequency is used to generate heat in nonconductive or dielectric materials.

Diemme   Membrane type filter press by Olds Filtration Engineering, Inc.

Diffusadome®   Coarse bubble air diffuser by Amwell, Inc.

Diffusair   Carbon dioxide diffuser system by Walker Process Equipment.

diffused aeration   The introduction of compressed air into water by means of submerged diffusers or nozzles.

diffuser   A porous plate or tube through which air, or another gas, is forced and divided into bubbles for diffusion in liquids.

Diffuserator   Carbon dioxide diffusion system by Walker Process Equipment.

diffusion   The movement of suspended or dissolved particles from a more concentrated to a less concentrated area.

diffusion dialysis   An ion exchange membrane process that separates ionic, non-ionic or colloidal species from solution based on differing diffusion rates, using concentration differentials across the membrane as the driving force.

diffusive air test   A high resolution test for detecting membrane integrity that measures the amount of air diffused through a wetted membrane.

Digesdahl   Digestion apparatus for lab sample preparation for crude protein and mineral analysis by Hach Co.

digester   A tank or vessel used for sludge digestion.

digestion   The biological oxidation of organic matter in sludge resulting in stabilization.

Digichem   Programmable titration analyzer by Ionics, Inc.

Digi-Flo®   Flow rate meter and totalizer by Blue-White Industries.

Dijbo   Hydraulically operated trash rake by Landustrie Sneek BV.

dike   An embankment or ridge of materials used to prevent the movement of liquid or sludges.

dilatant   Property of a liquid whose viscosity increases as agitation is increased.

diluent   Any liquid, solid, or gaseous material used for diluting or dissolving.

dilution   (1) Lowering the concentration of a solution by adding more solvent.
(2) The engineered mixing of discharged water with receiving water to lessen its immediate aesthetic and/or biochemical impact.

dilution factor   See "dilution ratio."

dilution ratio (1) The volumetric ratio of solvent to solute. (2) The ratio of the water volume in a stream to the waste volume introduced into the stream. The dilution ratio gives an indication of the capacity of a water body to assimilate a waste.

dimitic Lakes and reservoirs that freeze over and normally go through two stratification and mixing cycles a year.

Dimminutor® Open channel comminutor by Franklin Miller, Inc.

DIN Deutsche Industrie Normal.

dinocap A fungicide used primarily by apple growers to control summer diseases.

dinoseb A herbicide, also used as a fungicide and insecticide, that was banned by the EPA because it posed the risk of birth defects and sterility.

dioxin An aromatic halogenated hydrocarbon that is one of the most toxic compounds known.

Dioxytrol Centrifugal oxygenation system by Hazleton Environmental, Inc.

DIP® Dynamic inclined plane industrial skimmer by JBF Environmental Technology.

Dipair™ Static tube aerator by Infilco Degremont, Inc.

Diphonix™ Ion exchange resins by Eichrom Industries, Inc.

Di-Prime™ Automatic trash pump by Goodwin Pumps of America.

direct discharger A municipal or industrial facility which introduces pollution through a defined conveyance or system such as outlet pipes; a point source.

direct filtration Filtration process that does not include flocculation or sedimentation pretreatment.

direct reuse The beneficial use of reclaimed water with transfer from a reclamation plant to the reuse site.

Director™ Floating flow diversion baffle by Environetics, Inc.

DirecTube Eductor tube anaerobic gas mixing unit by Walker Process Equipment.

dirt holding capacity (DHC) The amount of contaminant fed to a filter on a weight basis that attains the filter's terminal differential pressure.

disc screen A screening device consisting of a circular disc fitted with wire mesh which rotates on a horizontal axis.

Discfuser® Coarse bubble air diffuser by USFilter/Diffused Air Products Group.

discharge The release of any pollutant, by any means, to the environment.

Discor Disc dryer by Andritz-Ruthner, Inc.

Discostrainer Fine screening device by Waterlink Separations, Inc.

Discotherm Thermal sludge processor by LIST, Inc.

Disc-Pak Cartridge filter by Alsop Engineering Co.

Discreen® Rotating disc screening device by Monoflo.

discrete particle settling Phenomenon referring to sedimentation of particles in a suspension of low solids concentration. Also called "type I settling."

Disc-Tube™ Membrane filtration system by Rochem Environmental, Inc.

Di-sep® Filtration systems by Smith & Loveless, Inc.

disinfectant A substance used for disinfection and in which disinfection has been accomplished.

disinfectant contact time The travel time, in minutes, for water to move from the point of disinfectant application to a point where the "residual disinfectant concentration" is measured. See also "C × T."

disinfection   The selective destruction of disease-causing microbes through the application of chemicals or energy.

disinfection byproduct (DBP)   A byproduct which occurs or is anticipated to occur from the addition of commonly used water treatment disinfectants including chlorine, chloramine, chlorine dioxide, and ozone.

disinfection byproduct formation potential (DBPFP)   An indirect measure of the amount of disinfection byproduct precursors in a sample.

disinfection byproduct precursor (DBPP)   A substance that can be converted into a disinfection byproduct during disinfection.

dispersant   An additive that prevents agglomeration of particulates or is used to break up concentrations of organic matter, such as oil spills.

dispersed suspended solids (DSS)   The suspended solids remaining in a sample's supernatant 30 minutes after settling.

dispersion model   A mathematical representation of the transport and diffusion processes that occur in the atmosphere.

dispersion rate   A diffusion parameter of gas plumes or stack effluents.

Disposable Waste Systems   Former name of JWC Environmental.

Dispos-A-Filter™   Metals testing filter capsule by Geotech/ORS Environmental Systems.

disposal   The discharge, deposit, injection, dumping, spilling, leaking, or placing of any liquid or solid waste on land or water so that it may enter the environment or be emitted into the air.

Disposorbs   Self-contained disposable adsorption systems by Calgon Carbon Corp.

dissociate   The process of ionization of an electrolyte or salt when dissolved in water, forming cations and anions.

dissolved air flotation (DAF)   The clarification of flocculated material by contact with minute bubbles causing the air/floc mass to be buoyed to the surface, leaving behind a clarified water. Use of a gas other than air is referred to as "dissolved gas flotation" or "DGF."

dissolved nitrogen flotation (DNF)   A variation of the DAF process where nitrogen, rather than air, is used to assist in the removal of suspended solids.

dissolved organic carbon (DOC)   The fraction of TOC that is dissolved in a water sample.

dissolved organic matter (DOM)   The fraction of organic matter in water that passes through a 0.45 micron filter.

dissolved oxygen (DO)   The oxygen dissolved in a liquid.

dissolved solids   Solids in solution that cannot be removed by filtration with a 0.45 micron filter. See "total dissolved solids (TDS)."

distill   See "distillation."

distillate   A liquid product condensed from vapor during distillation.

distillation   The process of boiling a liquid solution followed by condensation of the vapor for the purpose of separating the solute from the solution.

distributed control system (DCS)   A collection of modules, each having a specific function, interconnected to carry out an integrated data acquisition and control operation.

distributor See "rotary distributor."

diurnal (1) Occurring during a 24 hour period. (2) Happening during the day time, rather than the night time.

diurnal fluccation A daily fluctuation in flow or compostion that is of a similar pattern from one 24 hour period to another.

diversion chamber A chamber used to divert all or part of a flow to various outlets.

diversion dam A dam constructed to divert part of all of the water from a stream away from its main course.

diversity index A mathematical expression that depicts the diversity of a species in quantitative terms.

DLO™ Dynamic light obscuration technique used in a particle monitor by Chemtrac Systems, Inc.

DNA See "deoxyribonucleic acid (DNA)."

DNAPL See "dense, nonaqueous phase liquid (DNAPL)."

DNF See "dissolved nitrogen flotation (DNF)."

DNR Department of Natural Resources.

DO See "dissolved oxygen (DO)."

DOAS Differential Optical Absorption Spectrometry.

DOC See "dissolved organic carbon (DOC)."

doctor blade A scraping device used to remove or regulate the amount of material on a belt, roller, or other moving or rotating surface.

DOE See "Department of Energy (DOE)."

DOI U.S. Department of Interior.

Dokwed Back cleaned bar screen by USFilter/Hubert.

dolomite A natural mineral consisting of calcium carbonate and magnesium carbonate. Chemical formula is $CaMg(CO_3)_2$.

dolomitic lime Lime containing 30 to 50% magnesium oxide and 50 to 70% calcium oxide.

DOM See "dissolved organic matter (DOM)."

domestic wastewater Wastewater originating from sanitary conveniences in residential dwellings, office buildings and institutions. Also called "sanitary wastewater" and "domestic sewage."

domestic water use Water use within private homes.

Don-Press Screw-type solids/screenings press by Dontech, Inc.

DorrClone Grit classifier unit by GL&V/Dorr-Oliver, Inc.

Dorrco Brand name of GL&V/Dorr-Oliver, Inc. products.

dosage A specific quantity of a substance applied to a unit quantity of liquid to obtain a desired effect.

dose equivalent A measure of effective radiation dose that considers differences in biological effectiveness due to the radiation type and its distribution in the body.

dose-response relationship The relationship between the dose of a pollutant and its effect on a biological system.

dosfolat® A folic acid micronutrient solution to improve wastewater settling marketed by Bioprime, Ltd.

dosimeter An instrument used for measuring radiation exposure.

dosing siphon   A siphon that automatically discharges liquid onto a trickling filter bed or other wastewater treatment device.

dosing tank   A tank into which raw or partly treated wastewater is accumulated and held for subsequent discharge and treatment at a constant rate.

DOT   See "Department of Transportation (DOT)."

Double Dish   Pressure filter underdrain by USFilter/Warren.

Double Ditch   Oxidation ditch wastewater treatment process by USFilter/Krüger.

double salt   Any salt containing two different cations or anions in a crystalline matrix.

DoubleGuard™   Bar screen bar rack with two sets of offset bars by USFilter/Headworks Products.

double-suction pump   A centrifugal pump with suction pipes connected to the casing from both sides.

Dowex®   Ion exchange resin by Dow Chemical Co.

downcomer   A pipe directed downward.

downgradient   The direction that groundwater flows; similar to "downstream" for surface water.

downstream   Toward the mouth of a stream or river in the direction of flow.

dowsing   The scientifically unproven practice of locating groundwater or buried objects by walking above the area while holding a branch, wooden stick, or metal rods and noting where it appears to be drawn downward to a particular spot. Also called "witching."

DOX   Dissolved organic halogen.

DP   Differential pressure.

d-part®   Waste conditioner by Medina Products Bioremediation Division.

DPD   N,N-diethyl-p-phenylenediamine. An indicator which produces red gradations in the presence of chlorine residuals.

DPD-FAS   DPD-ferrous ammonium sulfate method of measuring all species of chlorine residuals in a water or wastewater.

dracunculiasis   A waterborne, parasitic disease caused by the *Dracunculus medinensis* worm, and the only known disease solely associated with unhealthy drinking water. Also called "guinea worm disease."

draft tube   A centrally located vertical tube used to promote mixing in a sludge digester or aeration basin.

Draft Tube Channel   Oxidation ditch process formerly offered by Lightnin.

drag tank   A rectangular sedimentation basin that uses a chain and flight collector mechanism to remove dense solids.

Drag-Star™   Rectangular clarifier sludge removal device utilizing a cable drive and scraper carriage by Smith & Loveless, Inc.

Draimad™   Solids dewatering system by Waterlink/Aero-Mod Systems.

drain tile   Short lengths of pipes laid in underground trenches to collect and carry away excess groundwater or to discharge wastewater into the ground.

drainage basin   The area of land that drains water, sediment, and dissolved materials to a common outlet at some point along a stream channel.

drainage water   Ground, surface, or stormwater collected by a drainage system and discharged into a natural waterway.

drainage well   A well drilled to carry excess water off agricultural fields. Because they act as a funnel from the surface to the groundwater below, drainage wells can contribute to groundwater contamination.

Drain-Dri®   Wet pit pump by Yeomans Chicago Corp.

Drainpac™   Storm drain filtration system by PacTec, Inc.

drawback   The reverse flow of water from the permeate side of a membrane to the feedwater or concentrate side as a result of osmosis.

drawdown   (1) The drop in the water table or level of water in the ground when water is being pumped from a well. (2) The amount of water used from a tank or reservoir. (3) The drop in the water level of a tank or reservoir.

DRE   See "destruction and removal efficiency."

dredge   To remove sediment or sludge from rivers or estuaries to maintain navigation channels.

dregs   Small solid particles in a liquid that settle to the bottom of a container.

Dresser/Jeffrey   Screening equipment product line acquired by Waste-Tech, Inc.

Drewcor®   Corrosion inhibitor chemical additive by Ashland Chemical, Drew Industrial.

Drewgard   Corrosion inhibitor by Ashland Chemical, Drew Division.

Drewpac®   Coagulant used in water and wastewater treatment by Ashland Chemical, Drew Industrial.

Drexelbrook   Former name of AMETEK Drexelbrook.

drift   (1) Water lost from a cooling tower as mist or droplets entrain in the circulating air. (2) Pollutants entrained in a plant's stack discharge.

drift barrier   An artificial barrier designed to catch driftwood or other floating material.

drift test   A part of an emissions certification process where the CEM must operate unattended for some period of time without the analyzers drifting out of calibration.

Driftor™   Drift eliminator by Kimre Inc.

drilling mud   A fluid, often containing bentonite, used to cool and lubricate a drilling bit and remove cuttings from the bit and carry them to the well's surface.

drinking water   Water safe for human consumption or which may be used in the preparation of food or beverages, or for cleaning articles used in the preparation of food or beverages.

Drinking Water Contaminant Candidate List (DWCCL)   A list of any chemical or microbial contaminant that is known or anticipated to occur in public water systems and which may be regulated in the future under the SDWA. The DWCCL replaces the Drinking Water Priorities List.

drinking water equivalent level (DWEL)   The lifetime exposure level at which adverse health effects are not anticipated to occur, assuming 100% exposure from drinking water.

Drinking Water Priority List (DWPL)   A 1988 list of drinking water contaminants which has been replaced by the "Drinking Water Contaminant Candidate List."

drip irrigation   A micro-irrigation water management technique used primarily for landscaping, where drips of water are emitted near the base of the plant.

drip proof   Designation for motor enclosure with ventilating openings so that drops of liquids or solids falling on the motor will not enter the unit directly or by running along an inwardly inclined surface.

drought   An extended period of dry weather which, as a minimum, can result in a partial crop failure or an inability to meet normal water demands.

drowned weir   See "submerged weir."

drum pulverizer   A rotating cylinder used to shred solid waste by the intermingling action of internal baffles acting on the wetted solid waste.

drum screen   A cylindrical screening device used to remove floatable and suspended solids from water or wastewater.

Drumm-Scale™   Drum weighing scale by Force Flow Equipment.

Drumshear   Rotating fine screen by Aer-O-Flo Environmental, Inc.

Drumstik   Feed tube for drum mounted chemical feed system by USFilter/Stranco.

dry bulb temperature   The air temperature measured by a conventional thermometer.

dry cleaning wastes   Wastewater from laundry cleaning operations that use nonaqueous chemical solvents to clean fabrics.

dry film thickness (DFT)   Thickness of a dried paint or coating usually expressed in mils.

dry ice   Solidified carbon dioxide frequently used as a refrigerant that vaporizes without passing through a liquid state.

dry weather flow (DWF)   The flow of wastewater in a sanitary sewer during dry weather. The sum of wastewater and dry weather infiltration.

dry well   (1) A dry compartment in a pumping station where pumps are located. (2) A well that produces no water.

Dry-All   Vacuum belt press by Baler Equipment Co.

DryCal®   Primary flow calibration instrument by BIOS International Corp.

drying bed   See "sludge drying bed."

DSS   See "dispersed suspended solids (DSS)."

DST™   Defined substrate technology. Reagent system designed to promote growth of target microbes by Environetics, Inc.

dual flow screen   A traveling water screen arranged in a channel so that water enters through both the ascending and descending wire mesh panels and exits through the center of the screen.

dual media filter   Granular media filter utilizing two types of filter media, usually silica sand and anthracite.

Dualator®   Gravity sand filter by Tonka Equipment Co.

dual-phase extraction   See "bioslurping."

Dubl-Safe   Measuring system for brine regeneration of softening unit by Culligan International Corp.

Duckbill®   Wastewater sampler by Markland Specialty Engineering Ltd.

duckweed   See "Lemnaceae."

duct   A tube or channel for gas or fluid flow.

dump   A site used to dispose of solid waste without environmental controls.

Dunkers   Flow balancing system by Munters.

Duo-Clarifier Wastewater treatment clarifier formerly offered by GL&V/Dorr-Oliver, Inc.

Duo-Deck Floating cover for anaerobic digester tank by USFilter/Envirex.

Duo-Filter Two-stage trickling filter formerly offered by GL&V/Dorr-Oliver, Inc.

Duo-Fine® Pleated media filter cartridge by USFilter/Filtration & Separation.

Duo-Flo Traveling mesh belt screen by Dontech, Inc.

Duo-Pilot® Coagulant control system using pilot filters by USFilter/Microfloc.

DuoReel Power cable reel system for clarifiers by Walker Process Equipment.

DuoSparj Coarse bubble diffuser by Walker Process Equipment.

DuoTherm® Biosolids production process by USFilter/Envirex.

Duo-Vac Inclined screen unit by the former Bowser-Briggs Filtration Co.

DuoVAL Automatically backwashed gravity sand filter by General Filter Co.

duplex pump A reciprocating pump having two side-by-side cylinders and connected to the same suction and discharge lines.

duplex stainless steel A high strength stainless steel containing two forms of iron, typically austenite and ferrite.

Duplex™ Air scrubber with two modules by USFilter/Davis Process.

Dupont® Brand name of E.I. DuPont De Nemours, Inc.

Dura-Disc Fine bubble membrane diffuser by Wilfley Weber, Inc.

Dura-Fuser™ Aeration piping system by USFilter/Davco.

Dura-Mix Mixer for aeration applications by Wilfley Weber, Inc.

Dura-Pac PVC sheet media by NSW Corp.

Dura-Pleat™ Fabric dust collector filters by Farr Co.

Durasan™ Reverse osmosis protective sleeve by Osmonics Desal.

Duratherm™ Reverse osmosis membranes for hot water sanitizing by Osmonics Desal.

Dura-Trac™ Sensor for streaming current transmitter by Chemtrac Systems, Inc.

Durco Filtration equipment by Aqua Care Systems.

Durex® Traveling water screen chain material by USFilter/Rex & Link-Belt Products.

Duriron Former name of Aqua Care Systems.

DuroFlow® Air blower by Gardner Denver Blower Division.

Durotex® Dewatering belts by Industrial Fabrics Corp.

Duroy Nonlubricated, metallic traveling water screen chain by USFilter/Rex & Link-Belt Products.

dust Fine-grained particles light enough to be suspended in air.

dust mask Face mask designed to provide relief from irritation of the nose, mouth, and throat caused by nontoxic dusts; not designed to be used when dust concentrations are greater than the PEL. Also called "nuisance mask."

dustfall jar An open container used to collect large particles from the air for measurement and analysis.

Dustube® Fabric filter by Wheelabrator Air Pollution Control, Inc.

DVB Divinylbenzene, the common crosslinking agent for making resin beads.

DVGW German Association for Gas and Water Plants.

DWCCL See "Drinking Water Contaminant Candidate List (DWCCL)."

DWEL See "drinking water equivalent level (DWEL)."

DWF See "dry weather flow (DWF)."

DWI   Drinking Water Inspectorate. Agency whose role is to enforce water quality/supply regulations for potable water which meets in England and Wales.

DWPL   See "Drinking Water Priority List (DWPL)."

DWS   Manufacturer of sewage grinders for JWC Environmental.

Dynabeads®   Anti-*Cyrptosporidium* by Dynal, Inc.

DynaBlend™   Polymer blending and feeding system by Fluid Dynamics, Inc.

DynaClear®   Packaged water treatment system incorporating contact filtration by Parkson Corp.

DynaCycle®   Regenerative catalytic oxidizer by Monsanto Enviro-Chem Systems, Inc.

DynaFloc Feedwell   Clarifier feedwell design to promote mixing of flocculants by GL&V/Dorr-Oliver, Inc.

Dyna-Grind   Screenings and sewage grinder formerly offered by USFilter/Headworks Products.

Dynalloy®   Filter cartridge with metal fiber media by USFilter/Filtration & Separation.

dynamic head   See "total dynamic head."

dynamic membrane   A transient membrane formed on the surface of an established membrane by solids filtered from the feed stream.

Dynamic Probe™   Filter control system by Roberts Filter Group.

Dynamic Systems   Product group integrated within USFilter/Control Systems.

DynaSand®   Continuously backwashed moving bed sand filter by Parkson Corp.

Dynasieve™   Self-cleaning rotary fine screen by Andritz-Ruthner, Inc.

Dynatherm   In-vessel composting system by Fairfield Service Co.

Dynatrol®   Measurement level switches and liquid measurement devices by Automation Products, Inc.

DynaWave®   Wet scrubbing system by Monsanto Enviro-Chem Systems, Inc.

DynaZyme™   Odor control biofilter system by Monsanto Enviro-Chem Systems, Inc.

dyne   A unit of measurement equal to the force that imparts an acceleration of 1 cm/s^2 to a 1 gram mass.

Dyneco   Former bar screen manufacturer acquired by Parkson Corp.

dysentery   A disease of the gastrointestinal tract usually resulting from poor sanitary conditions and transmitted by contaminated food or water.

Dystor   Gas holder system for anaerobic digesters by USFilter/Envirex.

dystrophic lakes   Acidic, shallow bodies of water that contain humus and/or other organic matter; contain many plants but few fish.

dystrophy   A disorder caused by defective nutrition or metabolism.

E

E. coli   See "*Escherichia coli (E. coli)*."

E.A. Aerotor   Packaged wastewater treatment plant by Lakeside Equipment Corp.

E/ONE Sewer™   Pressurized sewer system by Environment One Corp.

EA   Endangerment assessment; enforcement agreement; environmental action; environmental assessment; environmental audit.

EAD   Electro-acoustical dewatering.

EAF   Electric arc furnace.

EAFD   See "electric arc furnace dust (EAFD)."

EAG   Exposure assessment group.

EAP   Environmental action plan.

earthen dam   A dam made primarily of soil, sand, silt, and clay.

earthen reservoir   A reservoir constructed in earth using excavated materials to form embankments.

Earthtec®   Liquid copper water treatment product used as algicide/bactericide by Earth Science Laboratories, Inc.

easement   A legal right to the use of land owned by others.

EaseOut   Pivoting air header and drop pipe arrangement by Walker Process Equipment.

Eastern Econoline   Mixer product line by EMI, Inc.

EasyLogger   Stormwater data logger by Wescor, Inc.

EB   Emissions balancing.

ebb   (1) The flowing back of water brought in by the tide. (2) To recede from a flooded state.

ebb tide   Tide occurring at the ebb period of tidal flow.

EBCT   See "empty bed contact time (EBCT)."

e-beam   See "electron beam irradiation."

EBOD   See "effective BOD (EBOD)."

EC   Environment Canada; effective concentration.

E-Cell™   Modular electrodeionization process by E-Cell Corp.

ecesis   The successful establishment of a plant or animal in a new locality.

ECF   Elemental chlorine free.

Echo-Lock   Automatic compensating device for flow/level meters by Marsh-McBirney, Inc.

EchomaX®   Ultrasonic measurement transducer by Milltronics, Inc.

ECI   Environmental Conditioners, Inc., a former manufacturer of packaged wastewater treatment plants.

ECL   Electrochemiluminescence. A technology used for analyzing molecular and genetic material.

ECL   Environmental chemical laboratory.

Eclipse®   Dispersion polymers by Calgon Corp.

EcoCare™   Odor eliminator by Nature Plus, Inc.

Ecochoice   Catalytic oxidation system for dissolved organics by Eco Purification Systems USA, Inc.

Ecodenit   Biological nitrate removal process using ion exchange technology by USFilter/Krüger.

EcoDry   Sludge drying system by Andritz-Ruthner, Inc.

Ecodyne   Former name of Transunion product group now known as Graver Co.

ecoefficiency   Economically efficient use of resources by businesses.

Ecolo-Chief   Packaged wastewater treatment plants by Chief Industries, Inc.

***Eco-Logic*™**   Combination aeration and ultraviolet/ozonation system by Atlantic Ultraviolet Corp.

ecological indicator   A characteristic of the environment that, when measured, quantifies magnitude of stress, habitat characteristics, degree of exposure to stress, or ecological response to exposure.

ecology   The relationship of living things to one another and their environment.

Ecomachine   Belt filter press by WesTech Engineering, Inc.

***Econ-Abator*®**   Catalytic oxidation system by Huntington Environmental Systems, Inc.

Econex   Counterflow regeneration ion exchange by Ionics, Inc.

***Econ-NOx*™**   Selective catalytic reduction system by Huntington Environmental Systems, Inc.

***Econo Silo*™**   Enclosed composting system by USFilter/Davis Process.

***EconoBAY*™**   Composting system by USFilter/Davis Process.

Economixer   Solid bowl decanter centrifuge by Centrisys Corp.

economizer   A heat exchanger in a furnace stack that transfers heat from the stack gas to the boiler feedwater.

economy   In thermal desalination, the ratio of kilograms of distilled water produced per 2326 kJ of energy input.

economy-of-scale   The reduction of unit capital cost as the size of the unit increases.

Econopure   Reverse osmosis system by Osmonics, Inc.

EconoSep   Oil/water separator by Hydro-Flo Technologies, Inc.

***Econ-O-Star*®**   Batch wastewater treatment system by Alar Engineering Corp.

Econotreat   Package wastewater treatment plant by USFilter/Industrial Wastewater Systems.

Econ-O-Vap   Wastewater evaporation system by Fen-Tech Environmental, Inc.

Ecopure   VOC and particulate emission control product line by Dürr Environmental, Inc.

ecorock   A hard, dense rock produced from the ash of incinerated sludge and municipal solid waste suitable for use as a road aggregate.

***Ecosorb*®**   Odor control neutralizer by Odor Management, Inc.

ecosystem   The total community of living organisms together with their physical and chemical environment.

***EcoVap*™**   VOC control system by AMCEC, Inc.

ECRA   (1) Economic Cleanup Responsibility Act. (2) Environmental Cleanup Responsibilities Act.

ectotherm   An organism that cannot regulate its internal body temperature, and whose body temperature reflects the temperature of the environment. Fish and reptiles are ectotherms.

ED   (1) See "electrodialysis (ED)." (2) Effective dose.

EDAT   Environmental data acquisition telemetry.

EDB   See "ethylene dibromide (EDB)."

EDC   Endocrine-disruptive chemical.

eddy   A vortex-like motion of a fluid running contrary to a main current.

Eddyflow   High rate upflow clarifier by Gravity Flow Systems, Inc.

EDF   Environmental Defense Fund.

EDGE II™   Graded-cell catalyst by Alzeta Corp.

EDGE Plus™   VOC concentrator by Alzeta Corp.

EDGE QR™   Flameless VOC oxidizer by Alzeta Corp.

EDGE SB™   Thermal oxidizer by Alzeta Corp.

Edge Track   Drum filter cloth alignment mechanism by Baker Process.

EDI   Environmental Dynamics, Inc.

EDI   See "electrodeionization (EDI)."

EDL   Estimated detection limit.

EDR   See "electrodialysis reversal (EDR)."

EDS   European Desalination Society.

EDSTAC   Endocrine Distruptor Screening and Testing Advisory Committee.

EDTA   Ethylenediaminetetraacetic acid. A chelating agent.

Eductogrit   Aerated grit chamber by USFilter/Aerator Products.

Edur   Centrifugal pump for dissolved air flotation by Stanley Pump & Equipment Inc.

Edward & Jones   Filter press products marketed by USFilter/Asdor.

EDZ   Emission density zoning.

EEA   Energy and environmental analysis.

EECs   Estimated environmental concentrations.

EER   Excess emission report.

EERU   Environmental emergency response unit.

EESI   Environment and Energy Study Institute.

EESL   Environmental ecological and support laboratory.

EF   Emission factor.

effect   One of several units of an evaporator, each of which operates at successively lower pressures.

effective BOD (EBOD)   A measurement of wastewater strength which has been adjusted for elevated temperatures by increasing the standard BOD by 7% for each degree C rise above a temperature of 20°C.

effective rainfall   A rainfall that produces surface runoff.

effective size (ES)   A method of characterizing filter sand where the effective size is equal to the sieve size, in millimeters, which will pass 10%, by weight of the sand.

effective stack height   The sum of the plume rise and physical stack height at which particulates in a stack emission begin to settle to the ground after discharge.

Effizon®   Ozone production technology by PCI-Wedeco Environmental Technologies, Inc.

effluent   Partially or completely treated water or wastewater flowing out of a basin or treatment plant.

effluvium   Byproducts of food and chemical processes, usually in the form of wastes.

EFTC   European Fluorocarbon Technical Committee.

EG&G Biofiltration   Former name of AMETEK Rotron Biofiltration.

egestion   The act of discharging undigested or indigestible waste material from the body.

egg-shaped digester   Anaerobic digester characterized by an oval or egg shape which has been reported to promote efficient mixing.

EGL   See "energy grade line (EGL)."

EH   An expression of redox potential.

EHAP   Extremely hazardous air pollutant.

EHS   (1) Extremely hazardous substance. (2) Environmental health and safety.

EI   Emissions inventory.

EIA   Environmental impact assessment.

EIC   European International Contractors.

EIL   Environmental impairment liability.

Eimco®   Water and wastewater treatment product line by Baker Process.

EimcoBelt®   Continuous belt vacuum filter by Baker Process.

EimcoMet   Molded polypropylene components by Baker Process.

EIMIX®   Digester mixer by Baker Process.

EIR   Environmental impact report.

EIS   Environmental impact statement.

EIS/AS   Emissions inventory system/area source.

EIS/PS   Emissions inventory system/point source.

EJ   See "environmental justice (EJ)."

ejector   A device that uses steam, air, or water under pressure to move another fluid by developing suction through the use of a venturi. Sometimes referred to as an "eductor" or "jet pump."

EKA   See "electrokinetic analysis (EKA)."

Ekman water bottle   A tubular device used to sample water at selected depths.

EL   Exposure level.

El Niño   A climatic cycle resulting in warm, stormy weather in the Pacific caused by the warming of surface waters in the eastern Pacific Ocean.

El Niño/Southern Oscillation (ENSO)   The complete term for the "El Niño" climatic cycle which is thought to result from a "Southern Oscillation" circulation pattern in the atmosphere.

El Tor   A biotype of the *Vibrio cholerae* bacterium which takes its name from the El Tor quarantine camp in Sinai, Egypt where it was first isolated in 1906.

Elados®   Electronically controlled, electric motor-driven diaphragm pumps by PennProcess Technologies, Inc.

Elasti-Liner®   Containment liner material by KCC Corrosion Control Co.

elastomer   Synthetic material which is elastic or resilient and similar in structure, texture, and appearance to natural rubber.

Elastox®   Membrane air diffuser by Baker Process.

Elbac   Wastewater bioaugmentation product by Exceltec International Corp.

Elbow Rake   Hydraulically operated trash rake by the former Acme Engineering Co., Inc.

Elcat   Disinfection system product line by USFilter/Electrocatalytic.

Electraflote   Sludge thickener using electrolysis-generated bubbles formerly offered by Ashbrook Corp.

electric arc furnace dust (EAFD)   A byproduct of the production of steel using electric arc furnaces, usually containing heavy metals.

Electrocatalytic, Inc.   Former name of USFilter/Electrocatalytic.

electrochemical corrosion   Corrosion brought about by electrode reactions.

electrocoagulation   A wastewater treatment process where a direct current is used to precipitate heavy metals with ferrous hydroxides as metal hydroxides.

electrodeionization (EDI)   A water treatment process combining an electrodialysis membrane process with an ion exchange resin process to produce high purity, demineralized water.

electrodialysis (ED)   The separation of a solution's ionic components through the use of semipermeable, ion-selective membranes operating in a DC electric field.

electrodialysis reversal (EDR)   A variation of the electrodialysis process using electrode polarity reversal to automatically clean membrane surfaces.

electrokinetic analysis (EKA)   Method of measuring zeta potential at surface-liquid interfaces.

electrolysis   The passage of electric current through an electrolyte resulting in chemical changes caused by migration of positive ions toward the cathode and negative ions toward the anode.

electrolyte   A substance that dissociates into two or more ions when it dissolves in water.

Electromat   Electrodialysis equipment by Ionics, Inc.

Electromedia®   Processed mineral filter media by Filtronics, Inc.

electrometric titration   An acid or base titration where a pH meter is used for measuring end points.

electron beam irradiation   An oxidation process where compounds present in an aqueous solution are irradiated with high energy electrons produced by an electron accelerator to form water, carbon dioxide, and inorganic salts as end products. Also called an "e-beam."

electron microscope   A microscope that utilizes electromagnets as lenses and electrons instead of light rays to achieve a very high magnification.

electronic-grade water   Water used in the production of microelectronic devices which meets ASTM D-19 standards for resistivity, silica concentration, particle count, and other criteria.

electrophoresis   The movement of charged particles in a solution or suspension when an electric field is applied to them.

electrostatic precipitator (ESP)   Air cleaning system that imparts an electrical charge to airborne particles so that they can be removed by attraction to elements of opposite polarity.

electrotechnologies   Electrically-driven technologies used in waste source reduction, resource recovery, and end-of-pipe waste management and treatment.

elevated storage tank   A water storage reservoir supported by a column or tower.

Elf Atochem   Former name of Atochem North America, Inc.

Elf/Anvar   Oil/condensate coalescer-type oil/water separator by Graver Co.

ELI   Environmental Law Institute.

EloxMonitor™   On-line COD monitor by Anatel Corp.

Eltech   Former name of Exceltec International Corp.

eluant   A liquid used to extract one material from another.

elutriation   The process of washing sludge with water to remove organic and inorganic components to reduce chemical dosages required for additional treatment.

elutriator   An extension in an evaporator vapor body to thicken the solids slurry to minimize the loss of liquor.

EM   See "Enhanced Monitoring (EM)."

EMAS   Enforcement management and accountability system. Ecomanagement and auditing scheme.

EMC   Emission reduction credits.

emery   See "alumina."

Emery-Trailigaz   Ozone equipment supplier acquired by Praxair-Trailigaz Ozone Co.

Emerzone®   Ozone-generating systems by Praxair-Trailigaz Ozone Co.

EMI   Electromagnetic interference.

emission   Gas-borne particles or pollutants released into the atmosphere.

emission cap   A limit designed to prevent projected growth in emissions from existing and future stationary sources from eroding any mandated reduction.

emissions trading   U.S. EPA policy allowing a company to decrease pollution emissions from a facility while increasing levels at another as long as the total results are equal to, or less than the previous limits.

EMP   Electromagnetic pulse.

emphysema   A chronic, irreversible disorder of the lungs characterized by destruction of the alveoli and resulting in shortness of breath.

empty bed contact time (EBCT)   A measure of time that a water is in contact with the treatment medium calculated by dividing the empty volume in a contactor by the flow rate.

EMR   Environmental management report.

EMR™   Electrolytic metal recovery system by Kinetico Engineered Systems, Inc.

EMS   Environmental management system.

Emscher fountain   Another term for the "Imhoff tank," reflecting the part of Germany where this sewage treatment device was developed.

EMSL   Environmental monitoring support laboratory.

EMTS   (1) Environmental monitoring testing site. (2) Exposure monitoring test site.

emulsifying agent   An agent that aids in creating and maintaining an emulsion by altering the surface charge of droplets to prevent their coalescence.

emulsion   A heterogeneous mixture of two or more mutually insoluble liquids that would normally stratify according to their specific gravities.

emulsion breaker   A demulsifing agent that breaks an emulsion by neutralizing the surface charge of the emulsified droplets to allow their coalescence.

emulsion breaking   The use of heat, acids, oxidizing agents, or other chemicals to break an oil/water emulsion.

emulsion polymer   Dispersions of polymer particles in a hydrocarbon oil or light mineral oil used for conditioning sludge.

encapsulation   The complete enclosure of a waste in another material to isolate it from the external effects of air and water.

Encore®   Mechanical diaphragm metering pump by USFilter/Wallace & Tiernan.

encrustation   A covering, crust, or crust-like material on the surface of an object.

end point   The final state after a chemical or biological reaction is complete.

endangered species   Animals, birds, fish, plants, or other living organisms threatened with extinction by manmade or natural changes in their environment.

endemic   Restricted to a particular area or locality.

endocrine   Describing any gland that secretes hormones directly into the bloodstream rather than through a duct or directly into an organ.

endocrine disruptor   An external chemical substance or mixture that alters the structure or function of the endocrine system and causes adverse effects.

endogenous respiration   Bacterial growth phase where microbes metabolize their own protoplasm without replacement due to low concentrations of available food.

endospore   A bacterial spore formed within a cell and extremely resistant to heat and other harmful agents.

endotherm   An organism with the ability to maintain a constant body temperature. Mammals are endotherms.

endothermic   A process or reaction that takes place with absorption of heat.

endotoxin   A toxin, or poisonous substance, present in bacteria that is released during cell lysis.

endrin   a pesticide toxic to freshwater and marine aquatic life that produces adverse health effects in domestic water supplies.

Endura™   Wastewater treatment products by Aqua-Aerobic Systems Inc.

Endurance™   Conductivity and resistivity sensor by Rosemount Analytical, Inc.

Endurex   Coarse bubble diffuser by Parkson Corp.

Enelco   Former Environmental Elements Co. water treatment product line acquired by Infilco Degremont, Inc.

energy grade line (EGL)   The line joining the elevations of the energy heads.

Energy Mix   Rapid mix unit by Walker Process Equipment.

energy recovery   The retrieval of waste energy for some beneficial use.

energy recovery turbine (ERT)   A device used to recover pressure energy from reverse osmosis brine streams.

Engler   A scale for indicating viscosity.

enhanced coagulation   A modified coagulation process relying on the addition of excess coagulants to achieve increased removals of natural organic matter.

Enhanced Monitoring (EM)   Clean Air Act Amendment requirement for facilities to monitor emissions to certify compliance with permitted levels.

Enhanced Surface Water Treatment Rule (ESWTR)   An EPA drinking water regulation under development to include *Cryptosporidium* within the scope of the Surface Water Treatment Rule.

Enning ESD   Egg-shaped anaerobic digesters by CBI Walker, Inc. (U.S. licensee), Enning (German licensor).

enrichment   The addition of nutrients such as nitrogen, phosphorus, or carbon compounds from sewage effluent or agricultural runoff to surface water, which greatly increases the growth potential for algae and other aquatic plants.

ENRO™   Membrane elements by by Osmosis Technology, Inc.

ENSO   See "El Niño/Southern Oscillation."

EnSys   Immunoassay product line by Strategic Diagnostics, Inc.

enteric   Intestinal, or of intestinal, origin.

enteric bacteria   Bacteria that inhabit the gastrointestinal tract of warm-blooded animals.

Enterolert™   Reagent for *enterococci* detection by IDEXX Laboratories, Inc.

enterotoxin   A toxin or microbe that causes dysfunction in the human gastrointestinal tract.

enterovirus   A group of viruses including polio and hepatitis A viruses that replicates initially in the cells of the enteric tract.

Enterprise   Floating aspirating aerator by Air-O-Lator Corp.

Enterprise™   Regenerative thermal oxidizer by Megtec Systems, Inc.

enthalpy   The total heat content of a liquid, vapor, or body.

entrainment   (1) The incorporation of small organisms, including the eggs and larvae of fish and shellfish, into an intake system. (2) The carryover of droplets of water with vapor produced during evaporation.

entrainment separator   See "mist eliminator."

entropy   A measure of unavailable energy of an isolated thermodynamic system.

EnVessel Pasteurization™   Lime stabilization and pasteurization sludge treatment process to further reduce pathogens by RDP Technologies, Inc.

Enviro-Blend®   Specialty chemicals for heavy metal waste treatment by American Minerals, Inc.

EnviroDisc   Rotating biological contactor by Walker Process Equipment.

Envirofab   Former manufacturer of wastewater treatment equipment.

ENVIROFirst™   Sodium carbonate peroxyhydrate granules by Solvay America.

EnviroGard™   Chemical screening test kit by Strategic Diagnostics, Inc.

Enviromat   Wastewater treatment systems by Ionics, Inc.

environment   Water, air, and land and the interrelationship that exists among and between water, air, and land and all living things.

environmental audit   An independent assessment of the current status of a party's compliance with applicable environmental requirements or a party's environmental compliance policies, practices, and controls.

Environmental Elements   Equipment manufacturer whose Water Treatment Division product line was acquired by Infilco Degremont, Inc.

environmental impact assessment (EIA)   A method of analysis that attempts to predict probable repercussions of a proposed development on the social and physical environment of the surrounding area.

environmental impact statement   A detailed written report which identifies and analyzes the environmental impact of a proposed action.

environmental indicator   A measurement, statistic, or value that provides a proximate gauge or evidence of the effects of environmental management programs or the state or condition of the environment.

environmental justice (EJ)   The fair treatment and meaningful involvement of all people with respect to the development, implementation, and enforcement of environmental laws, regulations, and policies.

Environmental Protection Agency (EPA)   A U.S. agency with primary responsibility for enforcing federal environmental laws.

environmental racism   A belief that racism, whether intentional or unintentional, underlies variation in the distribution of environmental pollution.

Enviropac   Rotating biological contactor by Walker Process Equipment.

Enviropax   Tube settlers by Enviropax, Inc.

EnviroQuip International   Former name of USFilter/Diffused Air Products Group.

Enviro-Seal™   Valve packing system to prevent fugitive air emissions by Fisher Controls International, Inc.

Envirovalve   Telescopic valve by USFilter/Diffused Air Products Group.

enzyme   Organic catalysts that convert a substrate or nutrient to a form which can be transported into a cell.

EOP   End-of-pipe.

EOS   Predecessor company of USFilter/Operating Services.

EP   (1) Environmental profile. (2) European Pharmacopoeia.

EPA   See "Environmental Protection Agency (EPA)."

EPA200   The EPA 200 series of methods for chemical analysis (inorganic) of water and wastes.

EPA500   The EPA 500 series of methods for organic compounds in drinking water.

EPA600   The EPA 600 series of methods for analysis of pollutants under the Clean Water Act.

EPAA   Environmental Programs Assistance Act.

EPC   Engineer, procure, and construct.

EPCO™   Rotating biological contactors by USFilter Corp.

EPCRA   Emergency Planning and Community Right to Know Act.

ephemeral   Short-lived.

ephemeral streams   Streams in which no flow is common that only flow in response to precipitation.

EPI   Environmental Policy Institute.

EPIC™   Enhanced polymer control system by Norchem Industries.

epidemic   An outbreak of disease affecting many people at one time.

epidemiology   The study of the incidence, distribution, and control of disease in a population.

epilimnion   The upper layer in a stratified lake that results from varying water densities.

epm   See "equivalents per million (epm)."

EPNL   Effective perceived noise level.

EPRI   Electric Power Research Institute.

EPS   Eco Purification Systems USA, Inc.

Epsom salt   Hydrated magnesium sulfate having cathartic properties; also used in leather tanning and textile dyeing. Chemical formula is $MgSO_4 \cdot 7H_2O$.

EPTC   Extraction procedure toxicity characteristic.

EQ   See "equalization (EQ)."

EQL   See "estimated quantitation limit (EQL)."

equalization (EQ)   The process of dampening hydraulic or organic flow variations so that nearly constant conditions can be achieved.

equalization basin   A basin or tank used for flow equalization.

Equalizer®   Blower by Tuthill Pneumatics Group.

equator An imaginary circle around the earth that divides the earth's surface into the Northern and Southern Hemispheres and is equally distant at all points from the North and South Poles.

equivalent weight The weight of a compound that contains 1 gram atom of available hydrogen or its chemical equivalent, determined by dividing the molecular weight of a solute by the number of hydrogen or hydroxyl ions in the undissolved compound.

equivalents per million (epm) Ionic concentration determined by dividing an ion's concentration in ppm by its equivalent weight.

ER Energy recovery.

ERAMS Environmental radiation ambient monitoring system.

ERC (1) Emissions reduction credit. (2) Environmental Research Center.

ERDA Energy Research and Development Administration.

erg A unit of work or energy where 1 erg is equal to 10,000,000 joule.

ERL Environmental research laboratory.

Erlenmeyer flask A bell-shaped container used in laboratories to heat and mix chemicals.

ERNS Emergency response notification system.

erosion Wearing away of land by running water, waves, wind, or glacial activity.

erosion corrosion An attack on a material consisting of simultaneous erosion and corrosion through the effect of a rapidly flowing fluid.

ERT See "energy recovery turbine (ERT)."

ES See "effective size (ES)."

ES™ Electric straight-through thermal oxidizer by Thermatrix, Inc.

ESA Endangered Species Act.

escarpment A line of steep slopes or cliffs caused by erosion or faulting.

Escherichia coli (*E. coli*) Coliform bacteria of fecal origin used as an indicator organism in the determination of wastewater pollution.

ESD™ Egg-shaped anaerobic digesters by CBI Walker, Inc. (U.S. licensee), Enning (German licensor).

ESH Environmental safety and health.

ESP See "electrostatic precipitator (ESP)."

ESP® Sludge drying and pelletization system by Wheelabrator Water Technologies, Inc.

ESPA™ Low pressure polyamide membrane products by Hydranautics.

ESSD™ Engineered stainless steel design filter washtroughs by USFilter/General Filter.

ESSI EnviroSystems Supply.

esters Organic compounds produced by the reaction of an acid with an alcohol. Esters are often used in synthetic fragrances, industrial solvents, and the synthesis of plastics. Animal fats and vegetable oils are also esters.

estimated quantitation limit (EQL) The lowest concentration that can be reliably achieved within specified limits of precision and accuracy during routine laboratory operating conditions.

estuary A semi-enclosed coastal water body at the mouth of a river in which the river's current meets the ocean's tide, mixing fresh and salt water.

ESWTR   See "Enhanced Surface Water Treatment Rule (ESWTR)."

ESZ   Electrical sensing zone technology used in some particle counter sensors.

ET   (1) See "evapotranspiration (ET)." (2) Emissions trading.

ethanoic acid   See "acetic acid."

ethanol   An inflammable organic compound formed during the fermentation of sugars. Chemical formula is C_2H_5OH.

ethene   See "ethylene."

ethylene   A colorless, flammable, gas used in the manufacture of organic chemicals and plastics, and as a plant hormone to hasten the ripening of fruits. Chemical formula is $H_2C:CH_2$.

ethylene dibromide (EDB)   A toxic, carcinogenic chemical used as an agricultural fumigant and in certain industrial processes.

ethylene glycol   A colorless liquid used as an antifreeze and solvent. Chemical formula is $HOCH_2CH_2OH$.

etiologic agent   A viable microorganism, or its toxin, that causes, or may cause, human disease.

ETS   Environmental tobacco smoke.

E-Tube®   Wet electrostatic precipitator by Geoenergy International Corp.

ETV Program   Environmental Technology Verification Program by the U.S. EPA to identify promising, commercially ready pollution prevention technologies.

EU   Endotoxin units.

EUP   End-use product.

euphotic zone   The upper layer of water in a natural waterbody through which sunlight can penetrate.

Euroform   Mist eliminators by Munters.

eutectic   Easily melted.

eutrophic lake   A lake with an abundant supply of nutrients, excessive growth of floating algae, and an anaerobic hypolimnion.

eutrophication   Nutrient enrichment of water, causing excessive growth of aquatic plants and eventual deoxygenation of the water body.

Ev*ne/Scent™   Waterless vapor odor control system by Hinsilblon Laboratories.

Eva   Back-raked bar screen by Brackett Geiger.

Evap-O-Dry   Wastewater evaporation system by Fen-Tech Environmental, Inc.

evaporation   The process in which water is converted to a vapor that can be condensed.

evaporation pond   A natural or artificial pond used to convert solar energy to heat to accomplish evaporation.

evaporation rate   The mass quantity of water evaporated from a specified water surface per unit of time.

evaporator   A device used to heat water to create a phase change from the liquid to the vapor phase.

evaporimeter   A meteorological instrument used to measure natural rates of evaporation.

evaporite   A mineral produced as a result of evaporation.

evapotranspiration (ET)   Water withdrawn from the soil by evaporation and plant transpiration.

evapotranspiration treatment system   A wastewater treatment system utilizing surface evaporation and plant transpiration.

Evita®   Dissolved oxygen meter by Danfoss/Instrumark.

EVT   Belt filter press by Baker Process.

Eweson®   Compartmentalized rotary digester by Bedminster Bioconversion Corp.

EWPCA   European Water Pollution Control Association.

ex situ   Treatment or disposal methods that require movement of contaminated material.

Excel®   Cartridge filtration product line by Eden Equipment Co.

Excel®   High charge cationic flocculant by Cytec Industries, Inc.

excess lime-soda softening   The process of feeding excess lime and soda ash in addition to that required for lime-soda softening to further reduce hardness. Also called "railway softening."

exchange capacity   An ion exchanger unit's limited capacity for storage of ions.

excrement   Waste material from the bowels; feces.

excreta   Waste material excreted from the body.

excrete   To separate waste material from tissue or blood and eliminate it from the body.

excretion   The act or process of expelling waste material from the body in urine, feces, perspiration, or expired air.

excyst   To emerge from a cyst.

ExEx   Expected exceedance.

exfiltration   The wastewater leaking from breaks and cracks in a sewer line or manhole.

exhaustion   The condition that results when activated carbon, ion exchange resin, or other absorbents have depleted their capacity by using all available sites.

exothermic   A process or reaction which is accompanied by the evolution of heat.

Exotox®   Multi-gas detector by Neotronics of North America.

expanded metal   An open metal network produced by stamping or perforating sheet metal.

expansion joint   A joint installed in a structure to allow for thermal expansion or contraction.

explosion proof (XP)   Designation for a motor or electrical enclosure designed to withstand a gas or vapor explosion within the unit, and to prevent ignition of gas or vapor surrounding the unit by sparks, flashes, or explosions within the unit.

explosive limits   The amounts of vapor in the air that form explosive mixtures; limits are expressed as lower and upper and give the range of vapor concentrations in the air that will explode if an ignition source is present.

exposure   The amount of radiation or pollutant present in a given environment that represents a potential health threat to living organisms.

exposure assessment   Identifying the pathways by which toxicants may reach individuals, estimating how much of a chemical an individual is likely to be exposed to and the number likely to be exposed.

exposure indicator   A characteristic of the environment measured to provide evidence of the occurrence or magnitude of a response indicator's exposure to a chemical or biological stress.

ExpressClean™   Reverse osmosis cleaning service by Coster Engineering.

Expressor   Belt filter press by Baker Process.

extended aeration process   A variation of the activated sludge process with an increased detention time to allow endogenous respiration to occur.

Extendor   Detention tank for polymer mixing systems by Semblex, Inc.

extraction steam   Steam removed from a turbine at a pressure higher than the lowest pressure achieved in the turbine.

Extractor®   Horizontal belt press by Baker Process.

Extractoveyor   Composted sludge conveyor system by Fairfield Service Co.

Extreme Duty™   Mechanical digester sludge mixer by WesTech Engineering Inc.

extremely hazardous substances   Any of 406 chemicals identified by the EPA as toxic, and listed under SARA Title III. The list is subject to periodic revision.

Exxflow™   Crossflow microfiltration system by WesTech Engineering Inc.

E-Z™   Batch type centrifuge by Western States Machine Co.

E-Z Tray®   Air stripper by QED Environmental Systems, Inc.

EZ-BOD™   Sludge stability indicator by Bioscience, Inc.

F

F   See "Fahrenheit (F)."

F wastes   A class of hazardous wastes categorized from nonspecific sources as established by the EPA and identified in 40 CFR 261.31.

F.E.M.S.®   Relational database fugitive emissions management system by Enviro-Metrics.

F/M   See "food-to-microorganism ratio (F/M or F:M)."

fabric filter   A cloth device that catches dust particles from industrial emissions.

FAC   See "free available chlorine (FAC)."

face velocity   The linear velocity of air through a filter.

facultative bacteria   Microbes with the ability to survive with or without the presence of oxygen.

facultative lagoon   A lagoon or pond in which stabilization of wastewater occurs as a result of aerobic, anaerobic, and facultative bacteria.

facultative ponds   See "facultative lagoon."

Fahrenheit (F)   A temperature scale on which water's freezing point is 32° and boiling point is 212°.

Fairfield   In-vessel composting system by Fairfield Service Co.

falaj   System of surface or subsurface channeling of water fed by wells or springs to provide a community water supply (Arabic).

Falco   Catalytic oxidizer by Falmouth Products.

fall   A sudden change in the water surface elevation.

falling film evaporator   An evaporator with vertical heat transfer surfaces where liquor falling down the surfaces is heated by steam condensing on the other side of the surface.

fallout   Radioactive debris that settles to earth after a nuclear explosion.

Fallova™   Sewage shredder by ZMI/Portec Chemical Processing.

false positive   An erroneous test result indicating that something has a certain property when that property is not present.

famine   An acute food shortage in an area that leads to malnutrition and starvation.

Fan/Separator®   Fume scrubber by Tri-Mer Corp.

FAO   Food and Agriculture Organization.

Faraday's Law   A law stating that the amount of chemical change in an electrolysis process is proportional to the electrical charge passed.

Farm Gas   Former equipment manufacturer now part of Rosewater Engineering Ltd.

Fasflo   Floating oil skimmer by Vikoma International Ltd.

FAST®   Fixed activated sludge wastewater treatment plant by Scienco/FAST Systems.

Fast Flow®   Rotary strainer by Alar Engineering Corp.

Fastek™   Thin layer composite reverse osmosis membrane by Osmonics.

fathom   A marine unit of length equal to 6 feet (1.8 m).

fats   Triglyceride esters of fatty acids that are usually solid at room temperature.

fatty acid   Any of a class of lipids consisting of organic acids having the general formula R.COOH.

fauna   The animal life of an area or region.

Favair®   Dissolved air flotation product by USFilter/Warren.

FB®   Sodium percarbonate by Solvay America.

FBC   Fluidized bed combustion.

FBR   Fluid bed reactor.

FBS   Fine bar screen by Waste-Tech, Inc.

FBX   Fixed bed reactor.

FC   See "fecal coliform (FC)."

FC/FS   Ratio of fecal coliform to fecal streptococci that indicates if wastewater contamination results from animal or human wastes.

FDA   Food and Drug Administration.

FE   Fugitive emissions.

Fe³®   Liquid ferric sulfate by FE3, Inc.

fecal   Relating to feces.

fecal coliform (FC)   Coliforms present in the feces of warm-blooded animals, and considered indicators of fecal contamination of water.

Fecascrew   Screenings screw press by Hydropress Wallender & Co., AB.

Fecawash   Screenings washing and conveying unit by Hydropress Wallender & Co., AB.

feces   The excrement of humans and animals.

Federal Insecticide, Fungicide and Rodentcide Act (FIFRA)   1972 federal law requiring toxicity testing and registration of pesticides.

Federal Register   U.S. government daily publication detailing federal business including proposed and final rules and laws.

Federal Water Pollution Control Act (FWPCA)   The U.S. water control legislation of 1972 amended by the Clean Water Act of 1977.

feedlot wastes   Solid and liquid wastes from a facility where cattle or other animals are grown for market.

Feedpac   Coagulant feed system by Nalco Chemical Co.

feedwater heater   Heat exchangers in which boiler feedwater is preheated by steam extracted from a turbine.

feedwell   A circular, partitioned section within a clarifier or thickener that receives the feedwater and uniformly distributes it to the basin for sedimentation.

FEMA   Federal Emergency Management Agency.

fen   A wetland covered by shallow, stagnant, and usually alkaline water originating from groundwater sources.

fence-line concentration   Modeled or measured concentrations of air pollutants found at the boundaries of a property on which the pollution source is located.

Fenton's reagent   A mixture of iron and hydrogen peroxide used in oxidation, coagulation, or metal complexation reactions.

Feripac   Iron and manganese removal treatment plant by Vulcan Industries, Inc.

fermentation   The conversion of organic matter to carbon dioxide, methane, and other low molecular weight compounds.

Ferret™   Oil/water separating pump by QED Environmental Systems, Inc.

ferric   Relating to or containing iron in which the iron is trivalent or in a higher state of oxidation.

ferric chloride   An iron salt commonly used as a coagulant. The chemical formula is $FeCl_3$.

ferric sulfate   An iron salt commonly used as a coagulant. The chemical formula is Fe_2SO_4.

FerriClear®   Ferric sulfate product by Eaglebrook, Inc.

Ferri-Floc   Ferric sulfate by Boliden Intertrade, Inc.

Ferrosand®   Granular filter media used for removal of iron and manganese by Hungerford & Terry, Inc.

ferrous   Relating to or containing iron in which the iron is divalent or in a lower state of oxidation.

ferrous sulfate   An iron salt commonly used as a coagulant. The chemical formula is $FeSO_4$.

Ferrover   Reagent chemicals primarily used for iron analysis by Hach Co.

Ferrozine   Spectrophotometric reagents for iron and iron compounds by Hach Co.

Ferr-X   Iron removal process by Aquatrol Ferr-X Corp.

fertilizer   Materials usually containing nitrogen and phosphorous that are added to soil to provide essential nutrients for plant growth.

fetch   The distance traveled by a wind-generated wave before it reaches the coastline.

fetus   An animal embryo in the later stages of development; in humans from the beginning of the third month until birth.

FFFSG   Fossil-fuel-fired steam generator.

FGD   See "flue gas desulfurization (FGD)."

FGH   See "flue gas humidification (FGH)."

Fibercone Press   Cone-type dewatering press by Thermal Black Clawson.

FiberFlo™   Cartridge filters by Minntech Fibercor.

Fibrotex® Backwashable depth filter with steam pasteurization cycle by Smith & Loveless, Inc.

FID Flame ionization detector.

FIDIC Federation International des Ingenieurs-Conseils.

FIFRA See "Federal Insecticide, Fungicide and Rodentcide Act (FIFRA)."

filamentous growth Hair-like biological growth of some species of bacteria, algae, and fungi that results in poor sludge settling.

filamentous sludge Sludge characterized by excessive growth of filamentous bacteria that results in poor settling.

Filawound® Pressure vessel for reverse osmosis membranes by Spaulding Composites Co.

fill-and-draw Treatment process where a vessel is filled, the reaction occurs, and the contents are withdrawn. Typical of "sequencing batch reactors."

FilmShear Coarse and fine bubble diffusers by USFilter/Aerator Products.

Filmtec® Reverse osmosis membranes by Dow Chemical Co.

Filox® Iron, hydrogen sulfide, and manganese removal media by Matt-Son, Inc.

FiltaBand Continuous self-cleaning fine screen by Longwood Engineering Co., Ltd.

filter A device utilizing a granular material, woven cloth, or other medium to remove suspended solids from water, wastewater, or air.

Filter AG® Granular filter media used to adsorb tastes and odors by Clack Corp.

filter aid A polymer or other material added to improve the effectiveness of the filtration process.

filter area The effective area of a filter through which liquid passes, usually expressed in square meters (square feet).

filter backwash rate The volume of water per unit of time per unit of area required to flow backwards through a filter for cleaning.

filter bottom See "underdrain."

filter cake The layer of solids that is retained on the surface or upstream side of a filter.

Filter Cel® Diatomaceous earth filter media by Celite Corp.

filter cloth Cloth used as the filter media on a vacuum filter.

filter cycle The filter operating time between backwashes. Also called "filter run."

filter fly See "*Psychoda* flies."

filter gallery A passageway to provide access for installation and maintenance of underground filter pipes and valves.

filter loading, hydraulic The volume of liquid applied per unit area of the filter bed per unit of time.

filter loading, organic The pounds of BOD applied per unit area of the filter bed per unit of time.

filter press A dewatering device where water is forced from the sludge under high pressure.

filter ripening The gradual improvement in filter performance at the start of a filter run as solids deposition begins and serves to enhance filtration.

filter run See "filter cycle."

filter strip   Strip or area of vegetation used for removing sediment, organic matter, and other pollutants from runoff and wastewater.

Filterite®   Cartridge filter product line by USFilter/Filtration & Separation.

Filtermate   Filtration coagulant by BetzDearborn-Argo District.

FilterNet™   Net fabric composite by SLT North America, Inc.

Filterpak™   Plastic biological filter media by USFilter/General Filter.

Filter-Pak™   Gravity sand filter by Graver Co.

FilterSil®   Filtration sands by Unimin Corp.

filter-to-waste   An operational procedure in filtration where the filtrate is produced immediately after backwash is wasted.

Filterworx™   Automated filter control system by F.B. Leopold Co., Inc.

Filtomat®   Self-cleaning cooling water filters by Orival, Inc.

Filtra-Matic™   Pressure leaf filter by USFilter/Whittier.

FiltraPak   Packaged waste treatment and liquid-solids separation unit by Diagenex, Inc.

Filtrasorb®   Granular-activated carbon adsorption manufactured by Calgon Carbon Corp.

filtrate   Liquid remaining after removal of solids through filtration.

filtration rate   A measurement of the volume of water applied to a filter per unit of surface area in a stated period of time.

Filtroba®   Helical pressure filter element by Baker Process/Ketema.

Filtros®   Fine bubbles diffusers by Ferro Corp.

FIM   Friable insulation material.

Fimat   Electronic pulse control fish repelling system by Brackett Geiger.

final clarifier   See "secondary clarifier."

final closure   The closure of all hazardous waste management units at a facility in accordance with all applicable requirements.

final cover   Cover material that is applied upon closure of a landfill and is permanently exposed to the surface.

final effluent   The effluent from the final unit treatment process at a wastewater treatment plant.

fine bubble aeration   Method of diffused aeration using fine bubbles to take advantage of their high surface areas to increase oxygen transfer rates.

fine sand   Sand particles with diameters that usually range from 0.3 to 0.6 mm.

fine screen   A screening device usually having openings less than 6 mm.

FineAir   Ceramic fine bubble diffuser by Parkson Corp.

fines   Particles at the lower end of a range of particle sizes.

finished water   Water that has passed through a treatment plant in which all treatment processes are completed or "finished" and the water is ready to be delivered to consumers.

Fipro   Electronic pulse control fish repelling system by Brackett Geiger.

firetube boiler   Boiler in which the flame and hot gases are confined within tubes arranged in a bundle within a water drum.

first draw   The water that comes out when a tap is first opened, likely to have the highest level of lead contamination from plumbing materials.

first flush   Surface runoff at the beginning of a storm that often contains much of the pollution/solid matter washed from the streets and other surfaces.

first order reaction   A reaction in which the rate of change is directly proportional to the first power of the concentration of the reactant.

first-stage BOD   See "carbonaceous biochemical oxygen demand (CBOD)."

firth   A river estuary or wide inlet to the sea (Scottish).

Fischer & Porter   Former name of Bailey-Fischer & Porter.

fish ladder   A structure that permits fish to bypass a dam using a series of baffled chambers installed at progressively lower elevations to provide a velocity against which fish can more easily swim.

fish screen   (1) A screen at the head of an intake channel to prevent fish from entering. (2) A traveling water screen modified to remove impinged fish and return them to the water body.

fisheyes   A condition resulting from the improper mixing of dry polymer and water which results in the formation of lumps of undispersed polymer that resemble fisheyes.

fission   (1) The splitting of the nucleus of an atom into two or more nuclei with a concurrent release of energy. (2) A form of asexual reproduction where the parent organism splits into two independent organisms.

Fitch Feedwell   Clarifier feedwell with three horizontal chambers by GL&V/Dorr-Oliver, Inc.

FIX™   Heavy metals treatment system by ATA Technologies Corp.

fix, sample   A sample is "fixed" in the field by adding chemicals that prevent water quality indicators of interest in the sample from changing before laboratory measurements are made.

fixation   The stabilization or solidification of a waste material by involving it in the formation of a stable solid derivative.

fixed cover   Stationary anaerobic digester cover that allows a constant digester tank volume to be maintained.

fixed film process   Biological wastewater treatment process where the microbes responsible for conversion of the organic matter in wastewater are attached to an inert medium such as rock or plastic materials. Also called "attached growth process."

fixed matter   See "fixed suspended solids."

fixed solids   Inorganic content of suspended and dissolved solids in a water or wastewater sample, determined after heating the sample to 600°C.

fixed suspended solids   Inorganic content of suspended solids in a water or wastewater sample, determined after heating the sample to 600°C.

fixed-bed porosity   The ratio of void volume to total bed volume of a granular media filter.

fjord   A narrow inlet of the sea bordered by cliffs or highlands. Also called a "fiord."

flagellates   Microorganisms that move by the whipping action of a tail-like projection called a flagella.

flammable   Easily set on fire.

flammable liquid   Any liquid having a flash point below 38°C (100°F).

flange   A projecting rim or edge used for attachment with another object.

flap valve   A valve that is hinged on one edge and opens in the direction of normal flow and closes with flow reversal.

flare   A control device that burns hazardous materials to prevent their release into the environment; may operate continuously or intermittently, usually on top of a stack.

flash   The portion of a fluid converted to vapor when its pressure is reduced below the saturation pressure.

flash distillation   See "multistage flash evaporation (MSF)."

flash dryer   Sludge drying process that involves pulverizing sludge in a cage mill or by an atomized suspension technique in the presence of hot gases.

flash evaporator   A distillation device where saline water is vaporized in a vessel under vacuum through pressure reduction. See also "multistage flash evaporation."

flash mixing   Motor-driven stirring devices designed to disperse coagulants or other chemicals instantly, prior to flocculation.

flash point   The temperature at which a substance ignites.

flashing   The process of vaporizing a fluid by pressure reduction rather than temperature elevation.

Flashvap   Flash evaporator by Licon, Inc.

flashy   Characterization of rapidly changing conditions.

flavor profile analysis   An analysis profiling the matrix of odors in a water sample.

Fletcher   Filtration division of Edwards & Jones, Inc.

FlexAir™   Fine pore membrane diffuser by Environmental Dynamics Inc.

Flex-A-Tube®   Fine bubble aeration diffuser by Parkson Corp.

FlexDisc   Fine bubble membrane diffuser by USFilter/Diffused Air Products Group.

FlexDome   Fine bubble membrane diffuser by USFilter/Diffused Air Products Group.

Flexi Jet   Air sparging power mixer by USFilter/Aerator Products.

Flexi-Fabric   Filter press filter fabric by Baker Process.

Flexiflo   High speed surface aerator by USFilter/Aerator Products.

Flexi-Jet   Spray nozzles for sand filter rotary surface skimmer by F.B. Leopold Co., Inc.

Flex-i-liner®   Sealless self-priming rotary pumps by Vanton Pump & Equipment Corp.

Fleximix   High speed surface aerator by USFilter/Aerator Products.

Flexipak®   Submerged biofilm sewage treatment system by Gyulavari Consulting Kft.

Flexishaft™   Progressing cavity pump by MGI Pumps, Inc.

FlexKlear®   Inclined plate settler by Baker Process.

FlexKleen™   Filter underdrain nozzles by Baker Process.

FlexLine™   Nonbuoyant tubular diffuser by USFilter/Diffused Air Products Group.

Flexmate®   Skid-mounted water purification vessel by USFilter/Rockford.

Flexofuser®   Fine bubble diffuser with tube body and fine sheath by USFilter/Diffused Air Products Group.

Flexoplate™   Membrane type medium bubble diffuser by USFilter/Diffused Air Products Group.

Flex-O-Star®   Batch wastewater treatment and filtration system by Alar Engineering Corp.

FlexRO®   Skid-mounted reverse osmosis system by USFilter/Rockford.

FlexRO Mobile®   Mobile, skid-mounted reverse osmosis systems by USFilter/Rockford.

Flexscour®   Stainless steel filter underdrain for air-water backwash by Baker Process and Anthratech Western, Inc.

flight   (1) The horizontal scraper on a rectangular sludge collector. (2) The helical blade on a screw pump.

Flight Guide   Nonmetallic wear shoes for sludge collector flights by Trusty Cook, Inc.

Flint Rim   Cast sprocket with hardened chill rim by USFilter/Envirex.

FLM   Federal land manager.

float   The concentrated solids at the surface of a dissolved air flotation unit.

float switch   An electrical or pneumatic switch operated by a float in response to changing liquid levels.

floatables   See "flotables."

FloatAll   Dissolved air flotation system by Walker Process Equipment.

floating cover   Anaerobic digester tank cover free to move up or down to change the total internal capacity of the tank in response to sludge additions or withdrawals.

floating gas holder   Anaerobic digester tank cover that floats on a cushion of gas and moves up or down to change the total internal capacity of the tank in response to changes in gas volume.

Floating Plate®   Ozone generation technology by Pacific Ozone Technology, Inc.

Float-Treat®   Dissolved air flotation system by USFilter/Envirex.

Flo-Buster™   Organic solids agitator/separator by Enviro-Care Co.

floc   Small, gelatinous masses formed in water by adding a coagulant or in wastewater through biological activity.

Floc Barrier®   Inclined settling tubes for clarifiers by Graver Co.

Floc Tunnel®   Deep bed filter by Hazleton Environmental, Inc.

flocculant   An organic polyelectrolyte used alone or with metal salts to enhance floc formation and increase the strength of the floc structure.

flocculant settling   Phenomenon referring to sedimentation of particles in a dilute suspension as they coalesce or flocculate during sedimentation. Also called "type II settling."

flocculated suspended solids (FSS)   The suspended solids remaining in a sample's supernatant after 30 minutes of settling with flocculation.

flocculation   Gentle stirring or agitation to accelerate the agglomeration of particles to enhance sedimentation or flotation.

flocculator   A device used to enhance the formation of floc through gentle stirring or mixing.

Flocide®   Sanitizer additive by BioLab Inc.
Floclean   Membrane cleaning product by BioLab Inc.
Floclear   Solids contact clarifier by USFilter/Aerator Products.
Flocon®   Reverse osmosis feedwater additive by BioLab Inc.
Flo-Conveyor™   Screenings conveyor by Enviro-Care Co.
Flocor   Crossflow random fill media by the USFilter/Smogless.
Floc-Pac™   Compact flocculant packages by American Cyanamid Co.
Flocpress   Belt filter press formerly offered by Infilco Degremont, Inc.
Flocsettler   Combination flocculator/clarifier by Amwell, Inc.
Flocsillator   Horizontal oscillating flocculator by Baker Process.
FlocTreator   Coagulation and flocculation unit by the former PWT Americas.
Floctrol   Multiple stage, horizontal shaft, paddle flocculators by USFilter/Envirex.
Flo-Dar™   Noncontact area/velocity flowmeter by Marsh-McBirney, Inc.
Flodry®   Closed loop sludge drying process by USFilter/Smogless.
Flofilter™   Water treatment system combining flotation and filtration by Waterlink Separations, Inc.
FloGard™   Stormwater catch basin screen by KriStar Enterprises.
Flo-Lift™   Vertical screenings lift device by Enviro-Care Co.
Flo-Line   Fine mesh screening machine by Derrick Corp.
FloMag®   Magnesium-based granular, powder, and slurry water treatment products by Martin Marietta Specialties, Inc.
FloMaker™   Submersible mixer by ITT Flygt Corp.
Flo-Mate   Portable flow meter by Marsh-McBirney, Inc.
flood plain   The lowland area adjoining inland and coastal waters subject to a 1% or greater chance of flooding in a given year.
floor sweep   Capture of heavier-than-air gases that collect at floor level.
Flo-Poke   Portable flow rate instrument by Isco, Inc.
flora   Plants and plant life of a particular region or period.
Flo-Screen™   Reciprocating rake bar screen by Enviro-Care Co.
flotables   The floating material in a water or wastewater which must be removed prior to treatment or discharge.
flotation   A treatment process where gas bubbles are introduced to a water and attach to solid particles creating bubble-solid agglomerates that float to the surface where they are removed.
flotation thickening   Sludge thickening by means of dissolved air flotation.
Flotator™   Dissolved air flotation system by Baker Process.
Flotherm®   Sludge incineration process by USFilter/Smogless.
Flo-Tote   Computerized open-channel flow meter by Marsh-McBirney, Inc.
flotsam   Floating debris resulting from human activity.
Flottweg   Centrifuge product line by Krauss Maffei Corp.
flow control valve   A device that controls the rate of fluid flow.
flow equalization   Transient storage of wastewater for release to a sewer system or treatment process at a controlled rate to provide a reasonably uniform flow.
Flow Logger   Flow measuring device by Isco, Inc.
flow rate   The volume or mass of a gas, liquid, or solid material that passes some point in a stated period of time.

flow splitter   A chamber that divides incoming flow into two or more streams.

Flo-Ware   Data logging flow meter software by Marsh-McBirney, Inc.

FlowMAX™   Reverse osmosis system by USFilter.

Flowminutor™   Horizontal shaft comminutor by Enviro-Care Co.

FlowSorb™   Granular-activated carbon canisters by Calgon Carbon Corp.

Flowtrex   Pleated cartridge filter by Osmonics, Inc.

FLP   Flash point.

FLPMA   Federal Land Policy and Management Act.

flue   Any passage designed to carry combustion gases and entrained particulates.

flue gas   The gases and smoke released from an incinerator's chimney.

flue gas desulfurization (FGD)   The process of removing sulfur dioxide from exhaust gas, usually by means of a wet scrubbing process.

flue gas humidification (FGH)   A process to control SO_2 emissions by spraying a water/air mixture into flue gas.

FluePac™   Powdered activated carbon by Calgon Carbon Corp.

fluid   Any material or substance which flows or moves whether in a semisolid, liquid, sludge, or gaseous form or state.

Fluid Dynamics®   Engineered filter system by USFilter/Filtration & Separation.

Fluid Systems   Membrane products group of Koch Membrane Systems, Inc.

Fluidactor   Fluid bed sludge incinerator formerly offered by Walker Process Equipment.

fluidization   The upward flow of a gas or fluid through a granular bed at sufficient velocity to suspend the grains.

fluidized bed combustion   A method of burning particulate fuel, such as powdered coal, where the fuel is burned after injection into a rapidly moving gas stream.

fluidized bed furnace   An incinerator or furnace used to incinerate sludge by passing heated air upward through a fluidized sand bed.

Fluidizer-Minor   Flash dryer by Niro, Inc.

flume   A channel used to carry water.

Flump   Floating lagoon pump units by SRS Crisafulli, Inc.

Fluoretrack   Liquid water tracing dye by Formulabs, Inc.

fluoridation   The addition of fluoride to drinking water to aid in the prevention of tooth decay.

fluorimeter   An instrument used to measure the amount of fluorescent materials, dyes, or aromatic hydrocarbons in water.

fluorocarbons (FCs)   Any of a number of organic compounds analogous to hydrocarbons in which one or more hydrogen atoms are replaced by fluorine. FCs containing chlorine are called chlorofluorocarbons (CFCs).

fluorosis   A demineralization of tooth enamel caused by excessive ingestion of fluoride by children during the years of tooth calcification which is characterized by brown staining or mottling of the teeth.

fluorspar   Common source of commercially available fluoride compounds which may be used as a direct source of fluoride for water fluoridation.

Fluosolids   Fluid bed reactor for sludge disposal by GL&V/Dorr-Oliver, Inc.

flush tank   A tank in which water is stored for rapid release.

flush valve   A valve used to expel water and sediment from a pipeline.

Flush-Kleen ®   Nonclogging sewage ejector pump by Yeomans Chicago Corp.

fluvial   Produced by or found in a river or stream.

fluvial deposit   Sediment deposited by the actions of a river or stream.

flux   (1) Flowrate per unit area. (2) Heat transfer rate per unit area.

fly ash   The noncombustible particles in flue gas.

Flygt   Pump and mixer products by ITT Flygt Corp.

Flying Pig   Screenings compacting, dewatering, and conveying unit by Meurer Industries, Inc.

Flymitec ®   Submersible mixing concept by ITT Flygt Corp.

FMA   Free mineral acidity. See "mineral acidity."

FMC Process Additives   Former name of BioLab Inc.

FMC-MHS   Equipment product line acquired by USFilter/Envirex.

FmHA   Farmers Home Administration.

FML   Flexible membrane liner.

Foam Ban™   Foam control additive by Ultra Additives, Inc.

Foam-Abator   Foam abatement product by USFilter/Jet Tech.

Foamtex™   Foam control additive by Ultra Additives, Inc.

Foamtrol™   Foam control additive by Ultra Additives, Inc.

fodder crop   A crop grown principally for animal food.

FOG   Fats, oils, and grease.

fogging   The application of a pesticide by rapidly heating the liquid chemical so that it forms very fine droplets that resemble smoke or fog.

Folded Flow™   Enhanced flow scheme used in high rate water/wastewater treatment processes by USFilter/Envirex.

fomite   Objects that have been contaminated by pathogens from a diseased person and may serve in their transmission.

FONSI   Finding of no significant impact.

Font'n Aire ®   Floating decorative aerator/fountain by Air-O-Lator Corp.

food chain   A feeding hierarchy where food energy is passed from primary producers (plants) to primary consumers (herbivores) to secondary consumers (carnivores).

food chain crops   Crops grown for human consumption and for feed for animals whose products are used for human consumption.

food crop   Crops grown primarily for human consumption.

food poisoning   A gastrointestinal disorder caused by bacteria or their toxic products, occurring after consumption of contaminated food.

food waste   Organic residues generated by the handling, storage, preparation, cooking, and serving of food.

food web   The complex interconnected network of food chains in a community.

food-to-microorganism ratio (F/M or F:M)   The ratio of the influent BOD or COD to the volatile-suspended solids concentration in a wastewater treatment aeration tank.

foot-pound   Unit of measure of work performed by a force of 1 pound acting through a distance of 1 foot.

forage crop   Crops that can be used as feed by domestic animals by being grazed or cut for hay.

Forager™   Sponge containing a bonded polymer used to selectively remove heavy metals from water by Dynaphore, Inc.

force main   The pipeline through which flow is transported from a point of higher pressure to a point of lower pressure.

forced circulation evaporator   An evaporator in which circulation is maintained by pumping liquid through the heating element with relatively low evaporation per pass.

forced draft deaerator   Device to remove dissolved gases from solution by blowing an air stream through a packed column countercurrent to downflowing water.

forebay   A reservoir at the end of a pipeline or channel.

formaldehyde   A colorless, pungent, and irritating gas, CH_2O, used chiefly as a disinfectant and preservative and in synthesizing other compounds like resins.

formalin   A clear, aqueous solution of 37% formaldehyde and a small amount of methanol used as a sanitant.

FormAsorb™   Activated carbon product by Waterlink/Barnebey Sutcliffe.

formazin turbidity unit (FTU)   Unit of measure used in turbidity measurement based on a known chemical reaction that produces insoluble particulates of uniform size.

Fossil Filter™   Stormwater runoff filter by KriStar Enterprises.

fossil fuel   Natural gas, petroleum, coal, and any form of solid liquid or gaseous fuel derived from such materials for the purpose of creating useful heat.

fossil water   See "connate water."

Fotovap   Evaporator to treat photo processing lab wastewater by Licon, Inc.

fouling   Condition caused when bacterial growth, colloidal material, or scale forms a deposit on a filter, membrane, or heat transfer surface.

fouling factor   A design criteria used to allow for some variation of equipment performance due to fouling.

4-Beam™   Turbidity and suspended-solids measuring instrument by BTG, Inc.

4-log removal   The removal of 99.99% of a constituent.

four 9s DRE   Common term for an incinerator destruction and removal efficiency of 99.99%.

Fourier Transform Infrared Spectrometry (FTIR)   Continuous emissions monitoring system used to identify organic and inorganic compounds in liquids, solids, and gases.

Fox-Pac   Sanitation system for marine applications by Red Fox Environmental, Inc.

FP   Fine particulate.

FPA   (1) Federal Pesticide Act (FPA). (2) See "flavor profile analysis."

FPEIS   Fine Particulate Emissions Information System.

fpm   Feet per minute.

FPPA   Federal Pollution Prevention Act.

fps   Feet per second.

FQPA   Food Quality Protection Act.

FR   Federal Register.

fractionation   A distillation method used to separate a mixture of several volatile components of different boiling points in successive stages with each stage removing some proportion of one component.

Framco International   Product group of ABS Pumps, Inc.

frazil ice   Granular or spike-shaped ice crystals that form in supercooled water which is too turbulent to permit coagulation into sheet ice.

free available chlorine (FAC)   The amount of chlorine available as dissolved gas, hypochlorous acid, and hypochlorite ion that is not combined with ammonia or other compounds in the water.

free available chlorine residual   The concentration of residual chlorine remaining at the end of a specific contact time which is available as dissolved gas, hypochlorous acid, or hypochlorite ion, not combined with ammonia or in another less readily available form.

free carbon dioxide   The concentration of gaseous carbon dioxide present in water that is not combined in carbonates or bicarbonates.

free chlorine   See "free available chlorine residual."

free chlorine residual   See "free available chlorine residual."

free liquids   Liquids which readily separate from the solid portion of a waste under ambient temperature and conditions.

free mineral acidity (FMA)   See "mineral acidity."

free oil   Nonemulsified oil that separates from water, usually in 5 minutes or less.

free product   A regulated substance present as liquid not dissolved in water.

free radical   A chemical having an unpaired electron that makes it very reactive.

free settling   The settling of discrete, nonflocculant particles in a dilute suspension.

free water knockout (FWKO)   Gravity separation vessel used in an oil field to separate produced water from oil.

freeboard   The vertical distance between the normal maximum liquid level in a basin and the top of the basin that is provided so that waves and other liquid movements will not overflow the basin.

Free-Flow™   Ceramic diffuser plates for fine bubble aeration by USFilter/Davco.

Free-Slide   Wire mesh basket configuration for traveling water screens by USFilter/Rex & Link-Belt Products.

freeze concentration   A process involving the removal of heat from an aqueous mixture until one or more of the components crystallize.

freeze distillation   Production of distillate by freezing a saline solution and washing salts from the pure water crystals prior to melting.

Fre-Flo   Extruded cement underdrain for gravity filter by Infilco Degremont, Inc.

Freon®   Refrigerant compound by E.I. Dupont De Nemours Inc.

fresh water   Water that usually contains less than 1000 mg/L of dissolved solids.

freshet   A stream or current of fresh water that flows into the sea.

freshwater lens   An occurrence of fresh water found above salt water, usually found in coastal areas.

Freundlich isotherm   Graphical representation of data related to the removal of colloidal matter from water as a result of adsorption.

FREX   Freon extractable oil and grease.

friable   Material that may be easily crumbled, pulverized, or reduced to powder.

friction factor   A measure of the resistance to liquid flow that results from the wall roughness of a pipe or channel.

FrictionFlex®   Containment liner texturing process by SLT North America, Inc.

Fridgevap®   Package distillation/concentration system by Licon, Inc.

Frontloader   Reciprocating rake bar screen by Schreiber Corp.

Frontrunner   Reciprocating rake bar screen by Waste-Tech, Inc.

froth   A mass of bubbles in or on the surface of a liquid.

frothing   The formation of a thick layer of froth on an aeration basin.

FRP   Fiberglass reinforced plastic.

FRS   Formal reporting system.

fry   Juvenile fish.

FS   Feasibility study.

FSA   Food Security Act.

FSI®   Filter Specialists, Inc.

FSS   See "flocculated suspended solids (FSS)."

FTIR   See "Fourier Transform Infrared Spectrometry (FTIR)."

FTO   Flameless thermal oxidation.

FTU   See "formazin turbidity unit (FTU)."

Fuchs ATAD   Autothermal aerobic digestion system by USFilter/Krüger.

fuel economy standard   See "corporate average fuel economy standard (CAFE)."

fuel efficiency   The proportion of the energy released on combustion of a fuel that is converted into useful energy.

fugitive emission   Air pollutants emitted to the atmosphere other than those from chimneys, stacks, or vents.

fugitive source   Any source of emissions not controlled by an air pollution control device.

fuller's earth   A fine, clay-like substance also called "diatomaceous earth."

Full-Fit™   Membrane separator by Osmonics, Inc.

fulvic acid   Byproduct of decomposing organic matter that colors water and is a precursor of disinfection byproducts.

fume   Finely divided solids trapped in vapor in a gas stream.

fumigant   A pesticide vaporized to kill pests, commonly used in buildings and greenhouses.

fungi   Small, multicellular nonphotosynthetic organisms that feed on organic matter.

fungicide   A substance used to kill or inhibit the growth of fungi or molds.

furans   A family of toxic, chlorinated organic compounds present in minute amounts in the air emissions from hazardous waste incinerators.

FURS   Federal underground injection control reporting system.

fusion   An energy-producing nuclear reaction that results from combining nuclei of small atoms to form larger atoms.

Futura-Thane®   Potable water tank linings by Futura Coatings Inc.

future liability   Refers to potentially responsible parties' obligations to pay for additional response activities beyond those specified in the Record of Decision or Consent Decree.

Fuzzy Filter®   Upflow filter system by Schreiber Corp.

fuzzy logic   A process control system intended to replace a skilled human operator by using multi-level logic to adjust process operation based on a set of approximate, rather than exact, rules.

FWCA   Fish and Wildlife Coordination Act.

FWER   Final water effect ratio. The test procedure where a Water Effect Ratio test is repeated three times under different stream flow conditions. See also "water effect ratio (WER)."

FWKO   See "free water knockout (FWKO)."

FWPCA   See "Federal Water Pollution Control Act (FWPCA)."

FXB   Fixed bed reactor.

FY   Fiscal year.

G

g   See "gram (g)."

G value   See "velocity gradient (G value)."

G.T.M.™   Gas transfer membrane separation technology by Ecolochem, Inc.

GAAP   Generally accepted accounting principles.

gabion   A wire mesh container filled with rocks used to prevent soil erosion.

GAC   Granular-activated carbon.

GAC10   Granular-activated carbon with an empty bed contact time of 10 minutes and a carbon reactivation frequency of no more than 6 months.

GACT   Granular-activated carbon treatment.

gain output ratio (GOR)   A measure of evaporator performance which represents the ratio of mass flow of distillate to steam input.

gallon (U.S.)   A unit of volume for liquid substances equal to 231 cubic inches and approximately equivalent to 3.785 liters.

gallon, Imperial   A unit of volume for liquid substances approximately equal to 1.2 gallons (U.S.) or 4.546 liters.

gallons per flush   The number of gallons used to flush a toilet.

galvanic corrosion   Corrosion that occurs when two dissimilar metals are connected electrically and immersed in a conductive liquid.

galvanic series   The ranking of the relative nobility of different conducting materials in a certain environment.

galvanize   An electrolytic or hot dipping process to coat steel products with zinc to increase corrosion resistance.

gamma ray   A short wavelength, high energy form of electromagnetic radiation.

garbage   Solid wastes generated by the handling, storage, preparation, cooking, and serving of food.

GARD   Gravity-activated rotary trickling filter distributor by USFilter/General Filter.

Gardiner Equipment   Chemical induction system product line by USFilter/Stranco.

Gar-Dur   Ultra high molecular weight plastic products for chain and flight sludge collectors by Garland Manufacturing Co.

garnet   A dense mineral often used as media in a granular media filter.

Garra   Membrane bioreactor by USFilter/Memcor.

gas   One of the three states of matter having no fixed shape or volume and capable of expanding indefinitely.

gas blanket   The use of an inert gas in an enclosed tank's vapor space to minimize vapor formation or keep out unwanted air and moisture.

gas chlorination   The application of chlorine gas to water.

gas chromatography (GC)   An analytical technique used to determine the molecular composition and concentrations of various chemicals in water and soil samples.

gas tight   Operating with no detectable emissions.

gaseous emission   Volatile or uncondensed compounds discharged into the atmosphere.

gasification   See "coal gasification."

GasLifter   Anaerobic digester circulation and mixing system by Walker Process Equipment.

gasohol   A motor vehicle fuel containing 80-90% unleaded gasoline and 10-20% ethanol.

gasoline oxygenate   Combustible liquids containing oxygen that are added to gasoline to reduce atmospheric pollutants.

gastroenteritis   An inflammation of the stomach and intestinal tract that is usually accompanied by symptoms that include diarrhea, fever, and vomiting.

gastrointestinal   Related to the stomach or intestines.

gate valve   A valve with a disk-shaped closing element that slides over the opening through which water flows.

GBT   See "gravity belt thickener (GBT)."

GC   See "gas chromatography (GC)."

GC/MS   Gas chromatography/mass spectrometry.

GCL   See "geosynthetic clay liner."

GDT   Gas dissolution technology.

GDT Process™   Gas-degas treatment process for VOC removal by GDT Corp.

gear pump   A positive displacement pump where cavities created between the gear teeth of two meshing gears move from the suction to discharge side of the pump.

GEHO®   Piston-type pump for heavy sludges by Envirotech Pumpsystems.

Geiger counter   An instrument used to detect radiation.

gel polymer   High molecular weight monomers produced with gamma radiation.

gel zeolite   A synthetic sodium alumiosilicate ion exchanger.

Gelex   Standards used to standardize turbidimeters by Hach Co.

GelGester™   Bacterial additive for grease and odor applications by NuTech Environmental Corp.

Gemco   Spent liquor filter by Gauld Equipment Sales Co.

Gemini   Self-cleaning basket strainer by S.P. Kinney Engineers, Inc.

Gemini™   Granular-activated carbon contactors by Roberts Filter Group.

Gemini Polymaster   Emulsion and solution polymer blending equipment by Neptune Chemical Pump Co.

GEMS   Global environmental monitoring system.

Gen2®   Chemical feed system formerly offered by USFilter/Stranco.

General Filter   Former name of USFilter/General Filter.

generator   (1) Any person, group, or organization whose activities produce hazardous waste. (2) A rotating device used to produce electrical power.

Generon®   On-site gaseous nitrogen generator by MG Generon.
Generox®   On-site gaseous chlorine dioxide generator by MG Generon.
Genesis   Package gravity filter system by Roberts Filter Group.
Genesis   Sewage shredder by ZMI/Portec Chemical Processing.
genome   All the genetic material in an organism.
genotoxic   The ability of a chemical to cause adverse effects in the genetic material of a living organism.
Gen-Ozi   Ozone generator by Matheson Gas Products.
GeoCarb 6   Engineered granular carbon media by Geoenergy International Corp.
GeoCat™   Recuperative catalytic oxidizer by Geoenergy International Corp.
geodetic head   The total head without deducting velocity head or other losses.
geographic information system (GIS)   A computer system designed for storing, manipulating, analyzing, and displaying data in a geographic context.
Geoguard®   Groundwater sampling product by American Sigma, Inc.
geosmin   A trace organic chemical produced by certain algae which may impart an earthy/musty odor in surface waters. Chemical formula is $C_{12}H_{22}O$.
geosynthetic clay liner (GCL)   A landfill liner manufactured of bentonite clay supported by geotexiles.
Geothane®   Wastewater containment lining by Futura Coatings, Inc.
GeoTherm™   Regenerative thermal oxidizer by Geoenergy International Corp.
geothermal   Energy produced by the transfer of heat from the earth's interior and conducted to the surface by hot water or steam.
GEP   Good engineering practice.
germ   A disease-producing microbe.
germ cell   A reproductive cell; in humans, sperm or eggs.
germicide   Any compound that kills disease-causing microorganisms.
Gewe   Inclined plate settler by Waterlink Separations, Inc.
GFC   USFilter/General Filter.
GFF   Glass fiber filter.
GFS   Gravity Flow Systems, Inc.
ghanat   See "falaj."
GHG   See "greenhouse gases (GHG)."
Giardia lamblia   A protozoan parasite responsible for giardiasis.
giardiasis   A gastrointestinal disease caused by the ingestion of waterborne *Giardia lamblia*, often resulting from the activity of beavers, muskrats, or other warm-blooded animals in surface water used as a potable water source.
gill net   A vertical net used to capture fish of a certain size by allowing their heads to pass through the openings while catching their gills in the mesh.
Girasieve®   Externally fed rotating drum screen by Andritz Sprout-Bauer S.A.
GIS   See "geographic information system (GIS)."
glacial water   Bottled water originating from a glacier.
glacier   A large mass of accumulating ice and compacted snow that slowly moves down mountain valleys.
Gladiator   Groundwater remediation pump by Ejector Systems, Inc.
GLASdek   Synthetic media for evaporative cooling systems by Munters.
Glauber's salt   Sodium sulfate with ten waters of hydration.

GLC   Gas liquid chromatography.

GLERL   Great Lakes Environmental Research Laboratory.

GLI   See "Great Lakes Initiative (GLI)."

global warming   A theory that predicts the warming of the atmosphere as a result of an accumulation of atmospheric carbon dioxide.

globe valve   A valve where closure is accomplished by a spherical plug lowered onto a matching seat in the center of the valve.

GLP   Good laboratory practices.

GLWQA   Great Lakes Water Quality Agreement.

Glydaseal   Sluice gate by Rodney Hunt Co.

GM foods   Genetically modified foods made from plants that have been genetically altered.

GMCC   Global monitoring for climatic change.

Golfwater®   Aeration system for golf course ponds by Aeration Industries, Inc.

Gooch crucible   A heat resistant container fitted with a filter mat used for determination of suspended and total solids.

Goodfil   Rotary drum thickener by Gootech ASA.

GOP   General operating procedures.

GOR   See "gain output ratio (GOR)."

Gore-Tex®   Microporous membrane material by W.L. Gore & Associates, Inc.

Gould clarifier   A rectangular, secondary clarifier with dual chain and flight sludge collector mechanisms traveling toward a single sludge hopper at or near the middle of the tank.

GOX   Gaseous oxygen.

gpd   Gallons per day.

gpf   Gallons per flush.

gpg   See "grains per gallon (gpg)."

gpg imp   Grains per imperial gallon.

gpm   Gallons per minute.

GPS   Groundwater protection strategy.

gr   See "grain (gr)."

GR™   Gas recuperative thermal oxidizer by Thermatrix, Inc.

grab sample   A single water or wastewater sample taken at a time and place representative of total discharge.

Grabber   Tramp metal and heavy object catcher by Franklin Miller, Inc.

Grabber®   Reciprocating rake bar screen by Waterlink Separations, Inc.

Grace TEC   Former name of Megtec Systems, Inc.

grade   (1) The finished surface of a civil structure. (2) The inclination or slope of a surface or structure. (3) To rate according to a standard or size.

gradient   The rate of change of an elevation, velocity, pressure, temperature, or other parameter.

graduated cylinder   A glass cylinder with fine gradations used in a laboratory to measure liquid contents.

GrahamTek™   Filtration and desalination water treatment systems by MWD Technologies Ltd.

grain (gr)   A unit of mass equal to 0.0648 gram.

grains per gallon (gpg) A unit of measure where 1 gpg (U.S.) = 17.1 mg/L and 1 gpg (Imp) = 14.3 mg/L, frequently used in water hardness calculations.

gram (g) A unit of mass equal to the weight of 1 cubic centimeter (1 milliliter) of water or 0.03527 ounces.

Gram stain A staining procedure used to differentiate and categorize bacteria as either Gram negative or Gram positive.

granular activated carbon (GAC) A granular form of activated carbon used in filter beds or contactor vessels to absorb organic compounds.

granular media Grains of sand or other granular material used for filtering water or wastewater.

granular media filtration A tank or vessel filled with sand or other granular media to remove suspended solids and colloids from a water or wastewater that flows through it.

GRAS Generally regarded as safe.

Gravabelt Gravity belt thickener by Komline-Sanderson Engineering Corp.

gravel Rock fragments measuring 2 mm to 70 mm, often used as support material in granular media filters.

Gravex Zeolite softener by Graver Co.

Gravilectric Sludge wasting system using load cells to determine excess sludge accumulations by Patterson Candy International, Ltd.

Gravi-Merik™ Gravimetric belt feeder by Merrick Industries Inc.

gravimetric Pertaining to the measurement of the weight of samples or materials.

gravimetric feeder Dry chemical feeder that supplies a constant weight of chemical over a preset time period.

Gravipak Crossflow inclined plate clarifier by USFilter/Aerator Products.

Gravi-Pak Oil/water separator by the former Bowser-Briggs Filtration Co.

Gravisand™ Traveling bridge filter components by USFilter/Davco.

gravitational acceleration The acceleration of a free-falling body caused by the force of the earth's gravity, equal to 9.8 meters (32 feet) per second.

Gravitator™ Clarifier/thickener by DAS International, Inc.

gravity belt thickener (GBT) A sludge dewatering device utilizing a porous filter belt to promote gravity drainage of water.

gravity filter Granular media filter that operates at atmospheric pressure.

gravity spring See "seepage spring."

gravity system A hydraulic system that relies on gravity flow and does not require pumping.

gravity thickening A sedimentation basin designed to operate at high solids loading rates, usually with vertical pickets mounted to the revolving sludge scrapers to assist in releasing entrained water.

gray (Gy) An international standard dose unit of radiation equal to an absorption of 1 joule per kilogram; 1 Gy = 100 rads.

Gray Engineering Former equipment manufacturer.

gray water All nontoilet household water including the water from sinks, baths, and showers.

GRCDA Government Refuse Collection and Disposal Association.

grease   Common term used to include fats, oils, waxes, and related constituents found in wastewater.

Grease Grabber   Grease and oil skimmer by Abanaki Corp.

grease trap   A receptacle used to collect grease and separate it from a wastewater flow.

GreaseBurn   Grease and skimmings incinerator by Walker Process Equipment.

Great Lakes Environmental®   Equipment product line by Waterlink Separations, Inc.

Great Lakes Initiative (GLI)   Proposed guidelines to develop uniform water quality requirements for the U.S. Great Lake's basin.

Great Lakes Instruments   Former name of GLI International.

green belt   An undeveloped area immediately surrounding a town or development for the purpose of restricting indiscriminate outward expansion.

green liquor   The liquor resulting from dissolving molten smelt from the kraft recovery furnace in water.

greenhouse effect   The effect of CO_2 and other gases on the earth's atmosphere that is analogous to greenhouse glass as it restricts the outflow of radiative energy which results in the warming of the lower atmosphere.

greenhouse gases (GHG)   Gases including carbon dioxide, methane, nitrous oxide, and CFCs which have been recognized to contribute to the greenhouse effect.

Greenleaf Filter Control   Multiple cell rapid sand gravity filter using a central control and backwashing system by Infilco Degremont, Inc.

greensand   A filter sand containing glauconite with ion exchange properties, often used to remove iron or manganese from water.

GRGL   Groundwater residue guidance level.

Griductor®   Comminutor by Infilco Degremont, Inc.

Griffin Generator   Ozone generator by Ozonia North America.

Grifter®   Packaged pumping/grinding station by Ingersoll-Dresser Pump (U.S.) and H2O Waste-Tec (U.K.).

Grind Hog™   Mechanical shredding products by G.E.T. Industries, Inc.

grit   Sand, gravel, cinders, and other heavy solid matter that have settling velocities substantially higher than those of organic putrescible solids in wastewater.

grit chamber   A settling chamber used to remove grit from organic solids through sedimentation or an air-induced spiral agitation.

grit classifier   Mechanical device utilizing an inclined screw or reciprocating rake to wash putrescible organics from grit.

Grit King™   Grit removal unit by H.I.L. Technology, Inc.

grit removal   A preliminary wastewater treatment process to remove grit from organic solids.

Grit Snail™   Fine grit removal system by Eutek Systems, Inc.

grit washer   A device used to wash organic matter from grit.

GritLift   Air lift pump for grit removal by Walker Process Equipment.

Gritmeister™   Grit separator/screw conveyor by Waterlink Separations, Inc.

Gritreat   Aerated grit chamber by USFilter/Headworks Products.

Gritreator™   Grit treatment unit by Baker Process.

Gritt Mitt™   Grit removal and conveyor unit by WesTech Engineering Inc.

grizzly   A coarse screen used to prevent debris from entering a water intake.

Grizzly™   Grinder pump by F.E. Myers Co.

ground cover   Plants grown to keep soil from eroding.

groundwater   Subsurface water found in porous rock strata and soil.

groundwater infiltration (GWI)   The seepage of groundwater into shafts and tunnels.

groundwater recharge   The replenishment of a groundwater source, often by injection of tertiary treated wastewater.

grout   Fluid, or semi-fluid, cement slurry for pouring into joints of brickwork or masonry.

GRP   Glassfiber reinforced plastic.

grubbing   The process of removing tree stumps or roots.

GTO   Gas turbine oxidizer.

Guardian™   Strainer product line by Tate Andale, Inc.

guide vane   A device used to direct or guide the flow of a liquid or vapor.

guinea worm disease   See "dracunculiasis."

gunbarrel   A vertical settling tank used to separate free oil from water in the production of crude oil.

Gundline®   Containment liner by Gundle Lining Systems, Inc.

Gundnet   Drainage net by Gundle Lining Systems, Inc.

Gundseal   Geocomposite liner by Gundle Lining Systems, Inc.

Gutling   Former name of USFilter/Gutling.

GW   Groundwater.

GWDR   Ground Water Disinfection Rule.

GWI   Groundwater infiltration.

GWM   Groundwater monitoring.

GWUI   Groundwater under the direct influence of surface water.

gypsum   The mineral consisting primarily of fully hydrated calcium sulfate. Chemical formula is $CaSO_4\text{-}2H_2O$.

Gyrazur™   Softening clarifier by Infilco Degremont, Inc.

H

H. pylori   See "*Heliobacter pylori*."

H_2O   See "water."

H_2O_2   See "hydrogen peroxide."

H_2S   See "hydrogen sulfide."

HA   Health advisory.

ha   See "hectare (ha)."

HAA   See "haloacetic acid (HAA)."

HAA5   The sum of the concentration of five haloacetic acid compounds which include mono-, di-, and trichloroacetic acids and mono- and dibromoacetic acids.

HAA6   The sum of the concentration of six haloacetic acid compounds which include mono-, di-, and trichloroacetic acids, mono- and dibromoacetic acids, and bromochloroacetic acid.

HAAFP   Haloacetic acid formation potential.

habitat   The place where a human, animal, plant, or microorganism population lives, and its surroundings, both living and nonliving.

Hach One   pH Electrode by Hach Co.

hadal zone   The ecological zone of the ocean floor that lies in a deep trench below the abyssal zone, generally deeper than 6000 meters.

Halberg   Digester draft tube sludge mixer by Sterling Fluid Systems (USA).

half life   The time required for half of the atoms of a particular radioactive substance to transform or decay to another nuclear form.

half life, chemical   The time required for the concentration of a chemical being tested to be reduced to one half of its initial value.

halides   A compound containing a halogen.

haloacetic acid (HAA)   A type of disinfection byproduct formed during the chlorination of water containing natural organic matter.

halocline   A well-defined salinity gradient in an ocean.

halogen   One of the chemical elements of the group containing fluorine, chlorine, bromine, iodine, and astatine.

halogenated organic compounds (HOC)   Compounds having a halogen-carbon bond.

halon   Bromine-containing compounds with long atmospheric lifetimes whose breakdown in the stratosphere causes depletion of ozone.

halophyte   Plants capable of living in salt water or salty soil.

Hamburg Rotor   Surface aerator by Brackett Geiger.

HammerHead™   Groundwater pump by QED Environmental Systems, Inc.

hammermill   A device with hammer-like arms used to shred or grind solids to facilitate further treatment or disposal.

handler   A facility that accumulates and generates universal wastes or receives and/or sends universal wastes to a destination facility.

HAP   Hazardous air pollutant.

HAPEMS   Hazardous Air Pollutant Enforcement Management System.

HAPPS   Hazardous Air Pollutant Prioritization System.

Harbor Bosun   Dye tablets by Formulabs, Inc.

Harborlite®   Perlite material by Celite Corp.

hard water   Alkaline water containing dissolved salts that interfere with some industrial processes and prevent soap from sudsing.

HaRDE®   Electrostatic precipitator by Wheelabrator Air Pollution Control, Inc.

Hardinge   Manufacturer of traveling bridge filter whose product line was acquired by Infilco Degremont, Inc.

hardness   The total concentration of calcium and magnesium ions in water expressed as calcium carbonate.

hardpan   A compacted, impermeable layer of soil at or near the surface.

Hastelloy®   Corrosion resistant, nickel-based alloy by Haynes International, Inc.

HATREMS   Hazardous and Trace Emissions System.

HAV   Hepatitis A virus.

Hawker Siddeley Brackett   Former parent company of Brackett Geiger.

Hazardous and Solid Waste Amendments (HSWA)   The 1984 amendments to RCRA regulating underground tank storage and land disposal of certain hazardous wastes.

hazardous area, class 1   Locations where flammable gases or vapors may be present in the air in sufficient quantities to produce explosive or ignitable mixtures.

hazardous material   A solid, liquid, or gaseous material that is detrimental to human health.

Hazardous Ranking System (HRS)   The method used to evaluate the relative potential of hazardous substance releases to cause health or safety problems or ecological or environmental damage.

hazardous waste   Any waste or combination of wastes which pose a substantial present or potential hazard to human health or living organisms because they are nondegradable, persistent in nature, or may otherwise cause detrimental cumulative effects.

Hazen-Williams coefficient   A roughness coefficient related to the influence of pipe material on the velocity characteristics of a fluid.

HAZMAT   Hazardous materials.

HAZOP   Hazard and operability study.

HAZWOPER   Hazardous Waste Operations and Emergency Response.

HBV   Hepatitis B virus.

HC   (1) Hydrocarbon. (2) Hazardous constituents.

HCFCs   See "hydrochlorofluorocarbons (HCFCs)."

HCV   Hepatitis C virus.

HDPE   See "high density polyethylene (HDPE)."

HDT   (1) Hydraulic detention time. (2) Highest dose tested in a study.

HDXLPE   High density crosslinked polyethylene.

head   (1) A measure of the pressure exerted by a fluid expressed as the height of an enclosed column of the fluid which could be balanced by the pressure in the system. (2) The source or upper end of a system, e.g., headwater.

header   A pipe manifold fitted with several smaller lateral outlet pipes.

headloss   The difference in water level between the upstream and downstream sides of a treatment process attributed to friction losses. Sometimes called "pressure drop."

headwater   The source or upper reaches of a stream.

headworks   The initial structure and devices located at the receiving end of a water or wastewater treatment plant.

HEAL   Human exposure assessment location.

health advisory level   A nonregulatory health-based reference level of chemical traces in drinking water at which there are no adverse health risks when ingested over various periods of time.

health assessment   An evaluation of available data on existing or potential risks to human health posed by a Superfund site.

hearth   The bottom of a furnace upon which waste materials are exposed to the flame.

heartwood   The oldest, dense wood in the center of a tree which often contains an accumlation of resins and oils.

heat balance   An accounting of the distribution of a system's heat loss and heat gain.

heat exchanger   A device used to transfer heat from one substance to another. See also "shell-and-tube heat exchanger."

heat island   An urban phenomenon where air pollutants and the heat from a combination of tall buildings, concrete pavement, and other materials combine to create a localized haze dome that traps rising hot air resulting in higher temperatures and air pollution.

heat of condensation   The amount of heat released when a vapor changes state to a liquid.

heat of sublimation   The amount of energy required to convert ice directly to a vapor.

heat of vaporization   The amount of heat required to change a volume of liquid to a vapor.

heat pump   A device for transferring heat from a cooler reservoir to a hotter one by mechanical means involving the compression and expansion of a fluid.

heat recovery   The capture and use of heat that would otherwise be lost as waste heat.

heat sink   Any material that is used to absorb heat.

Heat Systems   Former name of Misonix, Inc.

heat tracing   The electrical or steam heating of piping and equipment to prevent freezing.

heat transfer   The transfer of heat from one body to another by means of radiation, conductance, or convection.

heat value   The quantity of heat that can be released from a sludge per unit mass of the sludge solids.

Heatamix   Heating and recirculation system for anaerobic sludge digesters by Simon-Hartley, Ltd.

heater treater   Oil field produced water treatment unit used to break water-in-oil emulsions with heat and chemicals.

HeatX   Anaerobic digester gas heating unit by Walker Process Equipment.

heavy metals   Metals that can be precipitated by hydrogen sulfide in an acid solution and which may be toxic to humans above certain concentrations.

heavy water   Water with a hydrogen isotope having an atomic weight of 2.

hectare (ha)   A unit of area equal to 10,000 square meters. One hectare is equal to 2.471 acres.

HEEB   High-energy electron beam irradiation.

HEI   Health Effects Institute.

Hela-Flow   Plastic laterals for water and wastewater treatment by Liquid-Solids Separation Corp.

Helaskim   Helical surface skimmer by Walker Process Equipment.

HeliCarb   Carbon dioxide contactor by CBI Walker, Inc.

Heliclean®   Open channel spiral screen with screenings washer by Waterlink Separations, Inc.

Helico   Screw type screenings press by Infilco Degremont, Inc.

Heliflow®   Spiral tube bundle heat exchangers by Graham Manufacturing Co.

helio-   A prefix referring to the sun or sunlight.

Heliobacter pylori   A bacterium that causes stomach ulcers and has been identified as an emerging waterborne health threat. Also known as *H. pylori*.

Heli-Press   Screenings compactor by Vulcan Industries, Inc.

Helisieve Plus™   Septage receiving station by Waterlink Separations, Inc.

Helisieve®   Open channel spiral screen by Waterlink Separations, Inc.

HeliSkim   Helical surface skimmer by Walker Process Equipment.

HeliThickener   Interrupted flight screw conveyor for sludge collectors by Walker Process Equipment.

Helixpress   Screenings dewatering press/conveyor by Waterlink Separations, Inc.

Hellmut Geiger   Former name of Brackett Geiger.

helminth   A parasitic worm.

HEM   Human exposure modeling.

hemi-hydrate   A crystalline compound having one molecule of water of crystallization per two molecules of compound.

Henry's Law   The weight of any gas that will dissolve in a given volume of a liquid at constant temperature is directly proportional to the pressure that the gas exerts above the liquid.

HEPA filter   See "high efficiency particulate air filter (HEPA)."

hepatitis   An acute viral disease which results in liver inflammation and may be transmitted by direct contamination of a water supply by sewage.

heptachlor   An insecticide banned for use on food products and seed treatment.

herb   A seed plant whose stem is not woody and whose leaves, stems, or roots are often used for seasoning, medicine, or fragrance.

herbicide   A synthetic organic compound used to control plant growth.

Hercules   Screening equipment product line of Moyno Industrial Products.

Hercules®   Pressure leaf filters by Liquid-Solids Separation Corp.

HES   Huntington Environmental Systems, Inc.

heterotrophic bacteria   Bacteria that derives it's cell carbon from organic carbon; most pathogenic bacteria are heterotrophic bacteria.

heterotrophic plate count (HPC)   A laboratory method of determining the level of heterotrophic bacteria in a sample. Formerly known as "standard plate count."

Hevi-Duty   Traveling water screen replacement parts products by USFilter/Rex & Link-Belt Products.

hexavalent chrome   A toxic form of chrome used in plating operations, usually reduced to the trivalent form and precipitated as a hydroxide.

HFCs   See "hydrofluorocarbons (HFCs)."

HGL   Hydraulic grade line.

HHC   Highly hazardous chemical.

HHE   Human health and the environment.

HHW   Household hazardous waste.

Hi-Cap®   High rate thickener by Baker Process.

Hidrostal   Screw/centrifugal impeller pump by Envirotech Pumpsystems.

Hi-Flo®   Depth filter and water softener products by Culligan International Corp.

HiFlo™   Thickener by WesTech Engineering, Inc.

Hi-GARD®   Rotary trickling filter by USFilter/General Filter.

high density polyethylene (HDPE)   A synthetic organic material often used as landfill liner because of its low permeability.

high efficiency particulate air filter (HEPA)   A filtering system capable of trapping and retaining at least 99.97% of all monodispersed particles 0.3 micrometer in diameter or larger.

High Flux Series™   Membrane cleaning products for restoring membrane flux by King Lee Technologies.

high performance liquid chromatography (HPLC)   Instrumental technique for measuring trace levels of organics by means of UV adsorption.

High Resolution Redox®   Disinfection and dechlorination control technology by USFilter/Stranco.

high velocity air filter (HVAF)   An air pollution control filtration device for the removal of sticky, oily, or liquid aerosol particulate matter from exhaust gas streams.

High-Flux TF™   Oil control chemical for use in membrane treatment system by King Lee Technologies.

high-level radioactive waste (HLW)   The highly radioactive waste material resulting from the reprocessing of spent fuel, the spent fuel itself, and much of the waste generated from nuclear weapons production, with activities measured in curies per liter.

high-test calcium hypochlorite   Calcium hypochlorite product containing at least 70% available chlorine.

high-to-low dose extrapolation   The process of prediction of low exposure risk to humans from the measured high exposure-high risk data involving rodents.

Hi-Iron   Closed pressure contact-aeration iron removal process by Aquatrol Ferr-X Corp.

Hi-Lift   Reciprocating rake bar screen by Longwood Engineering Co., Ltd.

Hi-Lucid   High rate coagulation and sedimentation process by Hitachi Metals America, Ltd.

hindered settling   Phenomenon referring to sedimentation of particles in a suspension of intermediate concentration where interparticle forces hinder the settling of neighboring particles. Also called "type III settling."

HIOPs   Heated iron oxide particles.

HiOx™   Aeration system by Parkson Corp.

HIP   High intensity belt press by Andritz-Ruthner, Inc.

Hi-Pass   In-line motionless mixer by Komax Systems, Inc.

Hiperfilter   Media filter by Axsia Serck Baker, Inc.

Hi-Rate Thickener   Circular gravity sludge thickening system with ancillary equipment by GL&V/Dorr-Oliver, Inc.

Hi-Tork   Portable mixer by Philadelphia Mixers.

HIV   Human immunodeficiency virus. See "AIDS (Acquired Immune Deficiency Syndrome)."

Hi-V®   Depth filter cartridge by USFilter/Filtration & Separation.

HLL   High liquid level.

HLW   See "high-level radioactive waste (HLW)."

HMIP   Her Majesty's Inspectorate of Pollution (U.K.).

HMIS   Hazardous Materials Information System.

HMR   Hazardous materials regulations.

HMTA   Hazardous Materials Transportation Act.

HMTR   Hazardous Materials Transportation Regulations.

HOA   Hand-off-automatic.

HOBr   See "hypobromous acid (HOBr)."

HOC   See "halogenated organic compounds (HOC)."

HOG™   Halogenated organic gas destruction process formerly offered by Quantum Technologies, Inc.

hogging ejector   A steam jet ejector which produces a vacuum in a vessel by evacuating the contents in a single stage to the atmosphere.

holding tank   A tank used to receive and store wastewater prior to its ultimate disposal.

Hollosave   Reverse osmosis system by Toyobo Co., Ltd.

Hollosep   Hollow fiber configuration reverse osmosis module by Toyobo Co., Ltd.

hollow fiber membrane   Type of reverse osmosis and ultrafiltration membranes formed into small diameter tubes.

homeowner water system   Any water system which supplies piped water to a single residence.

Homomix   Propeller type rapid mixer by Amwell, Inc.

HON   Hazardous organic NESHAP. A rule regulating emissions of listed organic chemicals from new and existing manufacturing sources.

honey wagon   Common term for vacuum truck used to remove accumulated septic tank solids.

hood capture efficiency   Ratio of the emissions captured by a hood and directed into a control or disposal device, expressed as a percent of all emissions.

hormone   A substance such as adrenaline or insulin which is formed in an organ of the body and which controls other organs or metabolic activities.

Hose pump   A peristaltic pump used for metering or fluid transfer.

host   Any plant or animal on, or in which, another lives for nourishment, development, or protection.

Hot Bottom®   Steam-powered evaporator by Lakeview Engineered Products, Inc.

Hot Box   Wastewater evaporator by Landa, Inc.

hot lime softening   Lime softening process conducted at temperatures of 104 to 125°C.

Hot Shot®   Electric powered evaporator by Lakeview Engineered Products, Inc.

Hot Tube®   Evaporator by Lakeview Engineered Products, Inc.

hot zone   The area immediately surrounding a hazardous materials accident site.

HOV   High-occupancy vehicle.

hp   Horsepower. A standard unit of power equal to 745.7 watts.

HPC   See "heterotrophic plate count (HPC)."

HPD   Former name of USFilter/HPD.

HP-Hybrid   Automatic filter press by Heinkel Filtering Systems, Inc.

HPI/CPI   Hydrocarbon processing industry/chemical processing industry.

HPLC   See "high performance liquid chromatography (HPLC)."

HPV   High priority violator.

HRB®   Solid waste baler by Harris Waste Management Group, Inc.

HRC™   Hydrogen release compound for bioremediation by Regenesis.

HRGC   High resolution gas chromatography.

HRMS   High resolution mass spectrometry.

HRS   See "Hazardous Ranking System (HRS)."

HRSG   Heat recovery steam generator.

HRT   See "hydraulic residence time (HRT)."

HRUP   High-risk urban problem.

HSC™   High pressure centrifugal pump for reverse osmosis by Pump Engineering, Inc.

HSDB   Hazardous Substance Database.

HSL   Hazardous substance list.

HSWA   See "Hazardous and Solid Waste Amendments (HSWA)."

HTA   High temperature additive.

HTC   Heat transfer coefficient.

HTC™   Hydraulic turbocharger for reverse osmosis pump system by David Brown Union Pumps Co.

HTH   High-test calcium hypochlorite product by Arch Chemicals.

HTO®   Regenerative thermal oxidizer by Dürr Environmental, Inc.

HTP   Heat-treated peat. A polymer spill encapsulation and cleanup product by American Products.

HTP   High temperature and pressure.

HTW   Hazardous toxic wastes.

Hubair®   Surface aerator by USFilter/Hubert.

Hubert   Former name of USFilter/Hubert.

Huisman   Oxidation ditch wastewater treatment system by USFilter/Envirex.

human equivalent dose   A dose that when administered to humans produces an effect equal to that produced by a dose in animals.

human exposure evaluation   Describing the nature and size of the population exposed to a substance and the magnitude and duration of their exposure. The evaluation could concern past, current, or anticipated exposures.

human health risk   The likelihood that a given exposure or series of exposures may damage the health of individuals.

Humboldt Decanter   Sludge dewatering equipment product line of Baker Process.

humic acid   Organic acids that are byproducts of decomposing organic matter which colors water.

humidity   The amount of water vapor within the atmosphere.

humidity ratio   The ratio of the mass of water vapor to the mass of dry air in the atmosphere.

humification   The decomposition of organic matter from plants and animals into soil.

Hummix™   Polymer blending system by Hinsilblon Laboratories.

humus   Dark or black decomposing organic matter in soil.

Hurox™   Spent caustic oxidation system by Huron Tech Corp.

Hurricane   Mixer by Franklin Miller, Inc.

Hurricane   Submersible aspirating aerator by Aeromix Systems, Inc.

Hurricane®   Combination centrifuge and cartridge filtration system by Harmsco Filtration Products.

HVAC   Heating, ventilating, and air conditioning.

HVAC/R   Heating, ventilating, air conditioning, and refrigeration.

HVAF   See "high velocity air filter (HVAF)."

HVIO   High volume industrial organics.

HVLC   High volume, low concentration.

HW   Hazardous waste.

HWCL   Hazardous Waste Control Law.

HWDMS   Hazardous Waste Data Management System.

HWF   Hazardous waste-derived fuel.

HWGTF   Hazardous Waste Groundwater Task Force.

HWIR   Hazardous Waste Identification Rule.

HWIS   Hazardous Waste Information System.

HWL   High water level.

HWLT   Hazardous Waste Land Treatment

HWM   Hazardous waste management.

HWTC   Hazardous Waste Treatment Council.

HX   (1) Heat exchanger. (2) Hydrogen halide.

Hy Flo Super-Cel®   Diatomaceous earth filter media by Celite Corp.

hyacinth   Floating aquatic plants whose roots provide a habitat for a diverse culture of aquatic organisms who metabolize organics in wastewater.

hybrid system   A system incorporating multiple processes or technologies.

Hycor®   Product line of Waterlink Separations, Inc.

Hydecat™   Hypochlorite destruction product by Synetix.

HYDRA   Sludge removal system by Hazleton Environmental, Inc.

Hydracap™   Ultrafiltration system by Hydranautics.

Hydradenser   Inclined screw thickener by Thermal Black Clawson.

Hydra-Mix®   Hydraulic mixer by Air-O-Lator Corp.

Hydrapaint   Ultrafiltration spiral membrane by Hydranautics.

Hydra-Press™   Hydraulic screenings compactor by Vulcan Industries, Inc.

Hydrasand®   Continuous cleaned moving bed sand filter by Andritz-Ruthner, Inc.

Hydrasep®   Oil/water separator by GNESYS, Inc.

Hydrasieve®   Static fine screen by Andritz-Ruthner, Inc. (Western Hemisphere) and Andritz Sprout-Bauer S.A. (Eastern Hemisphere).

hydrate   A compound formed by the union of water with another substance.

hydrated lime   The calcium hydroxide product that results from mixing quicklime with water. Chemical formula is $CaOH_2$.

hydration   The process of combining or uniting water with another substance.

Hydratower™   Static fine screen by Brandt Co.

Hydra-Tracker™   Self-tracking oil, grease, and sludge removal device by Dontech, Inc.

Hydraucone   Diffuser plate used with down flow mechanical aerator by Amwell, Inc.

hydraulic classification   A process of grading particles of the same specific gravity according to size by fluidization or backwashing.

hydraulic gradient   The slope of the hydraulic grade line which indicates the change in pressure head per unit of distance.

hydraulic jump   The sudden rise in water surface level which may occur when water flowing through an open channel at a high velocity is retarded.

hydraulic loading   Total volume of liquid applied per unit of time to a tank or treatment process.

hydraulic radius   The ratio of the area of a conduit in which water is flowing to its wetted perimeter.

hydraulic residence time (HRT)   Vessel volume (L) divided by the liquid removed (L/d).

hydraulics   The branch of science and engineering that deals with the mechanics of fluids.

hydrazine   A chemical compound used as an oxygen scavenger. Chemical formula is H_2NNH_2.

Hydro Grid™   Fine screening device by Schreiber Corp.

Hydro Group   Former name of Layne Christensen, Ranney Division.

Hydro Scour   Pulsed bed filter backwashing system by USFilter/Zimpro.

Hydro-Aerobics   Wastewater treatment equipment product line by Ashbrook Corp.

Hydrobend®   Water level control device by Grande, Novac & Associates, Inc.

hydroblast   The use of high pressure water jets to clean or remove debris.

Hydroburst™   Passive screen air backwash system by USFilter/Johnson Screens.

hydrocarbons   Organic compounds consisting predominantly of carbon and hydrogen.

HydroCeal   Mixed bubble diffuser by Ashbrook Corp.

Hydrocell®   Induced air flotation separators by USFilter/Whittier.

Hydro-Chek   Coarse bubble diffuser by Pollution Control Systems, Inc.

hydrochloric acid   An aqueous form of hydrogen chloride that is a strong corrosive agent. Chemical formula is HCl.

hydrochlorofluorocarbons (HCFCs)   Substances used as temporary alternatives to CFCs.

Hydro-Circ®   Nonmechanical sludge recirculation system by Graver Co.

Hydroclean®   Fine screening of combined sewer overflows by Grande, Novac & Associates, Inc.

Hydrocleaner   Dissolved air flotation aeration system by Baker Hughes Process Systems.

Hydro-Clear®   Pulsed bed gravity sand filter by USFilter/Zimpro.

HydroClor-Q™   Organic chlorine test kit by Dexsil Corp.

Hydro-Cone   Underdrain for sand filter by BIF.

hydrocyclone   A device used to separate grit and other solids from a liquid using centrifugal force.

Hydrodarco®   Activated carbon by Norit Americas Inc.

HydroDri™   Screenings press by Serpentix Conveyor Corp.

Hydrofilt™   Anthracite filter media by Lang Filter Media Co.

Hydroflo®   Metering pump product line by PennProcess Technologies, Inc.

HydroFlo™   Disposable in-line filter for groundwater monitoring by Schlicher & Schuell.

HydroFloat   Dissolved air flotation system by Hydro-Flo Technologies, Inc.

Hydro-Float   Flotation system for removal of fat, grease, and suspended solids by HydroCal, Inc.

Hydrofloc   Polyelectrolyte used to enhance liquid/solid separation by Aqua Ben Corp.

HydroFloc™   Rotating screen thickener by Klein America, Inc.

Hydrofluor Combo   *Cryptosporidium* and *Giardia* detection reagents by Meridian Diagnostics, Inc.

hydrofluoric acid   An aqueous form of hydrogen fluoride. Chemical formula is HF.

hydrofluorocarbons (HFCs)   Replacements for CFCs and HCFCs which contain no chlorine and have no ozone-depletion potential.

hydrofluosilicic acid   An acidic liquid used to fluoridate drinking water. Chemical formula is H_2SiF_6.

Hydroflush   Cable operated bar screen by Beaudrey Corp.

HydroForce™   Motorless polymer mixing system by USFilter/Stranco.

hydrogen ion concentration   See "pH."

hydrogen peroxide   An oxidizing agent used for odor control and disinfection. Chemical formula is H_2O_2.

hydrogen sulfide   A toxic gas formed by the anaerobic decomposition of organic matter containing sulfur. Chemical formula is H_2S.

hydrogenate   To combine hydrogen with a compound in a chemical reaction.

hydrograph   A graphical representation of a stream discharge at a single location.

Hydro-Grit™   Vortex degritter by Fluidyne Corp.

Hydrogritter   Grit washing system by Envirotech Pumpsystems.

Hydro-Jet   Self-cleansing screen by H.I.L. Technology, Inc.

Hydro-Lance™   Wet dust collector by Duall Division, Met-Pro Corp.

Hydro-Lift   Pre-fabricated steel lift stations by Ashbrook Corp.

Hydro-lite   Biological filter media by Ashbrook Corp.

hydrologic cycle   The natural cycle of continuous evaporation and condensation.

hydrology   The science dealing with the properties, distribution, and circulation of water.

hydrolysis   A chemical change or decomposition of matter produced by the combination with water.

Hydromation   Deep bed filter with walnut shell media by Filtra-Systems Hydromation.

hydrometer   An instrument used to measure the specific gravity of liquids.

Hydron®   Dissolved air flotation technology by Colloid Environmental Technologies Co.

hydronium ion   The hydrated hydrogen ion, H_3O^+.

Hydroperm®   Crossflow microfiltration system formerly offered by USFilter/Microfloc.

hydrophilic   Having an affinity for water.

hydrophobic   Having an aversion to water.

hydrophyte   A plant that grows in water or saturated soils.

Hydropillar™   Elevated water storage tank by Pitt-Des Moines, Inc.

hydropneumatic water system   A water system, usually small, in which a water pump is automatically controlled by the air pressure in a compressed air tank.

hydroponics   The science of growing plants in nutrient-rich solutions or moist inert material instead of soil.

Hydropress   Belt filter press by the former Clow Corp.

Hydro-Press   Belt filter press by HydroCal, Inc.

HydroPunch®   Groundwater sampler by QED Environmental Systems.

Hydrorake   Trash rake by Atlas Polar Co., Hydrorake Division.

HydroRanger   Ultrasonic level measuring system by Milltronics, Inc.

Hydro-Rotor   Brush type aerator by Amwell, Inc.

Hydro-SAFe   Biological aerated filter by Ashbrook Corp.

Hydroscreen   Static fine screen by Waterlink Separations, Inc.

HydroSeal®   Anaerobic digestion gasholder cover by Baker Process.

Hydroself®   Sewer and retention tank flushing device by Grande, Novac & Associates, Inc.

Hydrosep   Shallow basin, nonaerated grit removal system by USFilter/Aerator Products.

Hydroseparator   Industrial wastewater gravity thickener by GL&V/Dorr-Oliver, Inc.

Hydroshear   Aeration tank for low flow package treatment plant formerly offered by USFilter/Envirex.

Hydro-Shear   Internally fed rotary fine screen by Dontech, Inc.

Hydrosil   Static screen with brush/water jet cleaning device by Spirac.

Hydroslide®   Constant flow regulating device by Grande, Novac & Associates, Inc.

Hydro-Sock   Upflow cartridge filtration system by Ashbrook Corp.

hydrosphere   The aqueous environment of the earth including rivers, lakes, oceans, and glaciers.

hydrostatic pressure   The pressure exerted by water due to depth alone.

hydrostatic test   A test procedure in which a vessel or system is subjected to water pressure and examined for leaks, distortion, and/or mechanical failure.

hydrotest   The testing of piping, tubing, or vessels by the filling of water and the pressurization to test for integrity.

Hydrovex®   Flow control product line by John Meunier, Inc.

Hydrowash   Down flow recycling pump/aerator for grit removal by Amwell, Inc.

hydroxide alkalinity   Alkalinity caused by hydroxyl ions.

hydroxide ion   A negatively charged ion consisting of a hydrogen atom and an oxygen atom. Chemical formula is OH.

hydroxyl   A chemical group consisting of one hydrogen and one oxygen atom.

Hydro-Zap®   Ultra-violet wastewater disinfection systems by Ashbrook Corp.

Hydrozon®   Ozone system by Carus Chemical Co.

Hydrymax   Sludge dryer by D.R. Sperry & Co.

hyetograph   (1) A graphic representation of a rainfall that plots time vs. rainfall. (2) A rain recording gage.

HyFlo™   Continuous self-cleaning screenings belt by Waterlink Separations, Inc.

Hygene   Bacteriostatic filter media by Ionics, Inc.

hygrometer   An instrument for measuring the relative amount of moisture in the air.

hygroscopic   Readily absorbing moisture from the atmosphere.

hygroscopic water   Water in soil that is in equilibrium with the atmospheric water vapor and cannot be lost to evaporation or drain freely by gravity.

Hymergible®   Hydraulically driven submersible pump by Crane Pumps & Systems.

Hyper+Ion®   Cationic coagulants and flocculants by General Chemical Corp.

hyperfiltration   Filtration using a dense membrane that is often used synonymously with "reverse osmosis."

HyperFlex®   HDPE containment liner by SLT North America, Inc.

Hyperfloc®   Polyelectrolyte used to enhance liquid/solid separation by Hychem, Inc.

HyperNet™   Net fabric composite by SLT North America, Inc.

Hyperpress™   Combination belt filter/plate and frame press by Klein.

Hypersperse   Antiscalant for use in reverse osmosis systems by BetzDearborn-Argo District.

hypobromous acid (HOBr)   An effective biocide.

hypochlorite   Chlorine anion commonly used as an alternative to chlorine gas for disinfection. Chemical formula is OCl_3^-.

Hypo-Gen®   Sodium hypochlorite generation systems by Capital Controls Co.

hypolimnetic aeration   Aeration of water at the bottom of a lake.

hypolimnion   The lower layer of a stratified lake that results from varying water densities.

hypoxia   The condition that exists when a body of water has a very low dissolved oxygen level, usually less than 2 mg/L.

Hypress   Screenings dewatering press by Waterlink Separations, Inc.

Hy-Q   Flush bottom sluice gate closure by Rodney Hunt Co.

Hysep®   Decanter centrifuge by Westfalia Separator, Inc.

Hy-Speed®   Mixer by Alsop Engineering Co.

Hytrex®   Cartridge prefilters and prefilter housings by Osmonics, Inc.

Hyveyor   Troughing conveyor by Waterlink Separations, Inc.

I

I&C   Instrumentation and control.

I&M   Inspection and maintenance.

I/A   Innovative and alternative.

I/I   See "inflow/infiltration (I/I)."

I/O   Input/output.

IAF   See "induced air flotation (IAF)."

IAP   Indoor air pollution.

IAQ Indoor air quality.

IAS™ Induced air scour system for sand filters by USFilter/Davco.

IATDB Interim air toxics data base.

IBT Industrial biotest laboratory.

IBWA International Bottled Water Association.

IC See "ion chromatography (IC)."

ICBN International Commission on the Biological Effects of Noise.

ICE (1) Institute of Civil Engineers. (2) Internal combustion engine.

ice The solid, crystalline form of water.

ice apron A wedge-shaped structure for protecting a pier or intake from floating ice.

ice fog An atmospheric suspension of reflective ice crystals which affects visibility.

ICEAS® Intermittent cycle extended aeration system for wastewater treatment by ABJ product group of Sanitaire Corp.

Ice-Away® Ice melter by Air-O-Lator Corp.

ICOLD International Commission on Large Dams.

ICR See "Information Collection Rule (ICR)."

ICRE Ignitability, corrosivity, reactivity, and extraction.

ICRP International Commission on Radiological Protection.

ICS Intermittent control strategies.

ID Inside diameter.

ID$_{50}$ See "infectious dose 50 (ID$_{50}$)."

IDA International Desalination Association.

IDI® Infilco Degremont, Inc.

IDL See "instrument detection limit (IDL)."

IDLH See "immediately dangerous to life or health (IDLH)."

IDP Ingersoll Dresser Pump Co.

Idrex Pressure leaf filter by USFilter/Zimpro.

IDS Drumshear Rotating fine screen by Aer-O-Flo Environmental, Inc.

IE See "ion exchange (IX or IE)."

IEC International Electrotechnical Commission.

IEEE Institute of Electrical and Electronics Engineers.

IESWTR Interim Enhanced Surface Water Treatment Rule.

IFA Immunofluorescence assay.

IGCC Integrated coal gasification combined cycle.

IGF Induced gas flotation. See "induced air flotation (IAF)."

igneous rock A type of rock formed from cooled magma.

ignitability The characteristic of having a flash point less than 60°C.

ignition temperature The lowest temperature at which combustion of a material becomes self-sustaining.

IIA Incinerator Institute of America.

II-PLP® Double-pass reverse osmosis system with interstage chemical feeds for pH adjustment by USFilter/Rockford.

ilmenite A dense mineral often used as filter media in a granular media filter.

IMAC™ Ion exchange resins by Rohm & Haas, Co.

imbibition The absorption or adsorption of water by a solid or colloid that results in swelling.

IMF Protector™   Drinking water treatment system by Smith & Loveless, Inc.

Imhoff cone   Cone-shaped container used to determine the volume of settleable solids in a specific volume of water.

Imhoff tank   A two-story wastewater treatment tank developed by Karl Imhoff in which sedimentation occurs in the upper compartment and anaerobic digestion occurs in the lower compartment. Also called an "Emscher fountain."

immediately dangerous to life or health (IDLH)   The maximum environmental concentration of a substance from which one could escape from a 30 minute exposure without irreversible adverse health effects.

immiscible   Incapable of being mixed.

immunoassay   The identification of a substance based on its capacity to act as an antigen.

immunodeficiency   A lack of one or more immune functions.

Impac™   Packing media for air stripping towers by Lantec Products, Inc.

Impact™   Ion exchange resins by Sybron Chemicals, Inc.

impact fee   Fee assessed new connections to a water or sewer system intended to recover a portion of the capital cost of the system.

impeller   The rotating set of vanes in a turbine, blower, or centrifugal pump designed to cause rotation of a fluid mass.

Imperial gallon   See "gallon, Imperial."

impermeable strata   Layers of clay or dense stone in the earth through which water cannot penetrate in measurable quantities.

impervious   Not allowing the passage of water at ordinary hydrostatic pressure.

impingement   (1) The entrapment of fish and other marine life on the surface of an intake screen when a high water velocity prevents escape. (2) The striking of a surface by a moving fluid.

impoundment   A pond, lake, or reservoir created through the use of a structural barrier such as a dam, levee, or dike.

impressed voltage cathodic protection   The use of an impressed current to prevent or reduce the rate of corrosion of a metal in an electrolyte by making the metal the cathode for the impressed current.

Impulse®   Counter-current water softeners by USFilter/Lowell.

impulse turbine   An energy recovery device used to recover pressure energy from reverse osmosis brine streams.

impurity   A chemical substance which is unintentionally present with another chemical substance or mixture.

IMR   Infant mortality rate.

IMS   See "Ion Mobility Spectrometry (IMS)."

IMS®   Filter media support cap for sand filters by F.B. Leopold Co., Inc.

in situ   Treatment or disposal methods that do not require movement of contaminated material.

in vitro study   A laboratory study conducted in glassware.

in vivo study   A study conducted in a living organism.

in-channel storage   The water storage volume in a channel or canal above the minimum water level required for conveyance.

incidence of illness   The rate of occurrence of new cases of disease in a defined population over a specified period of time.

incidental recharge   Groundwater recharge occurring as a result of human activities such as irrigation which are unrelated to a recharge project.

incineration   The process of reducing the volume of a solid by burning of organic matter.

incinerator   A furnace or device for incineration.

inclined plate separator   A series of parallel inclined plates that can be used to increase the efficiency of clarifiers and gravity thickeners.

incompatible waste   A hazardous waste that may cause corrosion or decay of containment materials or is unsuitable for comingling with another waste under uncontrolled conditions because of the hazardous reactions that may result.

incrustant   Solids formed as a crust on the inside wall of a pipe.

incubate   To maintain optimum environmental conditions for growth and reproduction of viable microbes.

Incutrol   Biochemical oxidation measuring and temperature control apparatus for BOD incubation by Hach Co.

index organism   See "indicator organism."

indicator organism   Microbes that indicate the absence or presence of a specific pollutant.

indirect discharger   Water treatment plants that discharge pollutants to publicly owned treatment works.

indirect reuse   The beneficial use of reclaimed water after releasing it for storage or dilution into natural surface waters or groundwater.

indirect source   Any facility, building, property, road, or parking facility that attracts motor vehicle traffic and, indirectly, causes pollution.

induced air flotation (IAF)   The clarification of suspended material using dispersed air bubbles that attach to hydrophobic surfaces causing materials to collect as a froth on the surface.

induced draft cooling tower   A cooling tower in which the air flow through the tower is induced by means of an electrically operated fan.

industrial waste   Waste generated by manufacturing or industrial practices that is not a hazardous waste regulated under Subtitle C of RCRA.

industrial wastewater   Liquid wastes resulting from industrial practices or processes.

inert   Lacking active properties and unable to react with other substances.

inertial separator   A device that uses centrifugal force to separate waste particles.

infectious agent   Any organism that is capable of being communicated in body tissues and causing disease or adverse health impacts in humans.

infectious dose 50 (ID$_{50}$)   The microbe dose that will infect 50% of a population to which it is applied.

infectious waste   Equipment, instruments, pathological specimens, or other disposable wastes that may be contaminated from persons who are suspected to have a communicable disease or have been diagnosed with a communicable disease.

infiltration   (1) Water entering a sewer system through broken or defective sewer pipes, service connections, or manhole walls. (2) Wind-induced air movement into a building through openings in walls, doors, or windows.

infiltration gallery   A horizontal underground conduit of screens, perforated pipes, or porous material which collects percolating water, often under a river bed.

Infinity™   Continuous later filter underdrain by Roberts Filter Group.

inflammable   Easily set on fire.

inflow   Surface and subsurface water or stormwater discharged into a sewer system.

inflow/infiltration (I/I)   The total quantity of water from inflow and infiltration without distinguishing the source.

influent   Water or wastewater flowing into a basin or treatment plant.

Information Collection Rule (ICR)   An EPA rule requiring water utilities serving more than 10,000 customers to conduct monitoring which will aid in the gathering of data for use in developing the D/DPB Rule and ESWTR.

infrared furnace   A furnace using infrared radiant heat emitted from silicon carbide resistance heating elements to rapidly heat organic wastes to combustion temperatures.

infrared radiation   Low energy radiation with wavelengths longer than visible light and shorter than radio waves.

infrastructure   The fundamental network of facilities, installations, and utility systems serving a community.

infuse   (1) To pour a liquid into or upon. (2) To steep in water or liquid without boiling so as to extract soluble constituents.

inhalable diameter   The diameter of a particle considered to be less than 15 micrometers for humans which are capable of being inhaled and deposited anywhere within the respiratory tract.

inhalation LC_{50}   The concentration of a substance expressed as milligrams per liter of air which is lethal for 50% of the test population.

inhibitor   A chemical that interferes with a chemical reaction.

Inipol™   Oil slick dispersant by Elf Atochem North America, Inc.

initiator   A chemical that can cause the initial step in the carcinogenesis process.

injection well   A hole drilled below the ground surface into which wastewater or treated effluent is discharged.

inlet   (1) A surface connection to a drain pipe. (2) A structure at the diversion end of a conduit. (3) The upstream end of any structure through which water may flow. (4) A form of connection between the surface of the ground and a drain or sewer for the admission of surface or storm water. (5) An intake.

innovative technology   A process or technique which has not been fully proven under the circumstances of its contemplated use and which represents an advancement over the state of the art.

inoculum   (1) Bacterium placed in compost to start biological action. (2) A medium containing organisms that is introduced into cultures or living organisms.

inorganic carbon   The carbon present in an inorganic compound such as carbon dioxide or calcium carbonate.

inorganic compound   Compounds that contain no carbon or contain only carbon bound to elements other than hydrogen.

inorganic contaminant (IOC)   An inorganic substance regulated by the EPA for compliance with drinking water requirements.

inorganic matter   Substances of mineral origin that do not contain hydrocarbons and are not subject to decay.

insecticide   A pesticide compound specifically used to kill or prevent the growth of insects.

In-Situ Oxygenator™   Mechanical floating aerator by Praxair, Inc.

insoluble   A compound that has very low solubility.

InSpectra™   UV analyzer by Azur Enviromental.

Instant Ocean®   Mineral concentrate used to simulate seawater salinity by Aquarium Systems.

InstoMix   In-line and in-channel mixers by Walker Process Equipment.

instrument detection limit (IDL)   Lowest concentration of a chemical that can be detected by an instrument without correction for the effects of a sample matrix or method-specific parameters.

instrumentation   Use of technology to control, monitor, or analyze physical, chemical, or biological parameters.

intake   (1) The works or structures at the head of a conduit into which water is diverted. (2) The process or operation by which water is absorbed into the ground and added to the saturation zone. (3) The flow or rate of flow into a canal, conduit, pump, stack, tank, or treatment process before treatment.

Integra®   Bladder pumps by Solinst Canada Ltd.

integrated exposure assessment   Cumulative summation (over time) of the magnitude of exposure to a toxic chemical in all media.

Intellisieve   Rotary fine screen filter by Smith & Loveless, Inc.

IntensAer   Radial surface aerator formerly offered by Walker Process Equipment.

inter-   Prefix meaning between or among.

interceptor sewer   A sewer that receives flow from a number of other sewers or outlets for disposal or conveyance to a treatment plant.

intercondenser   A condenser used between stages to reduce steam consumption in the steam jet vacuum system in an evaporator system.

interconnecting piping   Piping, usually field-installed, that connects equipment skids or unit processes to one another.

interface   The common boundary between two substances such as water and a solid and water and a gas, or between two liquids such as water and oil.

interfacial tension   The tension that occurs at the interface between two fluids or a liquid and a solid.

intergranular corrosion   Corrosion at or near the grain boundaries.

Inter-Mix®   Slow speed mixer by Air-O-Lator Corp.

Internalift®   Enclosed screw pump by USFilter/CPC.

International Process System   Former supplier acquired by USFilter/CPC.

International Shredder   Sewage shredder by ZMI/Portec Chemical Processing.

Interox America   Former name of Solvay America.

Inter-Sep™   Rotary screen by Dontech, Inc.

interstate waters   Waters that flow across or form part of state or international boundaries, or coastal waters.

interstice   An open space in granular material that is not occupied by solid material.

interstitial monitoring   The continuous surveillance of the space between the walls of an underground storage tank.

interstitial water   (1) Water contained in the interstices of rocks. (2) Extracellular water between cells.

intra-   Prefix meaning within or inside.

Intracid®   Water tracing dye by Crompton & Knowles Colors, Inc.

inverse solubility   The characteristic attributed to a substance that becomes less soluble with increasing temperature.

inversion   The abnormal atmospheric condition that occurs when the air temperature increases with elevation.

invert   The lowest point of the internal surface of a drain, sewer, or channel at any cross section.

inverted siphon   A u-shaped pipe used to convey flow under a river, road, or other obstruction. Also called a "sag line."

in-vessel composting   Composting system with integral material handling and in-vessel mixing and aeration.

IOC   (1) See "inorganic contaminant (IOC)." (2) Inorganic chemicals.

iodinator   A mechanical device used to introduce iodine into water for sanitization purposes.

iodine   A nonmetallic element in the halogen group sometimes used as a disinfectant.

iodine number   A surrogate value indicating an activated carbon's ability to adsorb low molecular weight organics.

iodometric titration   See "Winkler titration."

ion   An electrically charged atom, molecule or radical.

ion chromatography (IC)   A technique for separating and measuring the quantities of different ions present in a sample based on their affinity for an adsorption medium.

ion exchange (IX or IE)   A chemical process involving the reversible exchange of ions between a liquid and a solid.

Ion Grabber   Electrolytic purification unit by Hoffland Environmental Inc.

Ion Mobility Spectrometry (IMS)   Continuous emissions analyzer used to measure pollutants in gases.

Ion Stick®   Electrostatic water treater for prevention of scale and fouling by York Energy Conservation.

Ionac®   Ion exchange resins by Sybron Chemicals, Inc.

ionic strength   A measure of solution strength based on both the concentrations and valences of the ions present.

ionization   The process by which an atom or molecule acquires a positive or negative electrical charge through the loss or gain of electrons.

ionization chamber   A device that measures the intensity of ionizing radiation.

ionizing radiation   Radiation that can strip electrons from atoms.

ionosphere   The upper level of the earth's atmosphere beginning at an altitude of approximately 80 km.

Ionpure®   High purity water treatment products and services by USFilter/Lowell.

Iopor   Low pressure ultrafiltration system formerly offered by GL&V/Dorr-Oliver, Inc.

IP   Inhalable particulates.

IPA   Isopropyl alcohol.

IPCC   Intergovernmental Panel on Climate Change.

IPLP®   Double-pass reverse osmosis system by USFilter Corp.

IPP   Independent power producer.

IPS Composting System   In-vessel composting system by USFilter/CPC.

IQS/3™   Programmable controller used to operate and monitor water treatment systems by Culligan International, Corp.

iron   A common naturally occurring metallic element found in dissolved form in most water supplies. Chemical formula is Fe.

iron bacteria   Bacteria capable of metabolizing ferrous iron often from water or steel pipes. See also "crenothrix polyspora."

Iron Humate®   An iron source for agricultural soils and turf grasses produced from water treatment plant sludges by Kemiron.

Iron Remover   Contact bed type iron removal system by Walker Process Equipment.

iron salt   An iron-based coagulant used in water and wastewater treatment.

irradiated food   Food subject to brief radioactivity, usually gamma rays, to kill insects, bacteria, and mold, and to permit storage without refrigeration.

irradiation   Exposure to radiation of wavelengths shorter than those of visible light for medical purposes, to sterilize foodstuffs, or to induce polymerization of monomers or vulcanization of rubber.

irreversible effect   An effect characterized by the inability of the body to partially or fully repair injury caused by a toxic agent.

irrigation   The artificial application of water to meet the requirements of growing plants or grass that are not met by rainfall alone.

irrigation efficiency   The amount of water stored in the crop root zone compared to the amount of irrigation water applied.

irritant   A substance that can cause irritation of the skin, eyes, or respiratory system. Effects may be acute from a single high level of exposure or chronic from repeated low-level exposures to such compounds as chlorine, nitrogen dioxide, and nitric acid.

ISA   Instrument Society of America.

ISCO   In-situ chemical oxidation.

ISEP®   Ion exchange systems by Advanced Separation Technologies.

ISF™   Induced air flotation unit by Baker Hughes Process Systems.

ISO   International Organization for Standardization.

ISO 9000   Certification conferred to a manufacturer that has demonstrated the capability of running an integrated business from initial design through manufacture.

ISO 14000   A series of guidance standards to provide businesses with a structure for managing varied environmental concerns.

ISO 14001   Environmental management system standards for manufacturing and service industries.

isobar   A line on a weather map that joins all points of equal barometric pressure.

isobath   A line on a map connecting all points of equal depth above the surface of a water-bearing formation or aquifer.

isochrone   A line on a map connecting all points having the same time of travel for contaminants to move through the saturated zone and reach a well.

isohaline   A line on a map connecting points having equal amounts of salinity.

isohyet   A line on a map connecting points that receive equal average amounts of precipitation.

isomer   A chemical compound that has the same molecular formula but different molecular structure, as another compound.

isopleth   A line on a map connecting points at which a certain variable has the same value.

isotherm   A line on a weather map connecting points that have the same temperature.

isothiazalon   A high molecular weight non-oxidizing biocide used to control membrane biofouling.

isotopes   Atoms with the same atomic number but different atomic weights.

isthmus   A narrow strip of land bounded by water on both sides and connecting two large land masses.

ITT Marlow   Former name of ITT A-C Pump.

IU   Industrial user.

IWEM   See "CIWEM."

IWPF   Industrial wastewater pretreatment facility.

IWS   USFilter/Industrial Wastewater Systems.

IWT®   Illinois Water Treatment product line of USFilter/Rockford.

IX   See "ion exchange (IX or IE)."

IXPER®   Calcium peroxide by Solvay America.

J

J   See "joule (J)."

J&A   Jones and Attwood product division of Waste-Tech, Inc.

JAC Oxyditch   Oxidation ditch treatment system formerly offered by Chemineer, Inc.

Jackbolt™   Aluminum clarifier cover by Enviroquip, Inc.

jacking   A method of installing pipe by forcing it into a horizontal opening with horizontal jacks.

JackKnife   Pivoting air header and drop pipe arrangement by Walker Process Equipment.

Jackson turbidity unit (JTU)   An obsolete unit of measure used to quantify the water turbidity by observing the outline of a candle flame viewed through a water sample. Replaced by "NTU."

Jacquelyn™   Corrosion and zebra mussel resistant coating by Cook Screen Technologies, Inc.

JAPCA   Journal of Air Pollution Control Association.

jar test   A test procedure using laboratory glassware for evaluating coagulation, flocculation, and sedimentation in a series of parallel comparisons.

Jayfloc   Polyelectrolyte used to enhance liquid/solid separation by Vulcan Performance Chemicals.

J-Belt   Belt filter press by USFilter/Dewatering Systems.

Jeffrey   Screening equipment product line offered by Jones & Attwood, Inc.

Jeffrey®   Chain, sprocket, and related component products by Jeffrey Chain Corp.

JelClear™   Granular filtration media having a coagulant bonded directly to the media grains by BetzDearborn-Argo District.

jet   A stream of pressurized liquid or vapor from a nozzle or orifice.

jet aeration   Wastewater aeration system using floor-mounted nozzle aerators that combine liquid pumping with air diffusion.

Jet Breaker™   Screenings washer/compactor by Headworks, Inc.

Jet Plant   Package wastewater treatment plant by Jet, Inc.

Jet Shear   Continuous mixing system using jet nozzles by Flo Trend Systems, Inc.

jet stream   A strong, thermally driven, high altitude wind.

Jet Tech   Former name of USFilter/Jet Tech.

Jet Tray   Deaerator by Cochrane Inc.

Jet VIP™   Pulsed fabric filter by Wheelabrator Air Pollution Control, Inc.

JETA   Vortex type grit collector by Waste-Tech, Inc.

Jeta-Matic   Spray jet pressure leaf filter cleaning system by USFilter/Whittier.

Jet-Chlor®   Tablet chlorine disinfection system by Jet, Inc.

JETIII®   Fabric filter by Wheelabrator Air Pollution Control, Inc.

JetMix™   Vortex mixing system for sludge storage basins by Liquid Dynamics Corp.

jetsam   Floating jettisoned material.

Jet-Tex   Leach bed filter fabric by Jet, Inc.

jetty   A structure that extends into an open body of water to influence currents or tides or protect a harbor.

Jet-Wet™   Dry polymer feeding system by Fluid Dynamics, Inc. and Ciba Specialty Chemicals.

J-Flow   Gravity belt thickener by USFilter/Dewatering Systems.

JH™   Series of air scrubbers for odor control by USFilter/Davis Process.

Jigrit   Screw-type grit washer by Jeffrey Chain Corp.

J-Mate®   Sludge volume reduction system by USFilter/Dewatering Systems.

Johnson Filter   Former name of USFilter/Johnson Screens.

Johnson Screen   Wedgewire screen media by USFilter/Johnson Screens.

Jones and Attwood   Environmental product division of Waste-Tech, Inc.

joule (J)   An SI unit of energy or work equal to the work done by the application of 1 newton acting through a distance of 1 meter.

Joule™   Electrochemical flocculation system by Ecoloquip Inc.

journal bearing   A cylindrical bearing that supports a rotating shaft.

J-Press®   Plate and frame filter press by USFilter/Dewatering Systems.

J-Spin   Centrifuge by USFilter/Dewatering Systems.

J-Track™   Nonmetallic return track for chain and flight sludge collectors by USFilter/Envirex.

JTU See "Jackson turbidity unit."

JUD Belt tracking system for belt filter press by Klein America, Inc.

junction box An enclosure within which electrical cables are connected and/or terminated.

J-Vap™ Mechanical sludge dewatering and drying system by USFilter/Dewatering Systems.

JWI Dewatering equipment product line by USFilter/Dewatering Systems.

K

K soil horizon A soil horizon in which the grains are cemented with $CaCO_3$ to form a hardpan layer.

K wastes A class of hazardous wastes categorized from specific sources as established by the EPA and identified in 40 CFR 261.32.

K² Modular™ Volumetric screw feeder by K-Tron North America.

Kaldnes System Biological wastewater treatment system by Waterlink Biological Systems.

kame A ridge or hill of stratified drift deposited by glacial meltwater.

KAMET Municipal effluent treatment system by Krofta Engineering Corp.

Kan-Floc™ Wastewater flocculant/coagulant by Kem-Tron.

kaolin A type of fine white clay material.

karst A geologic formation of irregular limestone deposits with sinks, underground streams, and caverns.

katabatic wind A localized wind that flows down valley or mountainous slopes — usually at night — caused by the descent of cold air as the valley slopes undergo rapid nocturnal cooling.

Katec® Thermal oxidizer by Megtec Systems, Inc.

Kat-Floc™ Wastewater flocculant/coagulant by Kem-Tron.

Katox Catalytic oxidizer by Adwest Technologies, Inc.

KD-HF™ Deionization system by Kinetico Engineered Systems, Inc.

Kebab™ Disc-type oil skimmer by Vikoma International Ltd.

Keene Former equipment manufacturer acquired by Amwell, Inc.

Kenics® Static mixer product line by Chemineer, Inc.

Kenite® Diatomite material by Celite Corp.

Ketema Product group of Baker Process/Ketema.

Key-Pac Package water/wastewater treatment plants by Pacific Keystone Technologies.

Keystone Former name of Pacific Keystone Technologies.

K-Floc™ Wastewater flocculant/coagulant by Kem-Tron.

K-Floor Suspended monolithic filter floor by the former PWT Americas.

kg See "kilogram (kg)."

kiln A heated enclosure for processing a substance by drying or burning.

kiln dust See "cement kiln dust."

kilocalorie   See "calorie."

kilogram (kg)   A unit of mass equivalent to 1000 grams or approximately 2.205 pounds.

kilopascal (kPa)   A unit of pressure equal to 1000 pascals, 0.3346 feet of hydraulic head, or 0.145 psi.

kilowatt (kW)   A measure of power equal to 1000 watts; 1 hp equals 0.746 kilowatts.

kilowatt-hour (kWh)   A unit of energy equal to that expended by 1 kilowatt in 1 hour.

kinematic viscosity   A fluid's absolute viscosity divided by its mass density.

kinesis   An organism's involuntary movement or reaction to an environmental stimulus.

kinetic energy   Energy that is possessed by a body of matter as a result of its motion.

kinetic head   The theoretical vertical height that a liquid may reach due to its kinetic energy.

KIWA   Netherlands Waterworks Testing and Research Institute.

Kiwi   Centrifuge conveyor by Alfa Laval Separation, Inc.

Kjeldahl nitrogen   The sum of the organic plus ammonia nitrogen in a water sample.

KL Series™   Cleaning powders for removal of membrane foulants by King Lee Technologies.

Klampress®   Belt filter press by Ashbrook Corp. (U.S.) and Simon-Hartley, Ltd. (U.K.).

Kleer Flow   Spiral wound reverse osmosis membranes by Great Lakes International, Inc.

Klenphos-300   Zinc phosphate corrosion inhibitor by Klenzoid, Inc.

Klensorb   Oil and grease absorbent by Calgon Carbon Corp.

Knight-Botec   Auger compost mixer by Knight Manufacturing Corp.

knot   A unit of speed equal to one nautical mile per hour (approximately 1.15 mph) used to describe the speed of winds or ships.

Koagulator   Solids contact clarifier by USFilter/Zimpro.

Koch Corrosion Control   Former name of KCC Corrosion Control Co.

KochKleen®   Reverse osmosis cleaning chemicals by Koch Membrane Systems, Inc.

KochTreat®   Antiscalants, stabilizers, and anti-microbial chemicals by Koch Membrane Systems, Inc.

Koflo®   In-line static mixers by Koflo Corp.

Komara™   Floating oil skimmer by Vikoma International Ltd.

Kompakt   Tubular block media by NSW Corp.

Kompress®   Belt filter press by Komline-Sanderson, Engineering Corp.

Komprimat   Fish/screenings separation system by Brackett Geiger.

Konsolidator   Wet scrubber solids filter by CMI-Schneible Co.

Kopcke   Former name of REKO Industrial Equipment B.V.

Koppers   Former manufacturer of traveling bridge filters whose product line was acquired by Infilco Degremont, Inc.

Koro-Z   PVC biological filter media formerly offered by B.F. Goodrich Co.

kPa   See "kilopascal (kPa)."

kraft   An alkaline chemical pulping process using salt cake as makeup.

Kraus-Fall   Peripheral feed clarifier by Smith & Loveless, Inc.

krill   A small, shrimp-like marine crustacean that is a major food source for whales, seals, and squid.

Krüger   Former name of USFilter/Krüger.

Kruger/Fuchs   Autothermal thermophilic aerobic digestion system by USFilter/Krüger.

KUBE³   Belt filter press by Klein America, Inc.

kW   See "kilowatt (kW)."

K-W Products   Former equipment manufacturer acquired by Smith & Loveless, Inc.

kWh   See "kilowatt-hour (kWh)."

Kyoto Protocol   A global agreement among 150 nations who met in Kyoto, Japan in 1997 to discuss a comprehensive plan to reduce greenhouse gas emissions in hopes of improving climatic conditions.

L

*L*ARO*   Reverse osmosis system by the former L*A Water Treatment Corp.

L/d   Liters per day.

L/d ratio   A simple parameter used to aid in sizing granular media filters based on the bed depth and effective size of media.

L-10 life   See "B-10 life."

La Niña   A climatic cycle that results in what climatologists believe is the most normal wind, pressure, and current patterns in the tropics.

LADD   Lowest acceptable daily dose.

LAER   Lowest achievable emission rate.

Lagco   Parshall flume by F.B. Leopold Co., Inc.

lagoon   An excavated basin or natural depression that contains water, wastewater, or sludge.

Lagoonerator™   Submerged, fine bubble aeration diffusion device by USFilter/Envirex.

laid length   The total length of a pipe or pipeline after it has been placed in position.

lake   A large, inland body of water, usually more than 200,000 square meters.

Lake Aid Systems   Former name of LAS International.

Lakos IPC   Self-cleaning pump intake screen by Claude Laval Corp.

LAL test   See "Limulus Amebocyte Lystate test (LAL test)."

lamel   A thin plate.

Lamella®   Gravity settler/thickener using inclined plates by Parkson Corp.

LamGard   Automated oxygen control system by Gardner Denver Blower Division.

laminar flow   A flow situation in which fluid moves in smooth parallel layers with essentially no mixing or turbulence, usually with a Reynolds number less than 2000.

LAMP   Lake Acidification Mitigation Project.

Lam-Pak®   Package treatment plant by Graver Co.

Lamson® Centrifugal blower products by Gardner Denver Blower Division.

Lanco Environmental™ Equipment product line of Waterlink Separations, Inc.

Lancom™ Flue gas monitoring system by Land Combustion.

Lancy™ Wastewater treatment product line offered by USFilter/Industrial Wastewater Systems.

land application The disposal of wastewater or municipal solids onto land under controlled conditions.

land ban RCRA provisions prohibiting land disposal of specific toxic materials unless they meet applicable treatment standards.

land disposal Application of municipal wastewater solids to the soil without production of usable agricultural products.

Land Disposal Restrictions (LDR) EPA-promulgated rules implementing the land ban.

landfarming Application of organic waste onto surface soil for the purpose of controlled biodegradation.

landfill (LF) A land disposal site that employs an engineering method of solid waste disposal to minimize environmental hazards and protect the quality of surface and subsurface waters.

Landox™ Aeration system by WesTech Engineering Inc. (North America) and Landustrie Sneek BV.

landscaping The enhancement of the appearance of land by changing its contours and planting decorative vegetation.

Landy™ Surface aerator product line by WesTech Engineering Inc. (North America) and Landustrie Sneek BV.

Langelier Saturation Index (LSI) A measure of the degree of saturation of calcium carbonate in water based on pH, alkalinity, and hardness. A positive LSI indicates that calcium carbonate may precipitate from solution to form scale.

Lanpac® Packing media for air stripping towers by Lantec Products, Inc.

lapse rate The rate at which temperature decreases as altitude increases.

Laran® An anaerobic wastewater pretreatment process by Lotepro Corp (Western Hemisphere) and Linde-KCA-Dresden GmbH.

large calorie (Cal) See "calorie."

large dam Criteria established by ICOLD to categorize dams whose height 15 meters or more; or whose height is between 10-15 meters with a crest length of at least 500 meters with a reservoir volume of at least 1 million cubic meters.

large quantity generator Person or facility generating more than 1000 kilos (2200 pounds) of hazardous waste per month and subject to all RCRA requirements.

large water system A water system that serves more than 50,000 persons.

large-quantity handlers (LQHs) Universal waste handlers who accumulate more than 5000 kilograms of wastes.

LAS See "linear alkyl sulfonate (LAS)."

Lasaire® Lagoon aeration system by A.B. Marketech, Inc.

latency period The time elapsing from the first exposure of a chemical until the appearance of a toxic effect.

latent heat The heat required to cause a change of state at constant temperature, such as the vaporization of water or the melting of ice.

lateral A secondary pipe that extends from a main water pipe or header.

Lateral Flow Sludge Thickener™ Gravity sludge thickener by Gravity Flow Systems, Inc.

lateral sewer A sewer which connects the collection main to the interceptor sewer.

launder A trough used to transport water.

laundering weir A v-notched overflow weir used to promote uniform flow rates.

laundry wastes Wastewater from industrial laundries that may be characterized by the presence of lint, fibers, oils, and greases.

Lavasol™ Liquid reverse osmosis membrane cleaners by Professional Water Technologies, Inc.

LC (1) Lethal concentration. (2) Liquid chromatography.

LC$_{50}$ See "lethal concentration."

LCL Lower control limit.

LCR Lead and Copper Rule.

LCRS Leachate collection and removal system.

LD See "lethal dose."

LD$_{50}$ See "lethal dose (LD)."

LDC London Dumping Convention.

LDCRS Leachate detection, collection, and removal system.

LDR See "Land Disposal Restrictions (LDR)."

LDS Leak detection system.

leach field The area of land into which a septic tank drains or wastewater is discharged.

leachate Fluid which percolates through solid materials or wastes and contains suspended or dissolved materials or products of the solids.

leachate collection system A system that gathers leachate and pumps it to the surface for treatment.

leaching The process by which soluble materials are washed out of soil, ore, or buried waste, and into a water source.

lead A trace element and cumulative poison that may be inhaled or ingested in food or water. Chemical formula is Pb.

Leadtrak Test kits used to determine lead content of water by Hach Co.

leaf filter A precoat filter with flat elements or leaves.

leakage (1) The presence of an ionic species in an ion exchanger effluent that usually indicates bed exhaustion. (2) The uncontrolled loss of water from a tank or aquifer.

Leakwise® Oil on water monitoring systems by Agar Corp.

L'eau Claire® Upflow filter products by USFilter/Whittier.

Lectra/San Marine sanitation system by Exceltec International Corp.

left bank The left hand bank of a river or stream when facing downstream.

Legionella A genus of bacteria, some species of which have caused a type of pneumonia called Legionnaires Disease.

LEL See "lower explosive limit (LEL)."

Lemna® Biological wastewater treatment system utilizing aquatic duckweed by Lemna Corp.

Lemnaceae   Floating aquatic plants that provide a habitat for aquatic organisms capable of metabolizing wastewater organics. Also known as "duckweed."

LemTec™   Biological wastewater treatment system and products by Lemna Corp.

lentic water   Standing or stagnant pond, swamp, or marsh water.

Leo-Lite   Fiberglass effluent and scum trough by F.B. Leopold Co., Inc.

Leopold®   Water and wastewater treatment products by F.B. Leopold Co., Inc.

Leo Vision   PC-driven graphics display of treatment plant operating conditions by F.B. Leopold Co., Inc.

LEPA   The designation for a low-energy, precision application irrigation sprinkler.

lethal   Causing, or being able to cause, death.

lethal concentration   The concentration of a substance which is fatal to a specified percentage of the population, usually expressed at the 50% level as LC_{50}.

lethal dose (LD)   The quantity of a substance which is fatal to a specified percentage of the population, usually expressed at the 50% level as LD_{50}.

LEV   Low emissions vehicle.

levee   A dike or embankment along a river built to prevent flooding of the surrounding land.

Level Bed Agitator   Agitator used in composting system by USFilter/CPC.

Level Mate™   Level measurement instrument by Ametek, PMT Products.

Level of Concern (LOC)   The concentration in air of an extremely hazardous substance above which there may be serious immediate health effects to anyone exposed for short periods.

levigation   The separation of fine particles from coarser ones by suspending them in a liquid.

Lewatit®   Ion exchange resin by Bayer Corp.

LF   See "landfill (LF)."

LFG   Landfill gas.

LFL   Lower flammability limit.

LHW   Liquid hazardous waste.

LI   See "Langelier Saturation Index (LSI)."

lichen   A sponge-like plant growing on wood, stone, or soil which is formed by an association of a fungus and alga, and often used as an air pollution indicator species.

life cycle cost   A method of comparing costs of various alternatives which considers capital, operations, and maintenance costs.

Lifeserver™   Built-in-place wastewater treatment plants by USFilter/Davco.

lifetime exposure   Total amount of exposure to a substance which a human would receive in a lifetime, usually assumed to be 70 years.

Lift Screen   Reciprocating rake bar screen by USFilter/Headworks Products.

lift station   A chamber that contains pumps, valves, and electrical equipment necessary to pump water or wastewater.

ligand   An atom, ion, or molecule bound to a central atom of a molecule to form a complex.

Lightspeed   Digital fiber optic flowmeter by Newport Electronics.

lignin   An organic substance that forms the chief part of wood tissue.

lignite   A type of coal with a low energy content; also called "brown coal."

LIMB   Limestone-injection multistage burner.

Limberflo   Precast filter bottom system by USFilter/Aerator Products.

lime   The term generally used to describe ground limestone (calcium carbonate), hydrated lime (calcium hydroxide), or burned lime (calcium oxide).

lime kiln   A unit used to calcine lime.

lime recalcining   Recovery of lime from water or wastewater sludge, usually with a multiple hearth furnace.

lime scale   Scale formed by hard water containing a high percentage of calcium carbonate.

lime slaker   A device used to hydrate quicklime.

lime softening   The addition of sufficient lime to raw water to achieve a reduction of carbonate hardness.

lime stabilization   The addition of lime to untreated sludge to raise the pH to 12 for a minimum of 2 hours to chemically inactivate microorganisms.

lime-and-settle   Common term for treatment technologies that utilize chemical precipitation and sedimentation processes.

lime-soda softening   The addition of sufficient lime and soda ash to raw water to achieve a reduction of carbonate and noncarbonate hardness.

limestone   A sedimentary rock composed primarily of calcium carbonate.

limestone scrubbing   Use of a limestone and water solution to remove gaseous stack-pipe sulfur before it reaches the atmosphere.

limit of detection (LOD)   The minimum concentration that can be detected by an analytical method. Generally the same as "instrument detection limit."

limnology   The study of fresh water lakes and their flora and fauna.

LIMS   Laboratory information management system.

Limulus Amebocyte Lystate test (LAL test)   A test used to determine presence of endotoxins in treated water, commonly used on water to be used for pharmaceutical purposes.

lindane   A pesticide that causes adverse health effects in domestic water supplies and is toxic to freshwater fish and aquatic life.

Lindox®   A pure oxygen, activated sludge wastewater treatment process by Lotepro Corp. (Western Hemisphere) and Linde-KCA-Dresden GmbH.

linear alkyl sulfonate (LAS)   A family of chemical compounds widely used as detergents, sometimes called "soft detergents" because they are readily degraded to simpler substances by biological action.

liner   (1) A barrier of plastic, clay, or other impermeable material which prevents leachate from contacting surface or subsurface water. (2) A protective, corrosion resistant layer attached or bonded to the inside of a tank.

Link-Belt®   Environmental equipment product line by USFilter/Envirex.

Linpor®   An activated sludge wastewater treatment process by Lotepro Corp. (Western Hemisphere) and Linde-KCA-Dresden GmbH.

LinX™   A spool valve used in reverse osmosis energy recovery devices by Desal Co. Ltd.

lipids   A group of organic compounds including fats that are water insoluble and important in the structure of cell walls and membranes.

lipophilic   Having an affinity for oil.

Liquaclone®   Hydrocyclonic solids separation unit for removal of granular solids from liquid discharges by Sanborn Environmental Systems.

Liquapac™   Solids removal unit for spent coolant/oil clarification applications by Sanborn Environmental Systems.

liquefaction   The process of making or converting a solid or gas to a liquid.

Liqui/Jector®   Liquid/gas coalescer by Osmonics, Inc.

Liqui-Cel®   Membrane contactors by Celgard LLC.

liquid   The state of matter between the solid and gaseous states in which matter possesses a definite volume and flows freely but has no definite shape.

Liquid A™   Sludge stabilization process by RDP Technologies, Inc.

Liquid Carbonic   Product line acquired by Praxair, Inc.

liquid chlorine   Chlorine compound that contains no water and results from gaseous chlorine under high pressure and which is stored in steel drums and cylinders.

liquid sludge   Sludge which contains sufficient water flow by gravity or permit pumping.

liquid-liquid extraction (LLE)   See "solvent extraction."

Liquid-Miser   Activated carbon odor absorbers by Westport Environmental Systems.

Liquidow®   Calcium chloride products by Dow Chemical Co.

LiquidPure   Small, low-cost activated carbon drum adsorber by American Norit Company, Inc.

Liqui-Fuge™   Internally fed rotary fine screen by Vulcan Industries, Inc.

LiQuilaz®   In-line sensor for particle measurement by Particle Measuring Systems, Inc.

Liquiphant   Liquid level indicator by Endress+Hauser.

Liqui-pHase®   Carbon-dioxide neutralization system by Praxair, Inc.

LiquiPro™   Metering pumps by Liquid Metronics, Inc.

Liquipure   Company acquired by USFilter/Lowell.

Liqui-Strainer   Externally fed rotating drum screen by Vulcan Industries, Inc.

Liquitron®   pH/ORP controllers by Liquid Metronics, Inc.

liquor   A aqueous solution of one or more chemical compounds.

listed hazardous waste   The designation for a waste material that appears on an EPA list of specific hazardous wastes or hazardous waste categories.

liter (L)   A unit of volume equal to 1000 cubic centimeters or 1.057 quarts. One liter of water weighs 1000 grams. Also called "litre."

lithology   The character or description of rocks in terms of their physical and chemical characteristics.

lithosphere   The solid portion of earth composed of rocks and soil.

Litmustik®   Pocket pH tester by Omega Engineering, Inc.

litre   See "liter (L)."

litter   Solid waste or garbage from human activity deposited indiscriminately.

Little Fox   Modular wastewater treatment plant for marine applications by Red Fox Environmental, Inc.

littoral zone   The area of the shore line between high and low tides where rooted water plants can grow.

live bottom bin   A storage bin in which controlled bottom discharge is facilitated by a vibrating device or other mechanical mechanism.

LLD   Lower limit of detection. Generally the same as "instrument detection limit."

LLE   Liquid-liquid extraction. See "solvent extraction."

LLL   Low liquid level.

LLQ   Lower limit of quantitation. Generally the same as "estimated quantitation limit."

LLW   See "low-level radioactive waste (LLW)."

LME®   Inclined plate separator by USFilter/Zimpro.

LMI   Liquid Metronics, Inc.

LNG   Liquefied natural gas.

Lo/Pro™   Packaged odor control system by USFilter/RJ Environmental.

Load Limitor   Automatic chain tensioning system for traveling water screens by USFilter/Rex & Link-Belt Products.

loading rate   The flow rate per unit area of treatment process through which water flows.

LOAEL   See "lowest-observed-adverse-effect level (LOAEL)."

loam soil   A rich soil consisting of organic material, sand, silt, and clay.

Lobe-Aire®   Rotary lobe blower by Spencer Turbine Co.

Lobeflo™   Rotary lobe pump by MGI Pumps, Inc.

Lobestar®   Mixing eductor by Vortex Ventures.

LOC   See "Level of Concern (LOC)."

local ventilation   The drawing off and replacement of contaminated air directly from its source.

localized corrosion   Corrosion taking place at a relatively high speed in limited sections of the area exposed to the corrosive medium.

Lo-Cat®   Hydrogen sulfide oxidation process for anaerobic bio-gas systems by USFilter/Gas Technologies.

loch   A lake or narrow body of water surrounded by land and stretching to the sea (Scottish).

lock   A short section of a canal equipped with gates at both ends so that the water level can be changed to raise or lower boats from one level to another.

LOD   See "limit of detection (LOD)."

LOEL   Lowest-observed-effect concentration.

log boom   A floating structure of logs or timber used to protect an intake, dam, or other structure by deflecting floating material.

log reduction   See "log removal."

log removal   A means of indicating the level of log_{10} removal, inactivation, or kill of pathogenic organisms through physical-chemical treatment of water. For example, 1-log removal equals a 90% reduction of the specified organism; a 2-log reduction equals a 99% reduction; and a 3-log reduction equals a 99.9% reduction.

log-death phase   Bacterial growth phase where the microbe death rate exceeds the production of new cells.

LogEasy™   Particle counter by Hach Co.

log-growth phase   Bacterial growth phase where cells divide at a rate determined by their generation time and their ability to process food.

Lo-Head™   Traveling bridge filter by Agency Environmental, Inc.

Lo-Hi™   Low pressure, high intensity ultraviolet lamps by PCI-Wedeco Environmental Technologies, Inc.

LOI   Loss on ignition.

long ton   A unit of weight equal to 2240 pound.

Longopac   Screenings bagging system by Spirac.

LoNox™   Combustion burner by John Zink Co.

LOOP   Package wastewater treatment process using oxidation ditch process by Smith & Loveless, Inc.

Loop Chain   Nonmetallic, filament wound sludge collector chain by USFilter/Envirex.

Lo-Pro™   Air stripper by Geotech/ORS Environmental Systems.

LOQ   Limit of quantitation. Generally the same as "estimated quantitation limit."

loss of head   A decrease in head energy that results from a bend, obstruction, or expansion in a channel or pipeline.

lotic water   Rapidly flowing water of a river or stream.

Love Canal   An industrial chemical waste site in Niagara Falls, NY which contaminated a residential area and contributed to public furor resulting in the 1980 enactment of Superfund.

low NOx burners   One of several combustion technologies used to reduce emissions of nitrogen oxides.

low sodium water   Bottled water containing 140 mg or less of sodium per serving.

low sulfur coal   See "compliance coal."

lower explosive limit (LEL)   The concentration of a compound in air below which the mixture will not catch on fire.

lowest-observed-adverse-effect level (LOAEL)   The lowest dose of a substance to cause an increase in the frequency or severity of an adverse effect in an exposed population.

low-flow toilet   A toilet that uses no more than 1.6 gallons of water per flush.

low-level radioactive waste (LLW)   Wastes less hazardous than most of those associated with nuclear reactor, usually generated by hospitals, research laboratories, and certain industries.

LOX   Liquid oxygen.

LP Block™   Low profile filter underdrain by Tetra Process Technologies.

lpf   Liters per flush.

LPG   Liquefied petroleum gas.

LQG   Large quantity generator.

LQHs   See "large-quantity handlers (LQHs)."

LSC™   Package spray-type deaerating heater by Graver Co.

LSI   See "Langelier Saturation Index (LSI)."

LSTK   Lump sum turnkey.

LTA   Low temperature additive.

LTESWTR   Long-Term Enhanced Surface Water Treatment Rule.

LUFT   Leaking underground fuel tank.

lumen   (1) The bore or axial hole through the center of a hollow fiber membrane or tubular structure. (2) A unit of light measurement equal to the light given off in a unit solid angle from a uniform point source of one candela.

LUST   Leaking underground storage tank.

LVHC   Low volume, high concentration.

LWL   Low water level.

LWT®   Liquid Waste Technology, Inc.

LX   Leachability index.

Lyco™   Wastewater treatment equipment product line by USFilter/Industrial Wastewater Systems.

Lynx®   Chain driven bar screen by Waterlink Separations, Inc.

lyse   To undergo lysis.

lysimeter   A device used to measure or obtain samples of water draining through soil.

lysis   The rupture of a cell that results in loss of its contents.

M

m³   See "cubic meter (m³)."

mA   Milliampere.

MacerAcer™   Screenings conditioning equipment by Brackett Geiger.

macerate   To chop or tear.

Macho Monster®   In-channel sewage grinder by JWC Environmental.

MacPac   Chemical feed package by Milton Roy Co.

Macro-Cat   Ion exchange resin by Sybron Chemicals, Inc.

macroencapsulation   Isolation of waste by embedding or surrounding it with a material that acts as a barrier between the waste and air, water, or other materials.

macrofloc   Destabilized floc particle that is too large to penetrate a granular media filter bed.

macrofouling   The biological fouling of a water system with macroorganisms including clams, barnacles, and mussels.

Macrolite®   Ceramic filter media by Kinetico Engineered Systems, Inc.

macroorganisms   All organisms larger than microscopic and visible to the unaided eye.

macrophyte   A type of macroscopic plant life.

macroporous resin   An ion exchange resin with a high resistance to oxidation and organic fouling used primarily in applications with high molecular weight organic matter.

macroreticular resin   An ion exchange resin having a pore structure even after drying.

macroscopic   Capable of being seen with the naked eye.

MACT   See "maximum achievable control technology (MACT)."

MACTherm®   Regenerative oxidizer by Applied Regenerative Technologies Co.

MADAM   Methacryloyl ethyl dimethyl amine.

MAF   Million acre feet.

Magicblock™   Fluid control system by Osmonics, Inc.

Magna   Rotor aerator by Lakeside Equipment Corp.

Magna Cleaner™   Liquid cyclone by Andritz-Ruthner, Inc.

Magnafloc®   Coagulant aid by Ciba Specialty Chemicals

Magnasol®   Inorganic and organic coagulant products by Ciba Specialty Chemicals

MagneClear®   Magnesium hydroxide water treatment products by Martin Marietta Specialties, Inc.

magnetite   A black, iron oxide mineral also known as "lodestone."

Magnifloc®   Polyelectrolyte used to enhance liquid/solid separation by Cytec Industries, Inc.

Magnum   Ultraviolet disinfection equipment by Atlantic Ultraviolet Corp.

Magnum®   Belt filter press by Parkson Corp.

Magnum™   Catalytic oxidizer by Megtec Systems, Inc.

Magox®   Magnesium oxide by Premier Chemicals.

Mahr™   Mechanically cleaned bar screen by Headworks, Inc.

makeup water   Fluid introduced in a recirculating stream to maintain an equilibrium of temperature, solids concentration, or other parameter(s).

malathion   A common organophosphate insecticide.

M-alkalinity   See "methyl orange alkalinity."

Mallard   Bridge-mounted clarifier scum removal system by NSW Corp. (U.S.) and Copa Group (U.K.).

malodor   An odor which causes annoyance or discomfort to the public and which has been determined to be objectionable.

Mammoth®   Brush aerator by USFilter/Zimpro.

mandatory recycling   Programs which by law require consumers to separate trash so that some or all recyclable materials are recovered for recycling rather than going to landfills.

manganese greensand   See "greensand."

Manganex Process   Manganese greensand filtration process by Roberts Filter Group.

Manhattan Process   High rate filtration process by Roberts Filter Group.

manhole   See "personnel access opening."

manifest   A form used to identify the origin, quantity, and composition of a hazardous waste during its transportation from the point of generation to the point of final treatment or disposal.

manifest system   Tracking of hazardous waste from generation through disposal with accompanying documents known as manifests.

mannich polymer   Sludge conditioning polymer produced by using a formaldehyde catalyst to promote a chemical reaction to create the organic compound.

Manning's formula   Formula used to measure flow in an open channel based on the cross-sectional area of the flowing stream and the hydraulic radius, slope, and roughness of the channel.

manometer A u-tube device filled with a liquid used to measure pressure differentials in liquids or gases.

Manor® Filter press by Simon-Hartley, Ltd.

Manu-Matic Manual pressure leaf filter cleaning system by USFilter Corp.

Manver Chemical composition used in analysis of water hardness by Hach Co.

Manville Filtration Former name of Celite Corp.

manway See "personnel access opening."

MARD™ Motor-actuated rotary distributor for trickling filters by USFilter/General Filter.

marine sanitation device Any equipment or process installed on board a vessel to receive, retain, treat, or discharge sewage.

Marox Pure oxygen wastewater treatment system by USFilter/Zimpro.

MARS Membrane-controlled biological wastewater treatment technology by USFilter/Krüger.

marsh gas Methane gas produced by the anaerobic decomposition of organic matter in wetland areas. Also called "swamp gas."

marshland An area of soft wet land vegetated by reeds and grasses.

masking The blocking out or covering of a sound or smell with another.

Maspac® Plastic packing media by Clarkson Controls & Equipment Co.

mass balance An analysis that delineates changes that take place in a reactor or system by quantifying system inputs and outputs. Also known as "material balance."

mass burn Solid waste incineration of garbage with a minimum of pretreatment or sorting.

mass loading The total amount of mass of a constituent flowing into a system.

mass spectrometer An instrument used for the analysis of organic materials in environmental samples by sorting ions according to their masses and electrical charges.

Master-Flo Bladder pump by American Sigma, Inc.

material balance See "mass balance."

Material Safety Data Sheet (MSDS) Data sheet containing descriptive information required by OSHA for hazardous materials.

materials recovery facilities (MRFs) A central facility where recycled materials are prepared and sorted.

maturation pond An aerobic waste stabilization pond used for polishing treated wastewater effluent.

MAX™ Reverse osmosis system product line by USFilter.

MaxAir™ Wide band coarse bubble diffuser by Environmental Dynamics Inc.

Maxflo™ Mixer by S&N Airoflo, Inc.

Maxi-Flo® Pressurized sand filter product line by USFilter Corp.

Maxim® Seawater conversion evaporator by Beaird Industries, Inc.

Maximizer Dewatering sludge press by Goodnature Products, Inc.

MaxiMizer® Solid bowl centrifuge by Alfa Laval Separation, Inc.

maximum achievable control technology (MACT) The level of air pollution control technology required by Clean Air Act.

maximum contaminant level (MCL) The maximum permissible level of a contaminant in water delivered to the free flowing outlet of the ultimate user of a public water system.

maximum contaminant level goal (MCLG)   The maximum level of a contaminant, including an adequate safety margin, at which no known or anticipated adverse effect on human health would occur.

maximum residual disinfectant level (MRDL)   The maximum level of disinfectant added for treatment that may not be exceeded at a consumer's tap without the possibility of adverse health effects.

Maxipress   Former name of J-Belt belt filter press by USFilter/Dewatering Systems.

Maxi-Rotor   Rotary brush aerator by USFilter/Krüger.

MaxiSep   Oil/water separator by Hydro-Flo Technologies, Inc.

Maxi-Strip®   Hydraulic venturi stripper for VOC removal by Hazleton Environmental, Inc.

Maxi-Tank™   Tank sump by Hazleton Environmental, Inc.

Maxi-Yield™   Polymer blending system by USFilter/Wallace & Tiernan.

Max-Load™   Cartridge filter by Ronningen-Petter.

Max-Pak™   Plastic packing media by Jaeger Products, Inc.

Maz-O-Rator   Solids grinder by Robbins & Myers, Inc.

MBAS   See "methylene blue active substance (MBAS)."

MBBR™   Moving bed biofilm reactor by Waterlink Biological Systems.

MBR   See "membrane bioreactor (MBR)."

MBS®   Molecular bonding system to stabilize heavy metals by Solucorp Industries Corp.

Mc®   Propeller flowmeter by McCrometer, Inc.

MCC   Motor control center.

McGinnes-Royce   Manufacturer of screening equipment whose product line was acquired by USFilter/Rex & Link-Belt Products.

MCL   See "maximum contaminant level (MCL)."

MCLG   See "maximum contaminant level goal (MCLG)."

m-ColiBlue24™   Laboratory coliform test broth by Hach Co.

MCRT   See "mean cell residence time (MCRT)."

MDI   Metered dose inhaler.

MDL   See "method detection limit."

MDTOC   Minimum detectable threshold odor concentration.

ME   Multiple effect. See "multiple effect distillation."

me   See "milliequivalent (me)."

mean cell residence time (MCRT)   The average time that a microbial cell remains in an activated sludge system. It is equal to the mass of cells divided by the rate of cell wasting from the system.

mean flow   The arithmetic average flow at a given point for a specified period of time.

mean sea level (MSL)   The average sea level for all stages of the tide.

mean velocity   The average velocity of a fluid flowing in a channel, pipe, or duct, determined by dividing the discharge by the cross sectional area of the flow.

meander   One of a naturally occurring series of bends or curves in a river, usually formed on a floodplain composed of unconsolidated alluvium.

meander belt   The outermost limits of a floodplain along which a stream meanders.

MEB   Multiple effect boiling. See "multiple effect distillation."

mechanical aeration   The mechanical agitation of water to promote mixing with atmospheric air.

mechanical coupling   A pipe coupling that does not require threading.

mechanical draft cooling tower   A cooling tower that depends on fans for introduction and circulation of its air supply.

mechanically emulsified oil   A classification of a free oil and water mixture subjected to severe turbulence where the oil droplets are in a range of 10 to 40 microns in size.

MECO®   Mechanical Equipment Co., Inc.

Mectan™   Grit chamber by John Meunier, Inc.

MED   See "multiple effect distillation (MED)."

media   Granular filtration or absorption material or ion exchange resin products used to form barriers to the passage of certain solids or molecules that are suspended or dissolved in water or wastewater. "Media" is the plural form of "medium."

medical waste   Any solid waste generated in the diagnosis, treatment, or immunization of humans or animals.

Medina®   Bioremediation products by Medina Products Bioremediation Division.

medium   The material used in a filter to form a barrier to the passage of certain suspended solids or dissolved molecules.

medium-size water system   A water system that serves 3300 to 50,000 persons.

Megacell™   Rectangular dissolved air flotation system by Krofta Engineering Corp.

megaliter (ML)   A unit of volume equal to 1 million liter.

Megatron™   Ultraviolet water purification system by Atlantic Ultraviolet Corp.

Megos®   Ozone generation system by Capital Controls Co.

MEK   Methyl ethyl ketone.

Mekor®   Corrosion inhibitor chemical additive by Ashland Chemical, Drew Industrial.

Mellafier   Inclined plate clarifier by Industrial Filter & Pump Mfg. Co.

melt water   Water derived from the melting of ice and snow.

meltdown   A defect in a nuclear reactor cooling system which results in an overheating of the reactor core and eventual melting of fuel rods.

melting point   The temperature at which a solid changes to a liquid.

Mem*Recon™   Reverse osmosis membrane reconditioning product by King Lee Technologies.

Membio®   Aerobic biological digester by USFilter/Memcor.

Membralox®   Ceramic membrane filters by USFilter/Rockford.

Membrana™   Cartridge filter housing by Osmonics.

membrane   A thin barrier that permits passage of particles of a certain size or of a particular physical or chemical property.

membrane bioreactor (MBR)   A modification of the activated sludge wastewater treatment process employing membrane filtration in place of conventional secondary clarifiers.

membrane contactor   A device that permits mass transfer between a gaseous phase and liquid phase of a material without dispersing one phase into another.

membrane diffuser   Fine bubble aeration diffuser with perforated flexible plastic membranes.

membrane filter   (1) A paper-like filter with small pore sizes that is capable of retaining bacteria for use in the laboratory examination of water. (2) A pressure-driven microfiltration or ultrafiltration membrane filter.

membrane processes   Processes including reverse osmosis, electrodialysis, and ultrafiltration that use membranes to remove dissolved material or fine solids.

membrane softening   A water softening process that utilizes semi-permeable nanofiltration or reverse osmosis membranes to remove hardness constituents such as calcium and magnesium from water.

MembraPro®   Ceramic membrane microfiltration system by USFilter Corp.

Membrastill™   Pharmaceutical water system by USFilter/Rockford.

Memclean®   Chemical solution used to clean microfiltration membranes and systems by USFilter/Memcor.

Memcor®   Continuous microfiltration systems by USFilter/Memcor.

Memexx™   Membrane filtration system by WesTech Engineering Inc.

Memloy   Microfiltration membrane product by USFilter/Memcor.

Memory-Flex™   Check valve by Val-Matic Valve & Manufacturing Corp.

Memstor   Membrane storage agent by King Lee Technologies.

Memtec   Microfiltration product line by USFilter/Memcor.

Memtek®   Microfiltration membrane products by USFilter/Industrial Wastewater Systems.

Memtrex™   Pleated filters by Osmonics, Inc.

Memtrol   Microfiltration control system by USFilter/Memcor.

meniscus   The curved upper surface of a column of liquid.

Mensch™   Reciprocating rake bar screen by Vulcan Industries, Inc.

MeOH   Common abbreviation for "methanol."

MEP   See "multiple extraction procedure (MEP)."

meq/L   See "milliequivalents per liter."

mercaptans   Organic compounds, or thioalcohols (thiols), containing sulfur and noted for their disagreeable odor.

Merco®   Centrifuge by Alfa Laval Separation, Inc.

mercury   A heavy metal element that when absorbed or ingested by humans is excreted from the body very slowly and can be lethal in very low concentrations.

Merlin®   Progressing cavity pump by MGI Pumps, Inc.

Mer-Made   Filter leaves for vacuum diatomite filters by Mer-Made Filter, Inc.

Mesa-Line®   Portable submersible pump by Crane Pumps & Systems.

mesh   The number of openings per lineal inch, measured from the center of 1 wire or bar to a point 1″ (25.4 mm) distant.

mesocosm   A physically confined, self-maintaining, multitrophic experiment for discerning processes involved in the fate and transformation of nutrients into organic matter.

mesophiles   Bacteria that grow best at temperatures between 25 and 40°C.

mesophilic digestion   Anaerobic sludge digestion within a mesophilic range of approximately 25 to 40°C.

mesophyte   A plant that grows under typical or moderate amounts of atmospheric water supply.

mesosphere   The level of the earth's atmosphere that exists above the stratosphere.

mesothelioma   A fatal form of cancer caused by exposure to asbestos.

mesotrophic lake   A lake between the oligotrophic and eutrophic stages that remains aerobic, although a substantial depletion of oxygen has occurred in the hypolimnion.

metabolism   The chemical and physical processes of living organisms which include the biological conversion of organic matter to cellular matter and gaseous byproducts.

metabolites   Any substances produced by biological processes, such as those from pesticides.

MetaGuard™   Powdered antiscalant and stabilizer for reverse osmosis systems by Professional Water Technologies, Inc.

metal   In general those elements that easily lose electrons to form positive ions.

metal finishing wastes   Wastewater from electroplating, galvanizing, and other metal finishing operations that may be characterized by the presence of acids, caustics, and metal contaminants.

metal salt coagulants   Salts of alum and iron commonly used as water treatment coagulants.

Metal-Drop™   Flocculant/coagulant by Kem-Tron.

metalimnion   The middle layer of a thermally stratified lake or reservoir.

MetalWeave®   Stainless steel fabric tank baffles by Baker Process.

metastasis   The spreading of disease from one part of the body to another.

meter   The basic SI unit of length equivalent to approximately 39.37 inches, or the distance traveled by light in a vacuum in 1/299,792,458 seconds. Also called "metre."

metering pump   A pump used to provide controlled injection of a chemical additive into a fluid flow.

metes and bounds   A description of the measurements and boundaries of a tract of land beginning at a given point in the boundary of the tract to be described, then the recitation of the courses (directions) and distances from point to point entirely around the tract.

Metex®   An anaerobic process for the removal of heavy metals by Lotepro Corp. (Western Hemisphere) and Linde-KCA-Dresden GmbH.

methane   A colorless, odorless combustible gas that is the principle byproduct of anaerobic decomposition of organic matter in wastewater. Chemical formula is CH_4.

methane formers   See "methanogens."

methanogens   Group of anaerobic bacteria responsible for conversion of organic acids to methane gas and carbon dioxide. Also known as "methane formers."

methanol   A solvent often used as a supplemental carbon source during denitrification. Chemical formula is CH_3OH.

methemoglobinemia   Disease occurring primarily in infants who ingest water high in nitrates. Also called "blue baby syndrome."

Method 18   An EPA test method which uses gas chromatographic techniques to measure the concentration of volatile organic compounds in a gas stream.

Method 24   An EPA reference method to determine density, water content, and total volatile content of coatings.

Method 25   An EPA reference method to determine the VOC concentration in a gas stream.

method detection limit (MDL)   The constituent concentration that when processed through the complete method produces a sample with a 99% probability that it is different from the blank.

methoxychlor   Pesticide that causes adverse health effects in domestic water supplies and is toxic to freshwater and marine aquatic life.

methyl   The monovalent hydrocarbon radical CH_3, usually existing only in combination with other atoms.

methyl orange   A color indicator used in acid and base titrations.

methyl orange alkalinity   A measure of the total alkalinity of an aqueous solution determined through titration with a methyl orange color indicator.

methylate   To replace one or more hydrogen atoms in a molecule with a methyl group.

methylene blue active substance (MBAS)   Anionic surfactants which react with methylene blue to form a chloroform-soluble complex.

Metito Arabia   Former name of USFilter/Metito.

metre   See "meter."

metric ton   A unit of mass equal to 1000 kilogram or approximately 2204 pounds. Also called "tonne."

Metrol™   Seawater injection product line by Baker Hughes Process Systems.

MEVA   Product line of Waterlink Inc.

MF   See "microfiltration (μF)."

MFS   Minimum functional specification.

MFT   Membrane filter technique.

mg   See "milligram (mg)."

mg/L   See "milligrams per liter."

mg/L as CaCO₃   See "calcium carbonate equivalent (mg/L as $CaCo_3$)."

mgd   Million gallons per day.

mgid   Million gallons (Imperial) per day.

mho   Unit of measurement for conductivity equal to the reciprocal of resistivity (ohm).

MHSA   Mine Health Safety Administration.

MHT®   Magnesium hydroxide by Dow Chemical Co.

miasma   Vapor rising from marshes, polluted water, or decaying organic matter which was once mistakenly thought to poison and infect the air causing malaria and other diseases.

MIB   Abbreviation for the trace organic 2-methylisoborneol which produces an earthy/musty odor in surface waters. Chemical formula is $C_{11}H_{20}O$.

MIC®   Mazzei Injector Corp.

Micrasieve™   Pressure-fed fine static screen by Andritz-Ruthner, Inc.

Micro Fine   Ultrafiltration system by USFilter/Memcor.

Micro/2000®   Residual chlorine, chlorine dioxide, and potassium permanganate analyzer by USFilter/Wallace & Tiernan.

microbe An organism observable only through a microscope. Also called "micro-organism."

microbial pesticide A microorganism that is used to control a pest, but of minimum toxicity to man.

Microbics Former name of Azur Environmental.

microbiocide See "biocide."

MicroBiotic™ Carbon biofiltration system by Geoenergy International Corp.

Microbloc Carbon bed VOC control system by USFilter/Westates.

Micro-Carbon® Filter cartridge with wound carbon batt by USFilter/Filtration & Separation.

Microcat® Microbial additive for use in biological wastewater treatment by Bioscience, Inc.

Microchem™ Tablet chlorinator by Mooers Products, Inc.

microclimate The localized climate conditions within an urban area or neighborhood.

MicroDAF Dissolved air flotation by the former Princeton Clearwater.

microelectronic water See "electronic-grade water."

microencapsulation Isolation of a waste material by mixing it with a material which then cures or converts to a solid, nonleaching barrier.

Microenfractionator™ Soil mixing technology by H&H Eco Systems, Inc.

microfauna Animals not visible to the naked eye.

MicroFIBR Skid-mounted biofiltration system by AMETEK Rotron Biofiltration.

MicroFID™ Portable FID by PerkinElmer Instruments.

microfiltration (MF) A low pressure (100-400 kPa, 15-60 psi) membrane filtration process which removes suspended solids and colloids generally larger than 0.1 micron diameter.

Microfloat® Dispersed air flotation system by Aeration Industries, Inc.

microfloc Destabilized floc particle that permits in-depth penetration of a granular media filter bed to optimize the filter's solid retention capacity.

Microfloc® Water treatment product line of USFilter/Microfloc.

microfouling The biological fouling of a water system with microorganisms including algae, fungi, and bacteria.

Microgap™ Ozonation equipment by Osmonics, Inc.

microgram (µg) A unit of mass equal to one-millionth of a gram.

micrograms per cubic meter (µg/m³) A measure of the concentration of particulate or gaseous matter in air, commonly used in reporting air pollution data.

micro-irrigation A water management irrigation technique using a micro-sprinkler or drip irrigation system to minimize water runoff.

Micro-Klean™ Wastewater treatment equipment product by Alar Engineering Corp.

MicroMass Dissolved air flotation unit by Komline-Sanderson Engineering Corp.

Micro-Matic® Rotating microscreen by USFilter/Industrial Wastewater Systems.

MicroMAX® Physical separation system by Micronair LLC.

Micromesh Strainer Microscreen by Lakeside Equipment Corp.

micrometer (µm) See "micron (µ)."

micromho A unit measure of conductivity equal to one millionth of a mho.

micron (μ)   A unit of length equal to one-millionth of a meter. Also called "micrometer."

micron rating   The term applied to a filter medium to indicate the particle size above which all suspended solids will be removed throughout the rated capacity.

Micronizer™   Fine bubble dissolved air flotation device by the former Microlift Systems, Inc.

microorganism   An organism observable only through a microscope. Also called "microbe."

Micro-Pi®   Pressure fed rotary screen by Andritz-Ruthner, Inc. (Western Hemisphere) and USFilter/Contra-Shear.

Micro-Polatrol®   Cathodic protection power units by Corrpro Waterworks, Inc.

MicroPore   Aeration mixing systems by Environmental Dynamics Inc.

MicroPurge®   Low-flow sampling by QED Environmental Systems, Inc.

microscope   An instrument used for visual magnification of small objects.

microscreen   A surface filtration device consisting of a rotating drum with a fine mesh screen fixed to its periphery. As water flows through the interior of the drum, solids are retained by the mesh for removal by a high pressure spray wash.

Microsep®   Ballasted Floc Reactor by USFilter/General Filter.

Micro-Sieve   Microscreen formerly offered by Passavant.

microsporidia   Spore-forming protazoan parasites recognized as pathogens of insects, fish, birds, and mammals.

microstrainer   See "microscreen."

Micro-T   Turbidimeter with remote station monitoring by HF Scientific, Inc.

Microtox®   Acute toxicity test by Azur Environmental.

Microtuff™   Microporous, vibrating diffuser by Clear-Flo International.

Microza™   Hollow fiber membrane filtration systems by Pall Corp.

midge fly   An insect that may infest a water system and whose larvae, known as "blood worms," feed on algae, protozoans, and decaying vegetation.

Midi-Rotor   Rotary brush aerator by USFilter/Krüger.

midnight dumping   The deliberate and illegal disposal of sludge or other waste materials at an unauthorized, nonpermitted location.

MightyPure™   Ultraviolet water purification system by Atlantic Ultraviolet Corp.

mil   A unit of measure equal to one-thousandth of an inch.

MIL   Military Specification.

milk of lime   A lime slurry formed by mixing water with calcium hydroxide.

mill   To grind or crush.

mill scale   An oxide coating formed on steel when heated in connection with hot working or heat treatment.

Millennium™   Regenerative thermal oxidizer system by Megtec Systems, Inc.

milliequivalent (me)   One-one thousandth of an equivalent weight.

milliequivalents per liter (meq/L)   An expression indicating the concentration of a solute which is calculated by dividing the concentration in milligrams per liter by the equivalent weight of the solute.

milligram (mg)   A unit of mass equal to one-thousandth of a gram.

milligrams per liter (mg/L)   A common unit of measurement of the concentration of a material in solution.

milliliter (mL)   A unit of volume equal to one cubic centimeter.

Milliscreen™   Internally fed rotary fine screen by Andritz-Ruthner, Inc. (Western Hemisphere) and USFilter/Contra-Shear.

mineral   A naturally occurring inorganic material having a definite chemical composition and structure.

mineral acidity   Acidity caused by the presence of hydrochloric, sulfuric, nitric, and other inorganic mineral acids.

mineral acids   Inorganic acids including hydrochloric, nitric, and sulfuric acid.

mineral water   Water containing a minimum of 250 mg/L of total dissolved solids and which come from a source tapped at one or more boreholes or springs originating from a geologically and physically protected underground source.

mineralization   The conversion of an organic material to an inorganic form by microbial decomposition.

Mini Monster®   Low wastewater flow solids reduction unit by JWC Environmental.

Mini Osec   Electrolytic chlorination system by USFilter/Wallace & Tiernan.

MiniBUS™   Ultrafiltration membrane cartridge by Cuno Separations Systems Division.

MiniChamp   Chemical induction unit by USFilter/Stranco.

MiniDisk™   Cloth media wastewater filter by Aqua-Aerobic Systems, Inc.

Minigas®   Multi-gas detector by Neotronics of North America.

Mini-Ject®   Pneumatic ejector lift station by Smith & Loveless, Inc.

Mini-Magna   Rotor aerator by Lakeside Equipment Corp.

Minimax   Dewatering pressure filter by Larox Inc.

Mini-Maxi   Dissolved air flotation unit by Tenco Hydro, Inc.

Mini-Milli™   Internally fed rotary fine screen by Andritz-Ruthner, Inc. (Western Hemisphere) and USFilter/Contra-Shear.

Mini-Miser™   Multiple feed dewatering system by Recra Environmental.

Minipure™   Ultraviolet water purification system by Atlantic Ultraviolet Corp.

Mini-Ring   Random packing media by USFilter/General Filter.

Mini-San   Tablet feeder disinfection system by Exceltec International Corp.

Miniseries™   Packaged desalination plant by Matrix Desalination, Inc.

minors   Publicly owned treatment works with flows less than 1 million gallons per day.

Minotaur™   Activated carbon product by Calgon Carbon Corp.

Miox®   Oxidant product line by Miox Corp.

miscible   Capable of being mixed.

MIST   Fine bubble diffuser by Aeromix Systems, Inc.

mist eliminator   A device used to remove entrained droplets of water from a vapor stream produced during evaporation.

Mist Pro™   Odor control system by NuTech Environmental Corp.

MistGard™   Oil mist collector by Monsanto Enviro-Chem Systems, Inc.

Mist-Master®   Mesh pad mist eliminators by ACS Industries, Inc.

mitigation   Measures taken to reduce adverse impacts on the environment.

mitochondria   Subcellular structures that contain genetic material and enzymes which produce ATP to provide energy for cell metabolism.

Mixaerator   Static mixing aerators by JDV Equipment Corp.

MixAirTech   MixAir Technologies, Inc.

Mixco   Batch mixer by Lightnin.

mixed bed demineralizer   Ion exchange demineralizer containing strong-acid and strong-base resins in a single vessel.

mixed liquor   The mixture of wastewater and activated sludge undergoing aeration in the aeration basin.

mixed liquor suspended solids (MLSS)   Suspended solids in the mixture of wastewater and activated sludge undergoing aeration in the aeration basin.

mixed liquor volatile suspended solids (MLVSS)   The volatile fraction of the mixed liquor suspended solids.

mixed low-level radioactive waste (MLLW)   Low-level radioactive waste that also contains hazardous constituents.

mixed media filter   Granular media filter utilizing two or more types of filter media of different sizes and specific gravities, usually silica sand, anthracite and ilmenite, or garnet.

Mixflo™   Aeration injection and dissolution system by Praxair, Inc.

mixing zone   Limited area where initial dilution of a discharge takes place, water quality changes may occur, and certain water quality standards may be exceeded.

Mix-Mate®   Multistage polymer mixing system by Neptune Chemical Pump Co.

mixotroph   Bacteria that do not grow in either anaerobic or highly oxygenated water.

ML   See "megaliter (mL)."

mL   See "milliliter (mL)."

MLLW   See "mixed low-level radioactive waste (MLLW)."

MLM™   Heat recovery media for regenerative thermal oxidation systems by Lantec Products, Inc.

MLSS   See "mixed liquor suspended solids (MLSS)."

MLTSS   Mixed liquor total suspended solids. See "mixed liquor suspended solids (MLSS)."

MLVSS   See "mixed liquor volatile suspended solids (MLVSS)."

mobile source   Any nonstationary source of air pollution such as cars, trucks, motorcycles, buses, airplanes, locomotives.

MobileFlow®   Mobile water treatment system by Ecolochem, Inc.

MobileRO   Trailer mounted reverse osmosis system by Ecolochem, Inc.

Mobius   Fine mesh belt screen by Pro-Ent, Inc.

modeling   A quantitative or mathematical simulation which attempts to predict or describe the behavior or relationships that result from a physical event.

MoDo-Chemetics   Former name of Kvaerner Chemetics.

Mod-U-Flo   Round bottom clarifier by Osmonics, Inc.

Moduflow   Concrete gravity sand filter by Smith & Loveless, Inc.

Modulab®   High purity water systems by USFilter Corp.

Modular Aquarius®   Modular water treatment plant by USFilter/Microfloc.

Modulozone   Skid-mounted ozone generator by Praxair-Trailigaz Ozone Co.

Modu-Plex   Wet well pump station by Smith & Loveless, Inc.

ModuStor™   Bolted steel tank by ModuTank, Inc.

MOE   Margin of exposure.

molality   The number of moles of solute per kilogram of solvent.

molarity   The number of moles of solute per liter of solution.

mole   (1) The molecular weight of a substance containing an Avogadro's number of atoms or molecules. (2) A massive harborwork, breakwater, or jetty.

molecular weight   The weight of a molecule that may be calculated as the sum of the atomic weights of its constituent atoms.

molecular weight cutoff (MWCO)   The smallest compounds that are generally rejected in a membrane filtration process.

molecule   The smallest division of a compound that still retains or exhibits all the properties of the substance.

mollusk   A group of invertebrate animals with a soft, unsegmented body protected by a shell.

Molpure®   Hollow fiber filtration membranes by Celgard LLC.

Molsep®   Hollow fiber filtration membranes by Celgard LLC and Daicen Membrane Systems Ltd.

molten salt reactor   A thermal treatment unit that rapidly heats waste in a heat-conducting fluid bath of carbonate salt.

Molyver   Reagent chemicals used to determine molybdenum concentration in water by Hach Co.

MOM   Management, operating, and maintenance.

monel   A nickel alloy containing approximately 30% copper and having good mechanical and corrosion-resistant properties.

monitoring well   A well used to obtain samples for analysis or to measure ground-water levels.

Monkey Screen   Reciprocating rake bar screen by Brackett Geiger.

Mono®   Pump products by Monoflo.

Monobelt®   Filter press by Waterlink/Aero-Mod Systems.

Monoblock   Carbon bed VOC control system by USFilter/Westates.

Monocluster   Package water treatment plant by Graver Co.

Monod equation   A mathematical equation that describes the relationship between biomass production and the concentration of growth-limiting substrate.

Mono-Ferm   Iron and manganese removal gravity filter by Graver Co.

monofilament   A single synthetic fiber of continuous length used in woven mesh or cloth.

monofill   A solid waste disposal facility containing only one type or class of waste.

Monoflo   Progressing cavity pump by MGI Pumps, Inc.

Monoflo®   Screenings grinder by Monoflo.

Mono-Floc®   Gravity sand filter with coagulant feed system by Graver Co.

Monoflor®   Cast-in-place filter underdrain by Infilco Degremont, Inc.

Monolift®   Vertical, progressing cavity groundwater pump by Monoflo.

monolithic underdrain   A concrete filter underdrain whose piers and floor are poured in place at one time.

monomedia filter   A granular media filter utilizing a single size and type of filter media.

monomer   The basic molecule of a synthetic resin or plastic.

monomictic   Relatively deep lakes and reservoirs which do not freeze over during the winter months and undergo a single stratification and mixing cycle during the year.

Mono-Pak   Concrete gravity filter by Graver Co.

Mono-Pilot®   Coagulant control center using a pilot filter column by USFilter/Microfloc.

Monorake   Traveling bridge raking mechanism for rectangular clarifiers by GL&V/Dorr-Oliver, Inc.

Monoscour®   High solids gravity sand filter by Graver Co.

MonoSparj   Coarse bubble diffuser by Walker Process Equipment.

Monosphere™   Ion exchange resin by Dow Chemical Co.

Monovalve® Filter   Gravity sand filter by Graver Co.

Monozone®   Ozone generation system by Capital Controls Co.

Montreal Protocol   A 1987 international agreement to phase out CFCs and replace them with HFCs. The full agreement name is the "Montreal Protocol on Substances That Deplete the Ozone Layer."

moraine   A mass of stone and earth carried by a glacier and deposited on land after a glacier recedes.

morbidity   Rate of disease incidence.

mortality rate   The number of deaths per 100,000 people in a population.

MOS   Margin of safety.

most probable number (MPN)   Statistical analysis technique based on the number of positive and negative results when testing multiple portions of equal volume.

mother liquor   The concentrated solution that remains after evaporation or crystallization. See also "bittern."

motive steam   High pressure steam used to operate a steam-jet ejector or thermocompressor.

MotoDip   Motorized slotted skimmer pipe by Walker Process Equipment.

mouth feel   A series of sensations such as aftertaste, astringent, burning, chalky, metallic, and flat which are usually easier to taste than to smell but cannot be strictly described as "tastes."

moving bed filter   A granular media filter that continuously cleans and recycles filter media while the filter continues to operate.

Moyno®   Pump product line by Moyno Industrial Products.

MP   Melting point.

MPA   Microscopic particulate analysis.

MPN   See "most probable number (MPN)."

MPRox Process   Organic destruction process for spent caustic wastewaters by MPR Services, Inc.

MPVT™   Multi-purpose vertical turbine pump by Patterson Pump Co.

MQL   Method quantitation limit. Generally the same as "estimated quantitation limit."

MRDL   See "maximum residual disinfectant level (MRDL)."

MRDLG   Maximum residual disinfectant level goal.

MRF Press   Belt filter press by Idreco USA, Ltd.

MRFs   See "materials recovery facilities (MRFs)."

MRL   Maximum residue limit.

M-roy   Metering pump products by Milton Roy Co.

MS Diffuser   Medium bubble diffuser by Enviroquip, Inc.

MSBR   Modified sequencing batch reactor.

MSD   Musculoskeletal disorder.

MSDS   See "Material Safety Data Sheet (MSDS)."

MSF   See "multistage flash evaporation (MSF)."

MSF-BR   Multistage flash evaporation, brine recirculation.

MSF-OT   Multistage flash evaporation, once-through.

MSL   Mean sea level.

MSW   Municipal solid waste.

MSWLF   Municipal solid waste landfill.

MT®   Reverse osmosis membrane cleaners by B.F. Goodrich Co.

MTBE   Methyl-tertiary-butyl-ether.

MTBF   Mean time between failures.

MTD   Maximum tolerated dose.

MTP   Maximum trihalomethane potential.

MTS   Wastewater treatment equipment product link of Waterlink Biological Systems.

MTS®   Trailer-mounted mobile treatment system for industrial wastewater treatment applications by Graver Co.

MTTR   Mean time to repair/replace.

MTZ   Mass transfer zone.

MUC   Maximum use concentration.

muck soils   Earth made from decaying plant materials.

MUD   Municipal Utility District.

mud balls   Agglomerations of floc, solids, and filter media in a filter bed which may grow into a larger mass and reduce filtration efficiency.

mud flat   A muddy, flat low-lying tidal area.

mud valve   A valve used to drain sediment from the bottom of a sedimentation basin.

Muffin Monster®   Wastewater solids reduction unit by JWC Environmental.

muffler   A device used to reduce or deaden noise.

mulch   A protective ground covering of compost, wood chips, sawdust, or other organic matter.

Multdigestion   Two-stage digestion system formerly offered by Dorr-Oliver, Inc.

Multicell®   Multiple cell gravity filter by USFilter/General Filter.

Multi-Chem®   NOx destruction system by Tri-Mer Corp.

multiclone   A set of individual cyclone separators arranged in parallel to remove particulate matter from air emissions.

Multicoil   Indirect sludge drying system by Kvaerner Eureka USA.

Multicone   Aluminum induction cascade aerator used to strip gases or aerate water supply by Infilco Degremont, Inc.

Multicrete®   Monolithic filter underdrain system by USFilter/General Filter.

MultiDraw   Circular clarifier with pumped suction sludge removal system using multiple nozzles by Walker Process Equipment.

Multiflo   Flow distribution nozzle for rotary distributors by Amwell, Inc.

Multiflo®   Pump column used with jet aeration system by USFilter/Jet Tech.

MultiFlow   Skid mounted water treatment system by Ecolochem, Inc.

Multi-Flow   PVC biological filter media formerly offered by B.F. Goodrich Co.

Multi-Jet™   Submersible self-aspirating aerator by Waterlink Biological Systems.

Multilogger   Water instrumentation device for measuring multiple parameters by Stevens Water Monitoring Systems.

Multi-Mag™   Electromagnetic flowmeter by Marsh-McBirney, Inc.

multimedia filter   Granular media filter utilizing two or more types of filter media of different sizes and specific gravities, usually silica sand, anthracite and ilmenite, or garnet.

Multi-Pack™   Oil/water coalescers by Mercer International, Inc.

Multipass®   Sludge dryer by USFilter/Envirex.

Multiple Barrier Filtration   Filtration system by USFilter/General Filter.

multiple effect distillation (MED)   A thin film evaporation process where the vapor formed in a chamber, or effect, condenses in the next, providing a heat source for further evaporation.

multiple extraction procedure (MEP)   Procedure used to simulate the leaching a waste will undergo from repetitive precipitation of acid rain on a material.

multiple hearth furnace   A furnace or incinerator consisting of numerous hearths which is used to incinerate organic sludges or recalcinate lime.

multiple stage flash evaporation   See "multistage flash evaporation (MSF)."

Multi-Point   Level controller by Drexelbrook Engineering Co.

Multiport Valve™   Valve used to manipulate filter backwashing and rinsing by USFilter/Rockford.

MultiRanger   Level and volume measurement device by Milltronics, Inc.

multistage flash evaporation (MSF)   A desalination process where a stream of brine flows through the bottom of chambers, or stages, each operating at a successively lower pressure, and a proportion of it flashes into steam and is then condensed.

Multi-Tech®   Chemical feed, contact flocculation, and filtration process by USFilter/General Filter.

Multi-Turi®   Wet scrubber with high energy venturi by CMI-Schneible Co.

Multiwash®   Sand filtration process using combined air/water backwash by USFilter/General Filter.

Multi-Wash®   Wet scrubber by CMI-Schneible Co.

Multi-Zone   Anaerobic digestion system by USFilter/Zimpro.

Muncher®   Sewage grinder products by Monoflo.

Munchpump®   Packaged pump/grinder assembly by Monoflo.

municipal waste   The combined solid and liquid waste from residential, commercial, and industrial sources.

municipal wastewater treatment plant   Treatment works designed to treat municipal wastewater.

Muniflo   Positive displacement rotary lobe sludge pump by Envirotech Pumpsystems.

Munox   Bacterial innoculant for wastewater treatment, soil, and groundwater remediation by Osprey Biotechnics.

Münster   Trash rake cleaning mechanism by Landustrie Sneek BV.

muntz metal   A brass containing approximately 60% copper and 40% zinc.

muriatic acid   Chemical formula is HCl, also known as "hydrochloric acid."

MUS   Minimum ultimate strength.

Mushroom Ventilator   Cast iron air diffuser by the former Knowles Mushroom Ventilator Co.

mutagen   A material that causes genetic change when interacting with a living organism.

mutagenic   A chemical or agent with properties that cause mutation or disfiguring.

Mutrator®   Packaged pumping/grinding station by Monoflo.

MVC   Mechanical vapor compression. See "vapor compression evaporation (VC)."

MVR   Mechanical vapor recompression. See "vapor compression evaporation (VC)."

MW   Megawatt.

MWCO   See "molecular weight cutoff (MWCO)."

Mx®   Magmeter by McCrometer, Inc.

Mycelx®   Chemical coating with an affinity for organic compounds by Mother Environmental Systems, Inc.

mycotoxin   A toxin produced naturally by molds or fungus.

Mystaire®   Air scrubber systems by Misonix, Inc.

N

NAAQS   See "National Ambient Air Quality Standards (NAAQS)."

NACE   National Association of Corrosion Engineers.

Nadir®   Cross flow filtration membranes by Celgard LLC.

NAE   National Academy of Engineering.

NAFCO®   Fibrous precoat filter aid by Liquid-Solids Separation Corp.

NaHMP   See "sodium hexametaphosphate."

Nalclear®   Anionic/nonionic polymer for sedimentation and sludge conditioning by Nalco Chemical Co.

Nalmet®   Solution polymer for reducing soluble metal concentrations to low levels by Nalco Chemical Co.

nanofiltration (NF)   A specialty membrane filtration process which rejects solutes larger than approximately 1 nanometer (10 angstroms) in size.

nanometer   A unit of length equal to one billionth (10^{-9}) of a meter.

NAPAP   National Acid Precipitation Assessment Program.

NAPC   National Air Pollution Control Association.

NAPL   See "nonaqueous phase liquid (NAPL)."

nappe   (1) A sheet of water flowing over a weir or dam. (2) An arch-shaped sheet of rock forced over underlying rocks by internal stresses.

napthalene   A synthetic organic chemical used as a moth repellent and fungicide. Chemical formula is $C_{10}H_8$.

Nara   Paddle dryer/processor by Komline-Sanderson Engineering Corp.

NAS   National Academy of Sciences.

Nasty Gas™   Regenerative blowers to move noxious or other exotic gases by AMETEK Rotron Biofiltration.

natality   Birthrate.

natality rate   The number of births per thousand in a specific population.

national ambient air quality standards (NAAQS)   U.S. standards established under the Clean Air Act to set limits on criteria pollutant levels in ambient (outdoor) air.

National Contingency Plan (NCP)   U.S. federal regulations promulgated to implement CERCLA and CWA.

National Emission Standards for Hazardous Air Pollutants (NESHAP)   U.S. standards established under CAA to set limits on pollutants which may pose an immediate hazard to human health.

National Environmental Policy Act (NEPA)   A 1969 U.S. public law declaring a national policy that encourages productive and enjoyable harmony between people and their environment to enrich their understanding of ecological systems and natural resources.

National Hydro   Former equipment manufacturer acquired by Amwell, Inc.

National Oil and Hazardous Substances Contingency Plan (NOHSCP)   The federal regulation that guides determination of the sites to be corrected under both the Superfund program and the program to prevent or control spills into surface waters or elsewhere.

National Pollutant Discharge Elimination System (NPDES)   A U.S. program to issue, monitor, and enforce pretreatment requirements and discharge permits under the Clean Water Act.

National Priorities List (NPL)   A U.S. federal list of hazardous waste sites addressed by CERCLA.

National Water Quality Standards (NWQS)   These set minimum requirements for water quality and require states to set standards to achieve CWA's water quality goals.

natural attenuation   Naturally occurring processes in the environment that reduce the mass, toxicity, mobility, volume, or concentration of contaminants in soils or groundwater.

natural draft cooling tower   A cooling tower in which the air flow through the tower occurs naturally, rather than mechanically, as a result of tower design.

natural gas   A naturally occurring mixture of hydrocarbon gases found in geologic formations beneath the earth's surface whose principal constituent is methane.

natural organic matter (NOM)   Term used to described the organic matter present in natural waters.

natural resource   An area, material, or organism useful to man.

Nautilus®   Traveling bridge siphon sludge collection system by USFilter/Microfloc.

navigable waters   Traditionally, waters sufficiently deep and wide for navigation by all or specified vessels and which are protected by certain provisions of the Clean Water Act.

NAWQA   National Water Quality Assessment.

NBOD   See "nitrogenous oxygen demand (NOD)."

NBS   National Bureau of Standards.

NCASI   National Council of the Paper Industry for Air and Stream Improvement.

NCCLS   National Committee for Clinical Laboratory Standards.

NCG   See "noncondensable gas (NCG)."

NCH   See "noncarbonate hardness (NCH)."

NCP   See "National Contingency Plan (NCP)."

NCS   Northwest Cascade, Inc.

NCWS   See "noncommunity water system."

ND   None (not) detected.

NDO   Natural draft opening.

NDP   See "net driving pressure (NDP)."

NDT   Nondestructive testing.

NDWC   National Drinking Water Clearinghouse.

neat solution   Full strength, undiluted solution.

NEC   National Electrical Code.

needle valve   A valve that controls flow by means of a tapered needle which extends through a circular outlet.

NEETF   National Environmental Education and Training Foundation.

negative head   Filter operating condition that occurs when the pressure in the filter bed is below atmospheric pressure during a filter cycle.

negative pressure   A gauge pressure less than the atmospheric pressure.

NELAP   National Environmental Laboratory Accreditation Program.

NEMA   National Electrical Manufacturers Association.

nematocide   A chemical agent which is destructive to nematodes.

nematode   A long, unsegmented and often parasitic worm.

Nemo®   Progressive cavity pump and macerator product line by Netzsch, Inc.

neoprene   A synthetic elastomer that is chemically, physically, and structurally similar to natural rubber.

Neosepta®   Electrodialysis membrane stack supplied by Graver Co.

Neozone™   Ozone generator by North East Environmental Products, Inc.

NEPA   See "National Environmental Policy Act (NEPA)."

nephelometer   See "turbidimeter."

nephelometric turbidity unit (NTU)   Unit of measure used in the measurement of turbidity by instrumentation.

Neptune Microfloc   Former name of Microfloc Products group of USFilter/Microfloc.

NESHAP   See "National Emission Standards for Hazardous Air Pollutants (NESHAP)."

Nessler tubes   Color comparison tubes used in making colorimetric measurements.

net driving pressure (NDP)   The net feed pressure of a reverse osmosis system plus the osmotic pressure of the permeate, minus the permeate line pressure and osmotic pressure of the feedwater.

net head   The head available for production of energy in a hydroelectric plant after deduction of all frictional losses.

net positive suction head (NPSH)   The difference between the total pressure head and the vapor pressure of the liquid being pumped.

NETA   National Environmental Training Association.

Net-Waste   Screw press by Olds Filtration Engineering.

Neutral Process™   Heavy metals treatment system by Geo-Chem Technologies, Inc.

Neutralite   Filter media used to neutralize acidic waters by USFilter/Warren.

neutralization   The chemical process that produces a solution that is neither acidic nor alkaline.

Neutralizer Plus™   Media for pH adjustment by Matt-Son, Inc.

Neva-Clog®   Filter media of perforated metallic sheets by Liquid-Solids Separation Corp.

New Source Performance Standards (NSPS)   Standards established under the Clean Air Act to impose federal technology-based control requirements on emissions from new stationary sources of pollution.

new water   Water from any discrete source such as a river, creek, lake, or well which is deliberately brought into a plant site.

Newtonian flow   The flow of a fluid in which the viscosity is independent of the shear rate.

NF   See "nanofiltration (NF)."

NFPA   National Fire Protection Association.

NFR   See "nonfilterable residue (NFR)."

NFRAP   No further remedial action planned.

ng/L   Nanograms per liter.

Nibbler™   Wastewater pretreatment system by Northwest Cascade, Inc.

Nichols   Former furnace manufacturer acquired by Hankin Environmental Systems, Inc.

night soil   Human fecal wastes spread on fields as fertilizer.

NIH   National Institute of Health.

NIMBY   "Not in my backyard." A common expression that indicates a preference for waste disposal or treatment to occur at some distant location.

NIOSH   National Institute for Occupational Safety and Health.

Niro   Decanter product line by Centrico, Inc.

Nirosta®   Stainless steel product by Krupp Thyssen Nirosta GmbH.

NIST   National Institute of Standards and Technology.

Nitox®   Activated carbon adsorbers by TIGG Corp.

Nitra-Select™   Selective nitrate removal media by Matt-Son, Inc.

nitrate   A stable, oxidized form of nitrogen having the formula NO_3^-.

nitrate formers   See "*Nitrobacter.*"

Nitraver   Reagent chemicals used to determine nitrite concentration of solutions by Hach Company.

Nitrazyme™   Nitrate solution for controlling wastewater odors by Vulcan Performance Chemicals.

nitric acid   A strong mineral acid having the chemical formula HNO_3.

nitric oxide (NO)   A gas formed by combustion under high temperature and high pressure in an internal combustion engine, which changes into nitrogen dioxide in the ambient air and contributes to photochemical smog.

nitrification   Biological process in which ammonia is converted first to nitrite and then to nitrate.

nitrite   An unstable, easily oxidized form of nitrogen with the chemical formula NO_2^-.

nitrite formers   See "*Nitrosomonas*."

Nitrobacter   Nitrifying bacteria that convert nitrites to nitrates, also called "nitrate formers."

nitrogen   A colorless, odorless, gaseous element that makes up 78% of the earth's atmosphere and occurs as a constituent of all living tissues in combined form. Chemical formula is N.

nitrogen cycle   A graphical presentation of nitrogen's natural cycle from living animal matter through dead organic matter and back to living matter.

nitrogen dioxide   A reddish brown gas, one of the primary air pollutants, that usually results from a combustion process and which causes respiratory irritation and illness in relatively low concentrations. Chemical formula is NO_2.

nitrogen fixation   The conversion of atmospheric nitrogen into nitrogen compounds through biological activity.

nitrogen, nitrate   See "nitrate."

nitrogen, nitrite   See "nitrite."

nitrogen oxides (NOx)   Compounds formed and released primarily by the burning of fossil fuels.

nitrogenous biochemical oxygen demand (NBOD)   The portion of biochemical oxygen demand where oxygen consumption is due to the oxidation of nitrogenous material, measured after the carbonaceous oxygen demand has been satisfied. Also called "second-stage biochemical oxygen demand."

nitrogenous BOD   See "nitrogenous oxygen demand."

nitrogenous oxygen demand (NOD)   That portion of the oxygen demand associated with the oxidation of nitrogenous material, usually measured after the carbonaceous oxygen demand has been satisfied.

nitrophenols   Synthetic organopesticides containing carbon, hydrogen, nitrogen, and oxygen.

Nitroseed   Nitrifying toxicity screening test by Polybac Corp.

Nitrosomonas   Nitrifying bacteria that convert ammonia to nitrites under aerobic conditions and derive their energy from the oxidation. Also called "nitrite formers."

Nitrox™   Nutrient removal process by United Industries, Inc.

NMFS   National Marine Fisheries Service.

NMO   Nonmethane organic compound.

NMR   See "nuclear magnetic resonance (NMR)."

NNI   See "noise and number index (NNI)."

no detectable emissions An atmospheric discharge with a concentration less than 500 parts per million by volume as measured by an appropriate detection instrument.

no effect level See "no observed adverse effect level (NOAEL)."

no observed adverse effect level (NOAEL) The maximum dose of a substance which produces no observed adverse effects.

NOAA National Oceanic and Atmospheric Administration.

NOAEL See "no observed adverse effect level (NOAEL)."

Nocardia Bacteria that can accumulate to create a nuisance foam in aeration basins and secondary clarifiers.

No-Cling Traveling water screen media insert by Norair Engineering Corp.

NOD See "nitrogenous oxygen demand (NOD)."

nodulizing kiln See "calciner."

NOEC No observed effect concentration.

NOEL See "no observed adverse effect level (NOAEL)."

Nogcoflot Flotation treatment system by Noggerath GmbH.

NOHSCP See "National Oil and Hazardous Substances Contingency Plan (NOHSCP)."

noise Any unwanted sound, independent of volume.

noise and number index (NNI) An index for assessing air traffic noise based on the average perceived decibel level of air traffic and the number of aircraft heard.

noise-induced hearing loss A hearing loss, or permanent threshold shift, resulting from noise exposure rather than the normal loss attributed to age.

NOM See "natural organic matter (NOM)."

nonaqueous phase liquid (NAPL) A liquid mixed with water but having distinct boundaries and properties different from water. In an oil/water mixture, oil is the nonaqueous phase liquid.

noncarbonate hardness (NCH) The hardness in water caused by chlorides, sulfates, and nitrates of calcium and magnesium.

noncombustible refuse Solid wastes that will not burn in a conventional incinerator.

noncommunity water system (NCWS) A public water system that serves a nonresident population such as a campground, school, or factory.

noncondensable gas (NCG) Gaseous material not liquefied when associated water vapor is condensed in the same environment.

noncontact cooling water system A once-through cooling water system which does not come into contact with hydrocarbons or other wastewater and is not recirculated through a cooling tower.

nonfilterable residue (NFR) See "suspended solids."

nonionic polymer A polyelectrolyte with no net electrical charge.

nonmetal Elements that hold electrons firmly and tend to gain electrons to form negative ions.

nonpoint source (NPS) A source, other than a point source, associated with widespread activities such as agriculture, atmospheric deposition, erosion, or runoff that discharges pollutants into the air or water.

nonpotable reuse   The beneficial use of reclaimed water other than potable water supply augmentation.

nonpurgeable organic carbon (NPOC)   The fraction of total organic carbon removed after purging a sample with an inert gas.

nonputrescible   Material that cannot be decomposed by biological methods.

nonrenewable resource   A naturally occurring finite resource that cannot be renewed once it has been used.

nonsettleable solids   Suspended solids that remain in suspension, usually for more than one hour.

Nopol®   Disc diffuser system by WesTech Engineering Inc. and Nopon Oy.

Nopon®   Aeration system offered by WesTech Engineering Inc. and Nopon Oy.

Noramer®   Water treatment polymers by Rohm & Haas Co.

Nordic Water™   Product of Waterlink Separations, Inc.

Norit Roz   Steam-activated, peat-based carbon product by Norit Americas Inc.

NORM   Naturally occurring radioactive materials.

normal solution   A solution that contains one equivalent weight of a substance per liter of solution.

normality   A solution's relation to the "normal solution."

Nor-Pac®   Tower packing by NSW Corp.

Nortex   Side and boot seals for traveling water screens by Norair Engineering Corp.

North™   Internally fed rotating drum screen products by Voith Sulzer.

North Filter   Rotary fine screen by Voith Sulzer.

North-American Hercules®   Pressure leaf filter product line by Liquid-Solids Separation Corp.

Norton   Biological reactor packing media by Sanitaire Corp.

Norwalk-type virus   A waterborne pathogen that is the most common viral cause of gastroenteritis in adults.

Notim™   Organic iron and tannin removal media by Matt-Son, Inc.

Novex TLS   Inclined plate separator module by Gyulavari Consulting Kft.

Novus®   Emulsion polymers by BetzDearborn, Inc.

No-Wear™   Traveling bridge filter backwash shoe by USFilter/Davco.

No-Well   Pier-mounted traveling water screen design that does not require channel type intake by USFilter/Rex & Link-Belt Products.

NOx   See "nitrogen oxides (NOx)."

Noxidizer™   Incineration system by John Zink Co.

Noxon®   Decanter centrifuge product line by Waterlink Inc.

NOxOut   Nitrogen oxide reduction system by Nalco Chemical Co.

NOxOUT®   Noncatalytic reduction process to reduce nitrogen oxide emission levels by Wheelabrator Air Pollution Control, Inc.

Nozzle Air   Dissolved air flotation aeration system by Baker Hughes Process Systems.

NPCA   National Precast Concrete Association.

NPDES   See "National Pollutant Discharge Elimination System (NPDES)."

NPDWR   National primary drinking water regulation.

NPE   Nonyl phenol ethoxylates.

NPHAP   National Pesticide Hazard Assessment Program.

NPL   See "National Priorities List (NRL)."
NPOC   See "nonpurgeable organic carbon (NPOC)."
NPS   See "nonpoint source (NPS)."
NPSH   See "net positive suction head (NPSH)."
NPSHA   Net positive suction head available.
NPSHR   Net positive suction head required.
NPT   National pipe thread.
NRA   National Rivers Authority.
NREP   National Registry of Environmental Professionals.
NRWA   National Rural Water Association.
NSDWRs   National Secondary Drinking Water Regulations.
NSF®   NSF International.
NSPS   See "New Source Performance Standards (NSPS)."
NSSC   Neutral sulfite semichemical pulping process.
NTA   Nitrilotriacetic acid. An organic chelating agent.
NTIS   National Technical Information Service.
NTNCWS   Nontransient noncommunity water system.
NTP   National Toxicology Program.
NTR   National Toxics Rule.
N-Trak   Test kit to determine nitrogen content of water by Hach Co.
NTU   See "nephelometric turbidity unit (NTU)."
nuclear magnetic resonance (NMR)   An analytical technique used to detect and distinguish between nuclear particles in a sample using magnetic fields.
nuclear winter   Prediction by some scientists that smoke and debris rising from massive fires of a nuclear war could block sunlight for weeks or months, cooling the earth's surface and producing climate changes that could, for example, negatively affect world agriculture.
Nuclepore®   Membrane cartridge filter by Corning, Inc.
nuclide   A species of atom characterized by the number of protons, neutrons, and energy in the nucleus.
NUG   Nonutility generator.
nuisance mask   See "dust mask."
Nu-Notch Mushroom   Cast iron air diffuser by the former Knowles Mushroom Ventilator Co.
Nupac®   Random packing media by Lantec Products, Inc.
NuTralite®   Odor control product for neutralizing disulfide and other odors by NuTech Environmental Corp.
Nu-Treat   Flocculator/clarifier by USFilter/Envirex.
nutrient   Any substance that is assimilated by organisms to promote or facilitate their growth.
Nutrigest®   Clarifier by Smith & Loveless, Inc.
nutshell filter   A filtration device that uses ground walnut or pecan shells as granular filter media to remove hydrocarbons and other suspended solids from water.
NVCU™   Vapor control unit by NAO Inc.
N-Viro   Pasteurization and chemical fixation process to disinfect and stabilize sludge by N-Viro International Corp.

NVOC   Nonvolatile organic carbon.

NWPA   Nuclear Waste Policy Act of 1982.

NWQS   See "National Water Quality Standards (NWQS)."

NWSIA   National Water Supply Improvement Association. Former name of "American Desalting Association."

nylon   Plastic compound that offers excellent load-bearing capability, low frictional properties, and good chemical resistance.

O

O&M   Operation and maintenance.

O/W   Oil-in-water emulsion.

O2 Minimizer®   Process controller used to control oxygenation of mixed liquor by Schreiber Corp.

OASES®   Oxygen-activated sludge wastewater process by USFilter/Krüger.

oasis   A fertile or green vegetated area in a desert or wasteland that is supplied with water.

OB/OD   Open pit burning/open detonation.

obligate aerobes   Bacteria that can survive only in the presence of dissolved oxygen.

obligate anaerobes   Bacteria that can survive only in the absence of dissolved oxygen.

obligate pathogen   A pathogen that is unable to live outside a living host.

OBS®   Turbidity sensors by D&A Instrument Co.

OCA 19™   Odor neutralizer by Hinsilblon Laboratories.

OCC   Old corrugated containers.

occlusion   An absorption process where one solid material adheres to another, sometimes resulting in coprecipitation.

Occupational Safety and Health Administration (OSHA)   A U.S. agency responsible for overseeing workplace health and safety.

ocean   The volume of salt water that covers approximately 71% of the earth's surface and is divided into five principal geographic regions: the Antarctic, Arctic, Atlantic, Indian, and Pacific.

ocean disposal   The discharge or disposal of wastes or sludges in ocean water.

ocean dumping   Disposal of wastes in the ocean or seas.

Ocean Dumping Act (ODA)   Authorizes regulation of intentional ocean disposal of materials, as well as related research and the establishment of marine sanctuaries.

Ocean Dumping Ban Act   A U.S. law making it unlawful to discharge sewage sludge into the Ocean after 1991.

ocean incineration   The burning of wastes on ocean-going vessels in waters remote from land.

ocean thermal energy conversion (OTEC)   Electrical generation process which relies on the temperature differential between the upper and lower layers of the ocean to vaporize a fluid and power a turbine generator.

OCPSF   Organic chemicals, plastics, and synthetic fibers.

OD   Outside diameter.

ODA   See "Ocean Dumping Act (ODA)."

Odin   Packaged water treatment plant by USFilter/Davis Process.

Odophos®   Hydrogen sulfide and phosphorus removal product by USFilter/Davis Process.

Odor Buster®   Aeration system used to reduce odors at pump stations and plant headworks by United Industries, Inc.

odor threshold   See "threshold odor number (TON)."

odor unit   See "threshold odor number (TON)."

OdorGard™   Enhanced packed tower scrubbing process by Monsanto Enviro-Chem Systems, Inc.

OdorLok™   Hydrogen sulfide corrosion and odor control system by Eaglebrook, Inc.

OdorMaster™   Electrolytic gas scrubber type odor control system by Pepcon Systems, Inc.

Odor-Miser   Vapor phase, activated carbon odor absorbers by Westport Environmental Systems.

Odor-Ox   Multistage dry chemical air scrubber by Purafil, Inc.

ODP   See "open drip proof (ODP)."

ODS   Ozone depleting substances.

OEM   Original equipment manufacturer.

offal   Trimmings and viscera of butchered animals.

off-gas   The gaseous emissions from a process or equipment.

offset   The requirement for a proposed air pollutant generator to reduce emissions or obtain emission reductions from other facilities to compensate for new emissions.

off-site facility   A hazardous waste treatment, storage, or disposal facility located away from the generating site.

off spec water (OSP)   Product water that does not meet purity specifications.

OFR   See "overflow rate (OFR)."

OGWDW   The U.S. Office of Groundwater and Drinking Water.

OHL   Overhung load.

ohm   Unit of electrical resistance where a potential difference of one volt produces a current of one ampere.

Ohmicron   Immunoassay product line by Strategic Diagnostics, Inc.

oil   (1) Any of various greasy, combustible substances obtained from animal, vegetable, and mineral sources. (2) Any of various liquids extracted from petroleum. (3) A naturally occurring hydrocarbon in liquid form.

oil fingerprinting   A method of identifying sources of oil, allowing spills to be traced to their source.

Oil Grabber®   Oil skimming system by Abanaki Corp.

Oil Pollution Act (OPA)   A 1990 U.S. federal law that places liability on tank owners or operators for removal costs and damages if oil or other hazardous materials are spilled or discharged.

oil skimmer   A device used to remove oil from a water's surface.

oil spill   An unintentional discharge of oil into the environment, especially into a waterbody.

OilMaster   Oil/water separator by National Fluid Separators, Inc.

Oil-Minder   Submersible pump unit by Stancor Pump, Inc.

oils and grease   Common term used to include fats, oils, waxes, and related constituents found in wastewater.

Oilspin II   Hydrocyclone by Axsia Serck Baker, Inc.

oily wastewater   An oil-in-water emulsion in which oil is dispersed in the water phase.

OKI™   Submerged aerator mixer by WesTech Engineering Inc. and Nopon Oy.

old growth forest   A forest with a large percentage of old trees which have never been cut or have not been cut for many years.

Oleofilter™   Filter for removal of hydrocarbons from water by Aprotek, Inc.

oleophilic   A characteristic describing a strong affinity for oils.

olfactometer   Device used to measure odors.

oligohaline   A term describing water with a salinity of 0.5 to 5%.

oligotroph   Bacteria that grow in a medium containing <1.0 mg/L organic carbon.

oligotrophic lake   A deep lake deficient in organic materials whose waters contain a high degree of dissolved oxygen and low BOD.

OM&M   Operation, maintenance, and management.

Omega   Horizontal rotor aerator by Purestream, Inc.

Omega®   Lime slaker and feeder package by PennProcess Technologies, Inc.

Omnichlor   Sodium hypochlorite generator for marine applications by Exceltec International Corp.

Omniflo®   Sequencing batch reactor wastewater treatment system by USFilter/Jet Tech.

Omnipac®   Sequencing batch reactor wastewater treatment package plant by USFilter/Jet Tech.

Omnipure   Marine sewage treatment plant product line by Exceltec International Corp.

oncogenic   A chemical or agent with tumor-causing properties.

one-hundred-year flood plain   Land adjoining inland and coastal waters which, on the average, is likely to flood once every 100 years.

Onguard®   Instrumentation and control products by Ashland Chemical, Drew Industrial.

oocyst   An outer shell that protects an organism in the environment. Pronounced "oh-oh-cist."

OPA   See "Oil Pollution Act (OPA)."

opacity   The degree to which emissions reduce the transmission of a beam of light, expressed as a percent of the light which fails to penetrate a plume of smoke.

open burning   The combustion of solid waste without containment of combustion reaction in an enclosed device, control of the emission of the combustion products, or controlling combustion air to maintain temperature for efficient burning.

open channel   A natural or artificial channel in which fluid flows with a free surface open to the atmosphere.

open cycle cooling system   A cooling water system where cooling water is discharged to a receiving body of water without being recycled.

open drip proof (ODP)   Designation for electrical motor enclosure in which the ventilating openings are so constructed that successful operation is not interfered with when drops of liquid or solid particles strike or enter the enclosure at any angle from 0° to 15° downward from vertical.

open dump   A land disposal site where solids are disposed of in a manner that does not protect the environment and is susceptible to open burning and exposure to the elements, insects, and scavengers.

OPIM   See "other potentially infectious material (OPIM)."

opportunistic pathogen   A microbe that can cause disease in ill, very young, or elderly persons, but usually not in healthy individuals.

OptiClean™   Powdered reverse osmosis membrane cleaners by Professional Water Technologies, Inc.

Opti-Core   PVC biological filter media formerly offered by B.F. Goodrich Co.

Optimem™   Reverse osmosis product formerly offered by USFilter.

Optimer®   Cationic polymer for sedimentation and sludge conditioning by Nalco Chemical Co.

Optimum   Direct filtration water treatment plant by BCA Industrial Controls.

oral toxicity   Ability of a pesticide to cause adverse effects when ingested by mouth.

Orbal™   Oxidation ditch wastewater treatment system by USFilter/Envirex.

ORC®   Oxygen release compound to enhance natural attenuation by Regenesis.

ORE   Orbital rod evaporation.

Orec™   Ozone-generating systems by Osmonics, Inc.

Organagro®   Agricultural compost by Bedminster Bioconversion Corp.

organic   Relating to, or derived from, a living thing. A description of a substance that contains carbon atoms linked together by carbon-carbon bonds.

organic loading   The amount of organic matter applied to a treatment process.

organic matter   Substances containing carbon compounds, usually of animal or vegetable origin.

organic nitrogen   Nitrogen bound to carbon-containing compounds.

organic phosphorus   Phosphorus that is bound to carbon-containing compounds.

organoclay   Chemically modified bentonite clay used as an ion exchange media or absorbent.

OrganoGuard™   Organic fouling control additive for reverse osmosis systems by Professional Water Technologies, Inc.

organophosphates   Commonly used phosphorous-based organic pesticides which are relatively nonpersistent in the environment.

organotins   Chemical compounds used in anti-foulant paints to protect the hulls of boats and ships, buoys, and pilings from marine organisms such as barnacles.

Ori-Cast   Cast elastomer material used in nonmetallic rectangular clarifier products by Oritex Corp.

orifice plate   (1) Flow measurement device that indicates flow as a function of differential pressure across a flow-restricting orifice. (2) Flow-limiting device.

Ori-Plastic   Plastic material used in nonmetallic rectangular clarifier products by Oritex Corp.

ORM   See "other regulated material (ORM)."

ORNL   Oak Ridge National Laboratory.

ORP   See "oxidation-reduction potential (ORP)."

OSEC®   Electrolytic chlorination system by USFilter/Wallace & Tiernan.

OSHA   See "Occupational Safety and Health Administration (OSHA)."

Osmo®   Water purification systems by Osmonics, Inc.

osmoconformers   Organisms that rely on osmotic pressures to maintain internal ionic balance.

osmosis   Movement of water from a dilute solution to a more concentrated solution through a permeable membrane separating the two solutions.

Osmostill   Distillation unit by Osmonics, Inc.

osmotic pressure   Excess pressure that must be applied to a concentrated solution to produce equilibrium and prevent the movement of a more dilute solution through a semipermeable membrane into the more concentrated solution.

Osmotik®   Reverse osmosis membrane elements by Osmosis Technology, Inc.

OSP   See "Off spec water (OSP)."

OST   Office of Science and Technology. A U.S. EPA office.

OSW   Office of Saline Water.

OTA® Aerator   Rotor aerator by Scoti-Zahner, Inc.

OTC   Odor threshold concentration. See "threshold odor number (TON)."

OTE   Orbital tube evaporation.

OTEC   See "ocean thermal energy conversion (OTEC)."

other potentially infectious material (OPIM)   Body fluids or unfixed tissues and organs visibly contaminated with blood, other body fluids, HIV, or HBV.

other regulated material (ORM)   The U.S. Department of Transportation's hazard classification of a particular hazardous material to label it for transport.

OUR   Oxygen uptake rate.

outfall   The location where a storm or sanitary sewer or effluent is discharged into a receiving water body.

outhouse   See "privy."

outsourcing   The use of an outside vendor to provide operating, maintenance, or other services for a facility.

ova   Plural of "ovum."

overburden   The soil and rock overlying a mineral deposit that must be removed prior to the start of strip mining.

overdraft   The pumping of water from a groundwater basin or aquifer in excess of the supply flowing into the basin.

overflow rate (OFR)   An expression used to indicate the upward water velocity in a sedimentation tank expressed as flow per day per unit of basin surface area. Also called "surface loading rate."

overflow weir   A weir over which excess water or wastewater is allowed to flow.

overland flow   A land application technique that cleanses wastewater by allowing it to flow over a sloped surface where contaminants are absorbed and the water is collected at the bottom of the slope for reuse.

ovum   A mature egg ready to undergo fertilization. The singular form is "ova."

Owamat®   Oil/water separator by BEKO Condensate Systems Corp.

oxbow lake   A lake that forms in the abandoned channel or a cutoff meander on a river's floodplain.

oxic   A biological environment that contains molecular oxygen.

Oxidair™   Thermal oxidizer for soil remediation and off-gas treatment by EPG Companies, Inc.

oxidant   A chemical substance, such as chlorine or ozone, capable of promoting oxidation.

oxidation   (1) A chemical reaction in which an element or ion loses electrons. (2) The biological or chemical conversion of organic matter into simpler, more stable forms.

oxidation ditch   An extended aeration waste treatment process that occurs in an oval-shaped channel or ditch (also called a "race track") with aeration provided by a mechanical brush-aerator.

oxidation pond   An earthen wastewater basin in which biological oxidation of organic matter occurs naturally or with the assistance of mechanical oxygen transfer equipment.

oxidation-reduction potential (ORP)   The potential required to transfer electrons from an oxidant to a reductant which indicates the relative strength potential of an oxidation-reduction reaction.

Oxidator   Combination aeration, flocculation, and sedimentation unit by Baker Process.

oxide   A compound of an element with oxygen alone.

oxidize   To bring about oxidation.

oxidizing agent   Any substance that can contribute electrons to a reaction.

Oxifree®   Ultraviolet disinfection system by Capital Controls Co.

Oxigest®   Cylindrical package extended aeration waste treatment plant by Smith & Loveless, Inc.

Oxigritter   Primary sewage treatment unit by Baker Process.

Oxitech®   Resin conditioning process for TOC reduction by USFilter/Rockford.

OxiTop   System to measure BOD by WTW Measurement Systems, Inc.

Oxitrace™   Oxidant analyzer and monitor by Capital Controls Co.

Oxitron™   Fixed film wastewater treatment plant by USFilter/Krüger.

Oxy Flo   Mechanical aerator by Aqua-Aerobic Systems, Inc.

Oxy stream™   Oxidation ditch wastewater treatment system by WesTech Engineering Inc.

Oxycap   Device to improve energy efficiency and reduce noise and aerosol emissions from surface aerators by DHV Water BV.

Oxycat   Air pollution abatement catalyst by Met-Pro Corp.

OxyCharger   Static aerator by Parkson Corp.

Oxychlor   Chlorine dioxide generator by International Dioxide, Inc.

Oxyditch   Oxidation ditch treatment system formerly offered by Chemineer, Inc.

Oxy-Gard   Aeration control system that monitors dissolved oxygen level to control blower operation by Gardner Denver Blower Division.

oxygen   A chemical element that comprises approximately 20% of the earth's atmosphere and is essential for biological oxidation.

oxygen deficiency   The additional amount of oxygen required to satisfy the oxygen requirement of wastewater.

oxygen sag   The temporary decrease in the dissolved oxygen level in a stream or river which occurs downstream from a point source of pollution.

oxygen scavenger   A chemical used to supplement mechanical deaeration.

oxygen transfer   The exchange of oxygen between a gaseous and a liquid phase.

oxygen transfer rate   The mass of oxygen transferred per unit time.

oxygen uptake   The amount of oxygen used during biochemical oxidation.

oxygenases   Enzymes that catalyze the insertion of one or both atoms of an oxygen molecule into an organic compound, resulting in a chemical transformation.

Oxygun™   Sub-surface self-aspirating aerator by ABS Pumps, Inc.

Oxyrapid   Air diffusion and recycling system for activated sludge system by Infilco Degremont, Inc.

Oxytrace®   Dissolved oxygen measurement instrument by Industrial Analytics, Corp.

Oxytrace™   Chlorine residual analyzer by Capital Controls Co.

Oxytrap™   Wastewater aerator by DAS International, Inc.

OZ   Ozone generation equipment by Ozone Pure Water, Inc.

Ozat®   Compact ozone generator by Ozonia North America.

Ozofloat®   Ozone flotation process for water/wastewater treatment by USFilter/Krüger (North America) and OTV.

ozonation   The process of using ozone in water or wastewater treatment for oxidation, disinfection, or odor control.

ozonator   An ozone generator.

ozone   An unstable, gaseous oxidizing agent with disinfection properties similar to chlorine, also used in odor control and sludge processing. Chemical formula is O_3.

ozone byproducts   Compounds such as aldehydes and aldoacids formed when ozone is used to disinfect water.

ozone contactor   A device used to promote the efficient transfer of ozone to water or wastewater.

ozone destruct unit   An ozone destruction system used to guarantee removal of any unused ozone before the reactor off-gas is discharged to the atmosphere.

ozone generator   Device used to produce ozone by passing air/oxygen through an electric field.

ozone layer   The portion of the stratosphere, extending from an altitude of approximately 20 to 50 km, in which naturally occurring ozone protects life on earth by filtering out harmful ultraviolet radiation from the sun.

Ozone Research & Equipment Corp.   Equipment manufacturer acquired by Osmonics, Inc.

Ozonmat®   Ozone analyzer by Zellweger Analytics, Inc.

P

P&ID   Piping and instrumentation diagram.

P&Ps   Practices and procedures.

P2   Pollution prevention.

P2Rx   Pollution Prevention Resource Exchange; a coordinated effort of nine regional pollution prevention resource centers funded by the U.S. EPA.

PAC   Valve positioner and controller by F.B. Leopold Co., Inc.

PAC   See "powdered activated carbon (PAC)."

Pace®   Oil/water separator by Scienco/FAST Systems.

Pacer II™   Package water treatment plant by Roberts Filter Group.

Pacesetter   Liquid/liquid gravity separator by Baker Hughes Process Systems.

Pacific Decadal Oscillation (PDO)   A warming/cooling pattern occurring in the Pacific Ocean which affects world weather patterns.

Pacific Flush Tank   Former digestion equipment manufacturer acquired by USFilter/Envirex.

package plant   Factory-assembled treatment plant generally incorporated in a single tank, or at most, several tanks.

packed bed scrubber   An air pollution control device in which emissions pass through alkaline water to neutralize hydrogen chloride gas.

packed column   A vertical vessel filled with packing material usually used to strip gases or degasify liquids.

packing   The fill material in a fixed film reactor or stripping vessel that provides a large surface area per unit volume.

PACl   Polyaluminum chloride.

Pacpuri®   Sodium hypochlorite generation system by USFilter/Electrocatalytic.

PACT®   Powdered-activated carbon wastewater treatment process by USFilter/Zimpro.

Pactank™   Portable spill containment system by PacTec, Inc.

Paddle Dryer   Sludge dryer by Komline-Sanderson Engineering Corp.

paddle flocculator   A flocculation device utilizing rotating baffles to accomplish mixing.

PADRE®   Moving bed resin adsorption system for VOC removal/recovery by Thermatrix, Inc.

PAH   See "polynuclear aromatic hydrocarbons (PAH)."

paint filter test   Test to determine free water content of sludge sample.

PakTOR   Multi-cell packed bed reactor by USFilter/General Filter.

palatable water   Water at a desirable temperature that is free from objectionable tastes, odors, colors, and turbidity.

P-alkalinity   See "phenolphthalein alkalinity."

PallSep™   Vibrating membrane filtration by Pall Corp.

Palmer-Bowlus Flume   A portable, venturi-type flume used to measure water or wastewater flow.

PAN   See "peroxyacetyl nitrate (PAN)."

pandemic   A worldwide epidemic.

PAPR   Powered air-purifying respirator.

Para Cone   Internally fed rotary fine screen by Andritz-Ruthner, Inc. (Western Hemisphere) and USFilter/Contra-Shear.

Paraflash   Forced circulation evaporator by APV Crepaco, Inc.

Paraflow   Plate heat exchanger by APV Crepaco, Inc.

Paramax   Speed reducer product line by Sumitomo Machinery Corp.

paraquat   A herbicide resistant to microbial degradation that has been used to control marijuana and whose exposure can result in serious health effects or death.

parasite   An organism that lives either on or inside a larger host organism and where the presence of the parasite is usually harmful to the host organism.

parasitic bacteria   Bacteria that require a living host organism.

Para-Stat   Static screen by Dontech, Inc.

Paravap   High solids evaporator by APV Crepaco, Inc.

parenteral solution   A solution introduced into the body by a vein, muscle, or pathway other than the mouth.

Parkwood   Sewage treatment equipment product line by Longwood Engineering Co., Ltd.

Parshall flume   A fixed, venturi-type flume used to measure water or wastewater flow.

parthenogenic   Capable of reproduction by means of an unfertilized egg.

partial closure   The closure of a hazardous waste management unit at a facility that contains other active hazardous waste management units.

partial pressure   The pressure exerted by each gas in a mixture proportional to the amount of that gas in the mixture.

particle counter   Instrument used to measure the size and count the number of particles in water.

particle counting   A quantitative measurement of the number and size of individual particles in a water or wastewater sample.

particle size analysis   Determination of the amounts of different particle sizes in a sample.

particulate   Usually considered to be a solid particle larger than one micron or large enough to be removed by filtration.

particulate organic carbon (POC)   The portion of organic matter that can be removed by filtration through a 0.45 micron filter.

particulate organic matter (POM)   Material of plant or animal origin suspended in water and usually capable of being removed by filtration.

Partisol®   Air sampler by Rupprecht & Patashnick Co. Inc.

parts per million (ppm)   A common unit of measure used to express the number of parts of a substance contained within a million parts of a liquid, solid, or gas. Generally interchangeable with "milligrams per liter" in dilute solutions and water treatment calculations.

parts per thousand (ppt)   A unit of measure used to express the number of parts of a substance contained within a thousand parts of a liquid, solid, or gas.

Generally used to specify a water's salinity and commonly indicated by the symbol "0/00."

PASS® Poly-aluminum-silicate-sulfate coagulant by Eaglebrook, Inc.

Passavant Wastewater treatment equipment product line by USFilter/Zimpro.

PASS-C® Poly-aluminum-chloride coagulant by Eaglebrook, Inc.

passivation The changing of a chemically active surface of a metal to a much less reactive state. Usually done to stainless steel by immersion in an acid bath.

passive screen Intake screening device that does not employ mechanical cleaning.

passive solar heating A heating system that provides heat directly from the sun's rays and does not employ pumps, blowers, or water-filled pipes to transfer heat.

Pastel UV® Former name of InSpectra UV analyzer by Azur Enviromental.

pasteurization A process for killing pathogenic organisms by applying heat for a specific period of time.

pathogen Highly infectious disease-producing microbes commonly found in sanitary wastewater.

Pathwinder® Screenings conveyor by Serpentix Conveyor Corp.

Patriot Fluid recovery treatment system for spent coolant and oils by Waterlink Biological Systems.

PATS Pesticide action tracking system.

Paygro In-vessel composting system by Fairfield Service Co.

PC See "physical-chemical treatment (PC)."

PCB See "polychlorinated biphenyl (PCB)."

PCC Process Combustion Corp.

PCE See "perchloroethylene (PCE, also PERC)."

PCI Patterson Candy International, Ltd.

pCi See "picocurie (pCi)."

PCM Phase contrast microscopy.

PCP Progressing cavity pump.

PCV See "positive crankcase ventilation (PCV)."

PD Positive displacement.

PD Plus® Heavy duty blower by Tuthill Pneumatics Group.

PDC™ Polymer dosage control system by Andritz-Ruthner, Inc.

PDM® Water storage tank product line by Pitt-Des Moines, Inc.

PDO See "Pacific Decadal Oscillation (PDO)."

PE (1) Professional engineer. (2) See "population equivalent (PE)."

Peabody Floway Former name of Floway Pumps, Inc.

Peabody TecTank Bolted steel storage tank by A.O. Smith Engineered Storage Products.

Peabody Welles Former manufacturer whose product lines were acquired by USFilter/Aerator Products.

peak flow Excessive flows experienced during hours of high demand, usually determined to be the highest 2-hour flow expected to be encountered under any operational conditions.

peaking factor The ratio of peak to average flow.

Pearlcomb® Fine bubble diffuser by USFilter/Diffused Air Products Group.

Pearth Anaerobic digester gas mixing system by USFilter/Envirex.

peat Material formed by partial decay of marsh vegetation with a moisture content greater than 75%.

pebble lime See "quicklime."

PEF See "primary effluent filtration (PEF)."

PEL See "permissible exposure limit (PEL)."

pelagic Referring to the open sea at all depths.

Pelican Wall-mounted clarifier scum removal system by NSW Corp. and Copa Group (U.K.).

Pelldry Liquid-absorbing pellet by the former Sheldahl Industrial Absorbents.

Pelletech® Indirect sludge dryer and pelletizing unit by Wheelabrator Water Technologies, Inc.

Pelton wheel An impulse hydraulic turbine that may be used as an energy recovery device in high head applications such as seawater reverse osmosis.

PEMS Predictive emissions monitoring systems. This monitors critical operational parameters of the source and relates the results to the emissions level.

Penberthy Former manufacturer whose product line is now offered by Chemineer, Inc.

Penfield® Water treatment product line by USFilter/Rockford.

Penro Reverse osmosis systems by Penfield Liquid Treatment Systems.

penstock A pipe which transports water to a turbine for the production of hydro-electric energy.

PentaPure® Disinfecting resin by WTC Industries, Inc.

Pentech Former manufacturer whose product line is now offered by Chemineer, Inc.

per capita Per person.

PERC See "perchloroethylene (PCE, also PERC)."

perc test See "percolation test."

perched aquifer An unconfined aquifer separated from the underlying water table by an impermeable layer or unsaturated zone.

perched water Zone of unpressurized water held above the water table by impermeable rock or sediment.

perchlorate A rocket fuel ingredient emerging as an environmental threat particularly to some drinking water supplies. Chemical formula ClO_4^-.

perchloroethylene (PCE, also PERC) A chlorinated hydrocarbon used as an industrial cleaner or solvent, often used to dry cleaning clothing. Also called "tetrachloroethylene."

Percol® Polyelectrolyte used to enhance liquid/solid separation by Ciba Specialty Chemicals.

percolating filter See "trickling filter."

percolation The flow or trickling of a liquid downward through a contact or filtering medium.

percolation test Test used to determine the water-absorbing capacity of soil where the drop in water level in a test hole is measured over a fixed time period. Also called "perc test."

Perc-Rite® Filtration system by Waste Water Systems, Inc.

perfected water right A permitted water right indicating that anticipated uses were of beneficial use.

perfection   The process of meeting terms and conditions of a water right permitting process which results in a perfected water right.

performance ratio   A unit of measurement used to characterize evaporator performance expressed as the mass of distillate produced per unit of energy consumed.

peripheral feed clarifier   A circular sedimentation basin in which the influent flows from the perimeter toward the center of the unit.

periphyton   Microscopic underwater plants and animals that are firmly attached to solid surfaces such as rocks, logs, pilings, and other structures.

peristaltic pump   A type of positive displacement pump where the fluid is squeezed through a flow tube by external rollers.

Perma-buoy   Foam-filled fiberglass flight for chain and flight sludge collector by Jeffrey Chain Corp.

Permachem   Packages containing chemical compositions and reagents by Hach Co.

permafrost   Permanently frozen subsurface soil layer in the polar regions.

Permaglas®   Storage tank coating system by A.O. Smith Engineered Storage Products.

Permaklip   Belt seam for filter press belt by Sefar America, Inc.

Permalife   Chain and flight sludge collector components by Jeffrey Chain Corp.

permanent hardness   Hardness associated with sulfates, chlorides, and nitrates of calcium and magnesium which remain after boiling.

permanent threshold shift (PTS)   A permanent hearing loss for a certain sound frequency.

Permasep®   Reverse osmosis products by E.I. Dupont De Nemours, Inc.

permeability   The property of a filter medium to permit a fluid to pass through it under the influence of pressure.

permeate   The liquid that passes through a membrane.

permeator   A pressure vessel containing semi-permeable membranes.

PermeOx®   Solid peroxygen by FMC Corp., Hydrogen Peroxide Division.

permissible dose   The dose of a chemical that may be received by an individual without the expectation of a significantly harmful result.

permissible exposure limit (PEL)   OSHA-established workplace exposure limit for 600 industrial chemicals.

Permofilter   Horizontal multiple cell pressure filter by USFilter/Warren.

Permujet®   Clarifier by USFilter/Warren.

Permupak   Package water treatment plant by USFilter/Warren.

PermuRO   Reverse osmosis system by USFilter/Warren.

Permutit®   Water treatment product line by USFilter/Warren.

peroxone   A blend of ozone and hydrogen peroxide used for disinfection and odor control.

Perox-Pure®   UV-catalyzed hydrogen peroxide system by Calgon Carbon Corp.

Perox-serv™   Odor control services by Vulcan Performance Chemicals.

Perox-stor™   Hydrogen peroxide user service by Vulcan Performance Chemicals.

peroxyacetyl nitrate (PAN)   A secondary pollutant and major component of photochemical smog formed when reactive hydrocarbons and oxides of nitrogen combine in the presence of sunlight.

Perpac   Surface water treatment plant by Vulcan Industries, Inc.

Perrin Dewatering equipment product line by USFilter/Dewatering Systems.

personal protective equipment (PPE) Devices or clothing worn to help insulate a person from direct exposure to hazardous materials.

personnel access opening An opening in a vessel or sewer to permit human entry. Also called a "manhole" or "manway."

PERT Program evaluation review technique.

pervaporation (PV) A process in which membranes are used to remove volatile organic compounds from an aqueous stream.

pesticides A chemical used to kill undesired insects or animals.

PET Polyethylene terephthalate. A resin used to make plastic bottles.

***PET*™** Demineralizer system by USFilter Corp.

petri dish A covered dish containing an agar media used in the laboratory to cultivate bacteria.

petrochemicals Products or compounds produced by the processing of petroleum and natural gas hydrocarbons.

***Petro-Flex*®** Portable holding tank by Aero Tec Laboratories, Inc.

petroleum The crude oil removed from the earth and the oils derived from tar sands, shale, and coal.

Petrolux Ceramic membrane filter by USFilter Corp.

Petro-Pak Coalescing media for oil removal system by McTighe Industries, Inc.

***Petro-Screen*™** Oil coalescer screen by Highland Tank & Manufacturing.

***Petro-Xtractor*™** Water well oil skimmer by Abanaki Corp.

pezodialysis Membrane process to remove salt from solution where salt, rather than water, passes through the membrane.

PFBC Pressurized fluidized bed combustion.

PFCs Perfluorinated compounds.

PFD Process flow diagram.

Pfiesteria piscicida A waterborne organism toxic to fish.

PFO Power fail open.

PFR See "plug flow reactor (PFR)."

PFRP Process to further reduce pathogens.

PFT Former equipment manufacturer acquired by USFilter/Envirex.

PFU See "plaque-forming units (PFU)."

pH The reciprocal of the logarithm of the hydrogen ion concentration in gram moles per liter. On the 0 to 14 pH scale, a value of 7 at 25°C (77°F) represents a neutral condition. Decreasing values indicate increasing hydrogen ion concentration (acidity) and increasing values indicate decreasing hydrogen ion concentration (alkalinity). Full form is "potential of hydrogen."

phagotroph An organism that obtains nutrients through the ingestion of solid particles of food.

***Phantom*® 4** Odor-fighting product by NuTech Environmental Corp.

pharmaceutical-grade water See "USP-purified water."

pharmakinetics The dynamic behavior of chemicals inside biological systems, including uptake, distribution, metabolism, and excretion.

PharmMAX™   Reverse osmosis system pharmaceutical and biotech applications by USFilter.

phase   The state of a substance; solid, liquid, or vapor.

phased reversal   A technique employed in EDR systems to improve product recovery by staging electrical polarity reversal.

PHC   Principal hazardous constituent.

PHD   Peak hourly demand.

PhD²   Portable multiple gas detector by Biosystems, Inc.

phenolphthalein   A color indicator that changes from colorless to pink/red, used to measure alkalinity.

phenolphthalein alkalinity   Alkalinity determined by titration with sulfuric acid to pH 8.3, indicated by a color change of phenolphthalein and expressed as mg/L of calcium carbonate. Also called "P-alkalinity."

phenols   Organic pollutant also known as carbolic acid occurring in industrial wastes from petroleum processing and coal-coking operations.

pheromone   A substance secreted by an organism that influences the behavior or developmental response of other members of the same species.

Phoenix®   Odor control system by Calgon Carbon Corp.

Phoenix Press   Belt filter press by Phoenix Process Equipment Co.

Phoenix System   Filter air scour and underdrain systems by AWI.

phosphate   A salt or ester of phosphoric acid.

phosphoric acid   An acid produced from mined phosphate rock used in fertilizers and detergents. Chemical formula is H_3PO_4.

phosphorous   (1) A nonmetallic chemical element with the chemical symbol P. (2) A nutrient essential to all life forms whose overabundance can contribute to the eutrophication of a water body.

Phostrip   Biological system for phosphorus and BOD removal by Tetra Process Technologies.

Phosver   Reagent chemical used to determine phosphate concentration in water by Hach Co.

photic zone   The upper level of a waterbody into which light penetrates.

photocatalytic oxidation   A process where organic and inorganic contaminants are oxidized by free radicals generated from the interaction of UV radiation with a chemical or metal oxide catalyst.

photochemical smog   A form of air pollution which results in an atmospheric haze and is caused by the reaction of sunlight with VOCs, nitrogen oxides, and other pollutants produced by combustion processes.

photooxidation   The use of ultraviolet light to induce or supplement an oxidation reaction.

photosynthesis   The process of converting carbon dioxide and water to carbohydrates, activated by sunlight in the presence of chlorophyll.

phototrophs   Organisms that rely on the sun for energy.

Photovac™   Portable VOC monitor by PerkinElmer Instruments.

photovoltaic cell   A device that utilizes crystalline materials to convert light from solar radiation directly into electricity.

phreatic   Of, or relating to, groundwater.

phreatic surface   The free surface of groundwater at atmospheric pressure.

phreatophyte   Any long-rooted plant that is able to obtain its water from the water table. Excessive growths of phreatophytes are undesirable in some areas since they may consume large quantities of scarce water.

pHREEdom™   Cooling water treatment chemicals by Calgon Corp.

PHSA   Public Health Service Act.

phys-chem   See "physical-chemical treatment."

physical treatment   A water or wastewater treatment process that utilizes only physical methods such as filtration or sedimentation.

physical-chemical treatment (PC)   Treatment processes that are nonbiological in nature. Also known as "phys-chem."

phytoplankton   Algae that exist floating or suspended freely in a body of water.

phytoremediation   A remediation technique where specialized plants are used to take up specific soil contaminants into their roots or foliage prior to harvesting and treatment.

phytotoxic   Harmful to plants.

PIC   See "products of incomplete combustion (PIC)."

Picabiol®   Activated carbon purification process for potable water by Pica USA, Inc.

pickets   Vertical paddles used in a gravity thickener.

pickle liquor   Waste acid from steel pickling process.

pickling   A chemical or electrochemical method of removing mill scale and rust from steel by washing or immersing in an acid or salt solution.

picocurie (pCi)   A unit of radioactivity equal to 3.7×10^{-12} curie.

Pielkenrood   Former name of USFilter/Rossmark.

piezometer   An instrument consisting of a small pipe and manometer fitted to the wall of pipe or container to measure pressure head.

piezometric head   The elevation plus pressure head.

pig   A water-propelled internal pipe cleaner.

pigging   A pipeline cleaning procedure using water-propelled pigs to scour solids from the interior walls of a pipe.

Pilgrim   Former equipment manufacturer acquired by Andritz-Ruthner, Inc.

piling   Timbers, concrete posts, or other structural elements embedded into the ground to support a load.

pilot plant   A water or wastewater treatment plant smaller than full scale and used to test and evaluate a treatment process.

pilot tests   Testing a treatment technology under actual site conditions to identify potential problems prior to fullscale implementation.

PIMA   Photonic ionization, manipulation, and augmentation. A desalination technology that achieves molecular modification of seawater through the use of lasers and photonics.

PIN   Pesticide information network.

pin floc   Small floc particle.

Pinch Press   High pressure filtration and dewatering device by Waste-Tech, Inc.

pinch valve   A valve where sealing is achieved by one or more flexible elements that can be pinched to stop flow.

pink water   The process of wastewater stream resulting from the manufacture of explosives.

pintle chain   Chain extensively used for elevating and conveying consisting of one-piece links cast with two offset sidebars and coupled with steel pins.

pipe gallery   A passageway to provide access for installation and maintenance of underground pipes and valves.

pipe spool   A prefabricated section of pipe.

Pipeliner   In-line grinder/cutter device by Robbins Myers, Inc.

pipette   A calibrated glass tube used to deliver prescribed volumes of liquids, usually less than 10 mL.

Pista®   Vortex type grit removal system by Smith & Loveless, Inc.

piston pump   A reciprocating pump whose piston normally incorporates a sliding seal with the cylinder wall.

Pit Hog®   Sludge pumping system by Liquid Waste Technology, Inc.

pitch   (1) The length of one link of chain measured from pin centerline to pin centerline. (2) The distance between the centers of adjacent tubes.

pitot tube   Flow measurement device that measures the velocity head of a liquid stream as the difference between the static head and the total head. Also called a "pitot gauge."

Pittchlor   High-test calcium hypochlorite product by PPG Industries, Inc.

pitting   Localized corrosion-causing attacks over small surface areas which may reach considerable depths.

PIV   Positive infinitely variable.

pK   The reciprocal of the logarithm of the ionization constant of a chemical compound.

Planet   Rotary distributor for fixed film reactor by Simon-Hartley, Ltd.

plankton   Small, passively floating or weakly swimming animal and plant life of a body of water.

plaque-forming units (PFU)   A measurement of viral particles where one PFU is equivalent to approximately 20 to 300 viral particles.

PLASdek   Cooling tower fill by Munters.

plasma   A high temperature, partially ionized gas which is electrically conductive.

plasma arc furnace   A high temperature furnace utilizing a plasma arc heater, or torch, for the destruction of hazardous gaseous, liquid, and solid wastes.

plate count   The number of microbe colonies that develop on a laboratory test dish after a fixed incubation period.

plate settler   Clarifier with enhanced sedimentation through the use of steeply inclined plates.

plate tectonics   The scientific theory that the earth's surface consists of large slabs, or plates, whose constant motion explains continental drift, the formation of mountains, and other geological changes.

plate tower scrubber   An air pollution control device that neutralizes hydrogen chloride gas by bubbling alkaline water through holes in a series of metal plates.

plate-and-frame press   A batch process dewatering device in which sludge is pumped through a series of parallel plates fitted with filter cloth.

Plate-Pak®   Vane mist eliminators by ACS Industries, Inc.

Platetube   Porous diffuser plates by Walker Process Equipment.

PLC   Programmable logic controller.

PLIRRA   Pollution Liability Insurance and Risk Retention Act.

PLM   Polarized light microscopy.

plug flow   Flow conditions where fluid and fluid particles pass through a tank and are discharged in the same sequence that they enter.

plug flow reactor (PFR)   A reactor in which the residence time for a given input, or plug, is equal to the theoretical hydraulic retention time.

plugging factor   See "silt density index (SDI)."

plume   The measurable or visible impact of a discharge into the air or a body of water.

plunger pump   A reciprocating pump whose plunger does not contact the cylinder walls but enters and withdraws from it through packing glands.

Plus 5   Air diffuser by USFilter/Diffused Air Products Group.

Plus 150™   Laboratory water system by USFilter Corp.

plutonium   A radioactive metallic element chemically similar to uranium.

pluvial   Of, or having to do with, rain.

pluvial lake   A lake formed during a period of abundant rainfall.

pluviometer   A rain gauge.

PM   (1) Particulate matter. (2) Preventative maintenance.

PM$_{2.5}$ Airborne particulate matter having a diameter equal or smaller than 2.5 microns.

PM$_{10}$ Airborne particulate matter having a diameter equal to or smaller than 10 microns.

PM$_{15}$ Airborne particulate matter having a diameter equal or smaller than 15 microns.

PM-100®   Clay filtration media by Colloid Environmental Technologies Co.

PMD™   Pipe mounted diffuser by Environmental Dynamics Inc.

PMR   Pollutant mass rate.

PNA   Polynuclear aromatics.

pneumoconiosis   Lung disease that results from chronic exposure to dusts.

PO*WW*ER™   Wastewater treatment process to reduce residual solids by USFilter/Gas Technologies.

POC   (1) Particles of complete combustion. (2) See "particulate organic carbon (POC)."

Pocket Pal   Portable pH tester by Hach Co.

pocosin   A low, flat swamp on the coastal plain of the Southeastern U.S.

PODR   Point of diminishing returns.

POE   See "point-of-entry (POE)."

pogonip   Native American term for a dense winter fog containing frozen ice particles, occurring in deep mountain valleys of the western U.S., particularly in the Sierra Nevada mountains.

pOH   The negative logarithm of the hydroxyl ion concentration approximated by the relationship: 14 minus pH = pOH.

POHC   See "principal organic hazardous constituent (POHC)."

point source discharge (PS)   A pipe, ditch, channel, or other container from which pollutants may be discharged.

point-of-entry (POE)   Location of a water treatment or water quality device at the point drinking water enters a house or building for the purpose of reducing the contaminants in the drinking water distributed throughout the house or building.

point-of-use (POU)   Location of a water treatment or water quality device at a faucet in an individual household.

polar zone   The high altitude regions of the earth located from latitude 66°34′ north and 66°34′ south to the poles.

polder   An area of dry, low-lying land that has been reclaimed from a body of water and is maintained through the use of dikes. In the Netherlands, polders are created for agricultural land.

Pol-E-Z®   Emulsion polymer to enhance solids/liquid separation by Calgon Corp.

pollutant   A substance, organism, or energy form present in amounts that impair or threaten an ecosystem to the extent that its current or future uses are precluded.

Pollutant Standard Index (PSI)   Measure of adverse health effects of air pollution levels in major cities.

pollution   The presence of a pollutant in the environment.

Pollutrol   Former equipment manufacturer.

Polly Pig   Internal pipeline cleaner by Knapp Polly Pig, Inc.

Polyad™   Fluidized bed VOC emission control system by Weatherly.

Poly-Alum   Polymerized inorganic coagulant by Rochester Midland.

polyamide   A molecular chain polymer made of amide linkages used in the construction of thin film composite reverse osmosis membranes.

PolyBlend®   Polymer mixing/feeding products by USFilter/Stranco.

PolyBoss™   Instrument to check settling velocity in clarifier feedwell by WesTech Engineering Inc.

Polybrake   Cleaning product for removal of polymers by AquaPro, Inc.

polychaete worm   A small worm common in seas and estuaries often chosen for bioassays of coastal regions.

polychlorinated biphenyl (PCB)   Class of hazardous organic compounds considered probable carcinogens formerly used in the manufacture of electrical insulation and heating and cooling equipment.

polyelectrolyte   A compound consisting of a chain of organic molecules used as coagulants or coagulant aids. See also "polymer."

polyethylene   An inexpensive plastic with a low coefficient of friction and excellent abrasion, impact, and chemical resistance.

Poly-Filter   Plate and frame filter press by the former Clow Corp.

Poly-Fine   Absolute rated cartridge filters by USFilter/Filtration & Separation.

polyhaline   A term used to characterize water with a salinity of 18,000 to 30,000 mg/L due to ocean salts.

Polyjet   Flow control valve by Bailey Polyjet.

PolyKleen®   ABS filter underdrain by Baker Process.

Poly-Links   Nonmetallic sludge collector chain by NRG, Inc.

Polymair   Package polymer processing system by Acrison, Inc.

polymaleic acid   A high performance scale control additive.

Polymaster® Polymer mixing/feeding system by Neptune Chemical Pump Co.

PolyMate® Liquid polymer blending/feeding system by Neptune Chemical Pump Co.

PolyMax Polymer mixing/feeding products by Semblex, Inc.

polymer (1) High molecular weight compounds derived by the recurring addition of similar molecules. (2) Common term for "polyelectrolyte."

polymer activity The percent of a polymer's molecular weight available to react with and flocculate solids particles.

Polymer Piping & Materials Former equipment manufacturer whose products are now offered by Jaeger Products, Inc.

polymerization A chemical reaction in which simple molecules combine to form larger, more complex molecules.

Polymetrics Membrane products company acquired by USFilter.

Polymetron™ Water analyzer product line by Zellweger Analytics, Inc.

PolyMixer Polymer mixing/feeding system by Hoffland Environmental Inc.

PolyMizer® Centrifuge by Alfa Laval Separation, Inc.

polyna An expanse of water surrounded by ice.

polycyclic aromatic hydrocarbons See "polynuclear aromatic hydrocarbons."

polynuclear aromatic hydrocarbons (PAH) A group of aromatic compounds, many of which are carcinogenic, formed by industrial processes and during the burning of gasoline, coal, and other substances.

Polypak Polymer feeding system by BIF.

polypeptide A compound consisting of multiple linked amino acids.

polyphosphates Phosphate compounds used as sequestration agents to prevent formation of iron, manganese, and calcium carbonate deposits.

Poly-Pleat™ Filter cartridges by Harmsco Filtration Products.

PolyPress Plate and frame filter press by Star Systems, Inc.

PolyPro Dry polymer feed system by AquaPro, Inc.

Polypure Water treatment chemical group of Polydyne, Inc.

PolyRex Polymer feed system by Bran+Luebbe.

polysaline A term used to characterize water with a salinity of 18,000 to 30,000 mg/L due to land-derived salts.

Polyseed® Bacterial culture for BOD seeding by Polybac Corp.

Poly-Stage™ Modular air scrubber system by USFilter/Davis Process.

polysulfone A synthetic thermoplastic polymer used in the manufacture of ultra-filtration membranes and in thin film composite and charged polysulfone reverse osmosis membranes.

PolyThickener Waste-activated sludge thickener by Walker Process Equipment.

Polytox™ Biological toxicity test kit by Polybac Corp.

PolyTube Tubular air diffuser formerly offered by Walker Process Equipment.

polyurethane An elastomer with tensile strength and abrasion resistance greater than that of natural rubber and capable of being formulated for injection molding or casting.

polyvinyl chloride (PVC) A thermoplastic with excellent corrosion resistance that is widely used to manufacture pipe and biological filter media.

POM See "particulate organic matter (POM)."

pond A body of water smaller than a lake, often formed artificially.

pond scum   A mass of filamentous algae that forms a green scum on the surface of ponds.

ponding   See "pooling."

Pond-X®   Odor control product by NuTech Environmental Corp.

Pontoon   Floating cover for anaerobic digesters formerly offered by USFilter/Envirex.

pooling   The formation of pools of liquid on the surface of a clogged filter.

population   (1) A group of interbreeding organisms occupying a particular space. (2) The number of humans or other living creatures in a designated area.

population at risk   A population subgroup that is more likely to be exposed to a chemical, or is more sensitive to the chemical, than the general population.

population equivalent (PE)   The daily wastewater typically produced by one person expressed in terms of BOD; i.e., 60g of BOD per population equivalent.

Porcupine   Indirect contact sludge dryer by Bethlehem Corp.

PORI   Former name of USFilter Recovery Services.

Poro-Carbon   Automatic liquid filter system by R.P. Adams Co., Inc.

Poro-Edge   Automatic water strainer by R.P. Adams Co., Inc.

porosity   The ratio of void volume to total bulk volume.

Poro-Stone   Automatic liquid filter system by R.P. Adams Co., Inc.

porous disk diffuser   A circular, fine bubble aeration device of porous plastic or ceramic construction.

Porta Dike™   Portable spill containment system by Environetics, Inc.

Port-A-Berm™   Portable secondary containment system by Aero Tec Laboratories, Inc.

Portacel®   Gas chlorination system by F.B. Leopold Co., Inc.

Porta-Cleanse   Submersible mixer for wet well pump station by Flygt.

Porta-Feed®   Chemical handling system by Nalco Chemical Co.

Porta-Pac™   Powdered-activated carbon wet injection system by Norit Americas Inc.

PortaPump®   Portable, submersible pump by Warren Rupp, Inc.

Porta-Tank   Liquid storage system by Environetics, Inc.

Portland cement   Cement made by heating a slurry of crushed chalk or limestone and clay to clinker in a kiln before grinding and adding gypsum.

Posi-Clean   Wedgewire straining element by Tate Andale, Inc.

Posirake   Reciprocating rake bar screen formerly offered by USFilter/Zimpro.

positive crankcase ventilation (PCV)   A method of reducing automobile engine emissions by directing crankcase emissions into the cylinders for combustion.

positive displacement pump   A pump where liquid is drawn into a cavity and its pressure is increased, forcing the liquid through an outlet port into the discharge line.

Positive Seal   Rotary distributor for trickling filter by Walker Process Equipment.

post chlorination   Addition of chlorine after completion of other treatment processes.

post treatment   Treatment of finished water or wastewater to further enhance its quality.

post-closure   The time period following the shutdown of a waste management or manufacturing facility, which, for monitoring purposes, is often considered to be 30 years.

potable reuse   The beneficial use of highly treated reclaimed water toward augmentation of potable water supply.

potable water   See "drinking water."

potassium permanganate   A crystalline salt of potassium and manganese used for taste, odor control, and iron/manganese oxidation. Chemical formula is $KMnO_4$, also called "purple salt."

potential of hydrogen   See "pH."

potentially responsible party (PRP)   An individual or company identified as potentially liable for cleanup or payment for cost of cleanup of a hazardous waste site.

potentiation   The effect of one chemical to increase the effect of another chemical.

potentiometric surface   The level to which water will rise in cased wells or other cased excavations into aquifers.

POTW   See "publicly owned treatment works (POTW)."

POU   See "point-of-use (POU)."

POU/POE   Point-of-use/point-of-entry.

pounds per square inch, absolute (psia)   The total pressure in a system which is the sum of the gage pressure and atmospheric pressure and is expressed in terms of the pounds of force exerted per square inch.

pounds per square inch, gage (psig)   The pressure measured by a gage and expressed in terms of the pounds of force exerted per square inch.

pour point   The lowest temperature at which a liquid will pour.

Powder Pop®   Chlorine test dispenser by HF Scientific, Inc.

powdered activated carbon (PAC)   A powdered form of activated carbon fed as a slurry to water to absorb organics, particularly taste and odor-causing constituents.

Powdex®   Combination ion exchange and filtration unit by Graver Co.

Power Backwash™   Concurrent air and water filter backwash system by Tetra Process Technologies.

Power Brush™   Internal pipeline cleaner by Pipeline Pigging Products, Inc.

Power Mizer™   Multi-stage centrifugal blower by Spencer Turbine Co.

Power Units™   Magnetic pipe treatment unit by Aqua Magnetics International, Inc.

PoweRake   Reciprocating rake bar screen by the former EnviroFab, Inc.

Powerhouse®   Alkaline cleaner/degreaser by Pro Products, Corp.

Powermatic   Reciprocating rake bar screen by Brackett Geiger.

Powerpak™   Filter press by Flo Trend Systems, Inc.

PowerSewer™   Low pressure sewer by Interon Corp.

PoweRupp®   Electrically driven diaphragm pump by Warren Rupp, Inc.

Powrclean   Front cleaned multiple rake bar screen by USFilter/Aerator Products.

Powr-Trols   Custom engineered pump station motor control panel by Healy-Ruff Co.

Poz-O-Lite®   A lightweight aggregate by Conversion Systems, Inc.

Poz-O-Tec®   Scrubber sludge and stabilizing additive mixture that produces a stable, nonleaching, monolithic mass of low permeability by Conversion Systems, Inc.

pozzolonic   Finely divided materials such as fly ash that aid in forming compounds possessing cementitious properties.

ppb   Parts per billion.

PPE   See "personal protective equipment (PPE)."

ppm   See "parts per million (ppm)."

ppmv   Parts per million by volume.

ppmw   Parts per million by weight.

PPP   Public-private partnership.

ppt   Parts per thousand.

PQ®   Sodium silicate corrosion inhibitor by PQ Corp.

PQL   Practical quantitation limit. Generally the same as "estimated quantitation limit."

Praestol®   Polymer products by Stockhausen, Inc.

preaeration   A preliminary treatment process in which wastewater is aerated to remove gases, add oxygen, promote flotation of grease, and/or aid coagulation.

prechlorination   The application of chlorine before other treatment processes.

precipitate   A solid that separates from a solution.

precipitation   The phenomenon that occurs when a substance held in solution passes out of solution into a solid form.

Precipitator™   Package treatment plants by USFilter/Warren.

Precision   Fine bubble tube diffuser by USFilter/Diffused Air Products Group.

precoat filter   Filter using a thin layer of very fine material, such as diatomaceous earth, to coat the filter surface before filtration cycles.

precursor   A substance or compound from which another substance or compound is formed. Usually refers to an organic compound capable of being formed into a trihalomethane.

prefilter   A filtration device located upstream of the main filtration process.

PreFLEX®   Skid-mounted pretreatment system for reverse osmosis and demineralizer systems by USFilter/Rockford.

preliminary assessment   The process of collecting and reviewing available information about a known or suspected waste site or release.

preliminary treatment   Treatment steps including comminution, screening, grit removal, preaeration, and/or flow equalization which prepare wastewater influent for further treatment.

Prerostal   System of adjusting pumping volume to inflow rate by Envirotech Pumpsystems.

prescriptive   Water rights which are acquired by diverting water and putting it to use in accordance with specified procedures.

presedimentation   A pretreatment process used to remove sand, gravel, or other gritty material before subsequent treatment.

Preservol™   Reverse osmosis membrane preservative by Professional Water Technologies, Inc.

pressate   The liquid waste stream from a filter press.

Pressflex Dewatering screen fabric by Geschmay GmbH.

***PressMaster*™** Hydraulic sludge dewatering unit by Baker Process.

pressure drop See "headloss."

Pressure Exchanger Energy recovery device for seawater reverse osmosis unit by Energy Recovery, Inc.

pressure filter Filter unit enclosed in a vessel which may be operated under pressure.

pressure filtration The process of removing solids from a liquid using external pressure to force the liquid through a filter media.

pressure head The amount of energy in water due to water pressure.

pressure sewers A system of pipes in which water, wastewater, or other liquid is pumped to a higher elevation.

pressure-swing adsorption (PSA) A process using an adsorbent to separate components of a gas mixture, sometimes used to separate oxygen from air to form pure oxygen.

Pressveyor Hydraulic screenings press/conveyor by Waterlink Separations, Inc.

Prestex Filter belt for belt filter press by Sefar America, Inc.

Presto-Tek Flowmeter product line of Newport Electronics.

presumptive test The first of three steps in the analysis of water and wastewater for the presence of fecal bacteria. After a sample is inoculated and incubated, the presence of acid and gas is an indication of a positive test and the water is presumed to be contaminated. This test is followed by a confirmed test and a completed test.

***Pretreat Plus*™** Antiscalant/dispersant by King Lee Technologies.

***Pretreat SDP*™** Antiscalant for small capacity RO systems by King Lee Technologies.

pretreatment (1) The initial water or wastewater treatment process that precedes primary treatment processes. (2) The treatment of industrial wastes to reduce or alter the characteristics of the pollutants prior to discharge to a POTW.

prevalent levels Levels of airborne contaminant occurring under normal conditions.

prevention of significant deterioration (PSD) A provision of the Clean Air Act that establishes a minimum air quality baseline for particular pollutants in specific areas.

primary clarifier Sedimentation basin that precedes secondary wastewater treatment.

primary contaminant A drinking water contaminant with health-related effects.

primary effluent filtration (PEF) The use of granular media or synthetic material filtration to remove contaminants from primary effluent prior to discharge.

primary industry categories The 34 types of industrial facilities that require the best available technology for toxic water pollutants under the Clean Water Act.

primary MCL EPA-mandated maximum contaminant level in drinking water based on health effects.

primary pollutant A pollutant that exists in the environment in the same form as when it was released.

primary sedimentation Principal form of primary wastewater treatment utilizing clarifiers to reduce the solids loading on subsequent treatment processes.

primary sludge Sludge produced in a primary waste treatment unit.

primary standards   National ambient air quality standards designed to protect human health with an adequate margin for safety.

primary treatment   Treatment steps including sedimentation and/or fine screening to produce an effluent suitable for biological treatment.

Prima-Sep®   Specialty clarifier with tray separator by Graver Co.

Prime Aire   Trash pump by Gorman-Rupp Co.

Primox®   Oxygen injection system for primary sewage by BOC Gases.

Princeton Clearwater   Former equipment manufacturer.

principal organic hazardous constituent (POHC)   Hazardous organic compounds that may form products of incomplete combustion when incinerated.

prior appropriation   A doctrine of water law that allocates the rights to use water on a first-come first-serve basis.

priority pollutants   A list of approximately 126 chemicals identified as toxic pollutants by the Clean Water Act.

privately owned treatment works   A treatment works that is not owned or operated by a state or municipality. Not a "POTW."

privatization   The involvement of nonpublic and entrepreneurial interests in project development, ownership, and/or operation of municipal facilities such as water and wastewater treatment systems.

privy   A pit toilet, also called an "outhouse."

*PRO*ECOLITH*   Monolithic regenerative catalytic oxidation product for VOC abatement by Süd-Chemie Prototech Inc.

*PRO*HHC*   Halogenated hydrocarbon catalyst by Süd-Chemie Prototech Inc.

*PRO*PEL*   Bead-type air pollution control catalysts by Süd-Chemie Prototech Inc.

*PRO*RCO*   Metal coated regenerative catalytic oxidation product for VOC abatement by Süd-Chemie Prototech Inc.

*PRO*VOC*   Ceramic and metal monolithic catalyst products for VOC abatement by Süd-Chemie Prototech Inc.

Probiotics™   Lagoon sludge oxidation products by Bio Huma Netics, Inc.

ProBlend   Liquid polymer feed system by AquaPro, Inc.

process wastewater   Wastewater generated during manufacture or production processes.

process water   Water that is used for, or comes in contact with, an end product or the materials used in an end product.

Prochem®   Agitator and mixer products by Chemics, Inc.

produced water   All waters produced during the production of crude oil and gas.

product staging   Reverse osmosis process configuration where the product from one stage is used as feedwater on a subsequent stage to improve product water quality.

product water   Water produced as a result of treatment or desalination processes.

products of incomplete combustion (PIC)   Carbon monoxide, hydrocarbons, and other organic matter generated when organic materials are burned.

Professional Services Group   Former group merged into USFilter/Operating Services.

profile wire   Term used to describe specially shaped wire that is generally triangular or trapezoidal in cross section. Also called "wedgewire."

Profiler   Sludge blanket level and suspended solids monitor by Mt. Fury Co., Inc.

progressing cavity pump (PCP)   Pump used for viscous fluids, including sludge, that consists of a single-threaded shaft rotor rotating in a double-threaded rubber stator.

progressive cavity pump   See "progressing cavity pump (PCP)."

ProGuard   Backwashable cartridge filters by ProGuard Filtration Systems.

Promal   Pearlitic malleable iron chain material by USFilter/Envirex.

ProMix™   Polymer wetting and blending unit by BlenTech Inc.

ProPack   Random trickling filter media by the former Gray Engineering Co.

ProPak™   Sampling system by Isco, Inc.

propellant   Liquid in a self-pressurized pesticide product that expels the active ingredient from its container.

prophylaxis   The observance of procedures necessary to prevent disease.

proportional weir   A weir whose discharge is directly proportional to the head.

Proportioneer   Mixer by Lightnin.

Propulsair   Aspirating aerator by Baker Process.

ProScreen®   Trommel screen by Knight Manufacturing Corp.

ProSep   Membrane treatment system for industrial wastes by PTI Advanced Filtration.

Prosonic™   Ultrasonic level control system by Endress+Hauser.

Prosser/Enpro   Pump product line by Crane Pumps & Systems.

ProTec RO™   Powder antifoulant by King Lee Technologies.

ProTechtor™   Headworks grease and grit removal system by Hi-Tech Environmental, Inc.

Protect   Portable carbon absorber by Waterlink/Barnebey Sutcliffe.

protein   Any of a large class of complex nitrogenous compounds that contains one or more polypetide molecules and occur in all animal and vegetable matter and are essential to life.

Protista   The class of living organisms that includes algae, bacteria, and protozoa.

Protoc   Total organic carbon analyzers by Tytronics, Inc.

protoplast   The living substance of a plant or cell, excluding the cell wall.

Prototech   Former name of Süd-Chemie Prototech Inc.

protozoa   A group of microorganisms including amoebas, flagellates, and ciliates that feed on bacteria and other protists and reproduce by binary fission.

ProTwin®   Sludge spreader by Knight Manufacturing Corp.

PRP   See "potentially responsible party (PRP)."

PS   See "point source discharge (PS)."

PSA   See "pressure-swing adsorption (PSA)."

PSD   See "prevention of significant deterioration (PSD)."

PSES   Pretreatment Standards for Existing Sources.

PSEU   Pollutant Specific Emissions Unit.

pseudo-hardness   The action exhibited by sea, brackish, and other waters containing high concentrations of sodium that interferes with the normal behavior of soap.

pseudomonas   A common rod-shaped aerobic bacteria.

PSG   Professional Services Group, now a part of USFilter/Operating Services.

psi   Pounds per square inch.

PSI   See "Pollutant Standard Index (PSI)."

psia   See "pounds per square inch, absolute (psia)."

psig   See "pounds per square inch, gage (psig)."

PSM   (1) Point source monitoring. (2) Process safety management.

PSNS   Pretreatment Standards for New Sources.

PSRP   Process to significantly reduce pathogens.

PSU   Primary sampling unit.

***Psychoda* flies**   A small, dark-colored fly that creates a nuisance by breeding in trickling filter beds. Commonly known as "filter flies."

psychrometer   An instrument used to determine the relative humidity and vapor tension of the atmosphere.

psychrophiles   Bacteria that grow best at temperatures below 20°C.

PTE   (1) Potential to emit. (2) Permanent total enclosure.

ptomaine   Any of a class of alkaloid substances formed in decaying animal or vegetable matter by bacterial action on proteins.

ptomaine poisoning   Food poisoning caused by bacteria, erroneously thought to be caused by ptomaines.

PTS   See "permanent threshold shift (PTS)."

public utility   A private business organization subject to government regulation that provides an essential product or service such as water, electricity, transportation, or communication to the public.

public water system (PWS)   A system that pipes water for human consumption to at least 25 people or has 15 or more surface connections.

publicly owned treatment works (POTW)   Treatment works owned by a state or municipality, including sewers, pipes, or other conveyances used to convey wastewater.

PUC   Public Utilities Commission.

pug mill   A grinder used to reduce the size of solid waste or sludge to facilitate further treatment or disposal.

Pullman Power Products   Chimney and storage silo product line by Wheelabrator Air Pollution Control, Inc.

PullUp   Removable aeration header and drop pipe assembly by USFilter/Aerator Products.

***Pulsapak*®**   Potable water treatment plant by Infilco Degremont, Inc.

pulsation dampener   A device using air or other compressible gas to absorb pressure irregularities and induce uniform flow in pump suction or discharge lines.

***Pulsator*®**   Solids contact clarifier by Infilco Degremont, Inc.

***PULSAtrol*®**   Microprocessor based system controller by Pulsafeeder, Inc.

***Pulsatron*®**   Electronic metering pump by Pulsafeeder, Inc.

***Pulsemate*™**   Metering pumps and controls by USFilter/Wallace & Tiernan.

***Pulse-Mix*®**   Short-term backwashing process to regenerate sand filter media by USFilter/Zimpro.

pump   A mechanical device used to apply pressure to a fluid to cause its flow.

pump curves   A set of graphical pump characteristics that represent pump performance by comparing total discharge head, net positive suction head, and efficiency relative to capacity.

pump stage   The number of impellers in a centrifugal pump.

pump station   A chamber that contains pumps, valves, and electrical equipment necessary to pump water or wastewater.

Pumpak   Standard pump control systems by Healy-Ruff Co.

Purac   Product line of Waterlink Separations, Inc.

Purafilter™   Air filters by Purafil, Inc.

Puratex™   Cartridge filter by Osmonics, Inc.

PurCycle™   VOC removal system by Purus Corp.

pure oxygen process   Variation of the activated sludge process using pure molecular oxygen for microbial respiration rather than atmospheric oxygen.

Purelab Plus™   Laboratory water purification system by USFilter Corp.

Pureone®   Laboratory water system by USFilter Corp.

Puresep®   Resin separation process by USFilter/Warren.

Purgamix   Sludge mixing and heating system by Waste-Tech, Inc.

Purge Saver™   Groundwater analyzer by QED Environmental Systems, Inc.

Purifax®   System using wet oxidation process for sludge stabilization by BIF.

purified water   Bottled water that has undergone significant treatment prior to bottling. See also "USP-purified water."

Puritan   Fluid recovery treatment system for spent coolant and oils by Waterlink Biological Systems.

purl   To flow in ripples with a murmuring sound.

Purofine™   Ion exchange resin by Purolite Co.

PURPA   Public Utilities Regulatory Policies Act.

purple salt   See "potassium permanganate."

Purspring   Biofiltration system by Bioway America.

Pusher™   Dewatering screw press by USFilter/Dewatering Systems.

putrefaction   The decomposition of organic matter by bacteria, fungi, and oxidation which results in the formation of noxious products and/or foul-smelling gases.

putrescible   Organic matter that is likely to result in a rotten, foul-smelling product as it undergoes decay or decomposition.

Putzmeister   Sludge cake pump marketed by USFilter/Asdor.

PV   See "pervaporation (PV)."

PVC   See "polyvinyl chloride (PVC)."

PVDF   Polyvinylidene fluoride.

PWMP®   Pure water management program for outsourced water services by USFilter Corp.

PWS   See "public water system (PWS)."

PWSS   Public water supply system.

PWT   Professional Water Technologies, Inc.

Pyramed™   Direct retention filter underdrain system by WesTech Engineering Inc.

pyrite   A mineral containing sulfur and iron, frequently found in coal.

Pyrofluid®   Fluidized bed incinerator by USFilter/Krüger (North America) and OTV.

pyrogen   Cell material from bacteria that produces fevers in mammals.

Pyrolox™   Granular media for removal of iron, manganese, and hydrogen sulfide by Clack Corp.

pyrolysis   The chemical decomposition of a material by heating in oxygen-deficient conditions.

Pyrospout   Fluid bed combustor/incinerator by Process Combustion Corp.

Python Press   High pressure filtration and dewatering device by Waste-Tech, Inc.

Q

Q·E·D plus™   Mixer by Chemineer, Inc.

QA   See "quality assurance (QA)."

QC   See "quality control (QC)."

QL-1™   Ultraviolet disinfection system by Infilco Degremont, Inc.

QLS   Quick-lock sprocket by Budd Co.

Q-Pac™   Random plastic media by Lantec Products, Inc.

Q-Tracker™   Sewer collection flow monitor by Badger Meter, Inc.

Quadra Press   Plate and frame filter press by Aqua Care Systems.

Quadra-Clean™   Vertical basket batch type centrifuge by Western States Machine Co.

Quadra-Kleen   Sand filter backwash system by Culligan International Corp.

Quadramatic III®   Automatic batch type centrifuge by Western States Machine Co.

Quadrant™   Thermal oxidizer by Catalytic Products International.

Quadrasep   Integrated wastewater treatment system by Blace Filtronics, Inc.

Quadricell®   Mechanical gas induction flotation separator by USFilter/Whittier.

Quadrufil   Gravity filter system by Vulcan Industries, Inc.

quagga mussel   Freshwater mollusk that can foul water intake screens and piping by attaching itself to a solid structure, eventually restricting flow.

quaking bog   A wholly or partially floating bog of peat that shakes when walked upon.

qualitative   The general description of a substance without specifying its exact amount or concentration.

quality assurance (QA)   A definitive plan for operation that specifies the measures used to produce products or data of known precision and bias.

quality control (QC)   Set of measurements within an analysis methodology to assure that the process is in control.

Quanti-Cult™   Quality control test organisms by IDEXX Laboratories, Inc.

quantitative   Description of a substance in exact terms.

Quanti-Tray™   Water sample test kit by IDEXX Laboratories, Inc.

Quantum®   Floating aerator by Air-O-Lator Corp.

quart   A liquid and dry unit of measure equal to a quarter of a U.S. gallon or 2 pints.

quaternary ammonium   A basic chemical group [$N(CH_3)_3^+$] which provides the site of activity of certain anion exchange resins.

quench   To cool suddenly by immersion in water or oil.

quench tank   A water-filled tank used to cool incinerator residues or hot materials during industrial processes.

quicklime   A calcium oxide material produced by calcining limestone to liberate carbon dioxide, also called "calcined lime" or "pebble lime," commonly used for pH adjustment. Chemical formula is CaO.

Quick-Purge®   Soil and groundwater remediation technology by Integrated Environmental Solutions, Inc.

quicksand   A loose, wet, unstable sand deposit caused by the upward pressure of sand, and which may engulf an object.

Quickwipe™   Quartz lamp cleaning technology used in UV disinfection systems by Calgon Carbon Corp.

Quik-Clamp   Clamp-on spray nozzle by USFilter/Rex & Link-Belt Products.

Quik-Solid®   Granulated absorbent polymer by Colloid Environmental Technologies Co.

R

R&B Filtration   Sludge dewatering equipment product line of Baker Process.

R&D   Research and development.

RA   (1) Reasonable alternative. (2) Regulatory alternatives. (3) Regulatory analysis. (4) Remedial action. (5) Resource allocation. (6) Risk analysis. (7) Risk assessment.

rabble arms   Rotating rake arms used to scrape sludge in a multiple hearth furnace.

RAC   Radiation Advisory Committee.

race track   See "oxidation ditch."

RACM   (1) See "reasonably available control measures (RACM)." (2) See "regulated asbestos-containing material (RACM)."

Racod™   Biological sensor to provide a rapid indication of BOD by USFilter/Wallace & Tiernan.

RACT   See "reasonably available control technology (RACT)."

rad   Radiation absorbed dose. A measure of the absorbed dose of radiation.

Radial Filter   Tertiary filtration system by Waterlink/Aero-Mod Systems.

radial flow   Direction of flow from either the center to the periphery or the periphery to center.

Radial Plate Dryer   Sludge drying system by B-H Process Systems.

radial well   A well system where one or more cylindrical screens are driven horizontally into a water-bearing stratum, radiating from a central sump.

radiation   The transfer of energy by means of electromagnetic waves or high-speed particles.

radiation sickness   A sickness that results from overexposure to radiation whose symptoms may include nausea, bleeding, hair loss, and death.

radiation standards   Regulations that set maximum exposure limits for protection of the public from radioactive materials.

radical   A combination of atoms in a molecule that remains unchanged throughout most chemical reactions.

Radicator   Solid waste incinerator by Hitachi Metals America, Ltd.

radioactive   A description of an unstable atom that undergoes spontaneous disintegration and gives off radiant energy in the form of particles or rays.

radioactive material   A material that spontaneously emits ionizing radiation having a specific activity greater than 0.002 microcuries per gram.

radioactivity   The spontaneous decay or disintegration of an atomic nucleus that is accompanied by radiation.

radioisotopes   Chemical variants of an element with potentially oncogenic, teratogenic, and mutagenic effects on the human body.

radionuclide   An atom that spontaneously undergoes radioactive decay.

radius of vulnerability zone   The maximum distance from the point of release of a hazardous substance in which the airborne concentration could reach the level of concern under specified weather conditions.

radon   A radioactive gas produced from the decay of radium which may be inhaled when the gas is released from groundwater, often during showering, bathing, or cooking.

radon daughters   Short-lived radioactive compounds produced during the decay of radon that decay into longer-lived lead isotopes. Also called radon progeny or radon decay products.

RADS   See "reactive airways dysfunction syndrome (RADS)."

radwaste   Radioactive waste.

raffinate   The extracted waste stream containing contaminants in a solvent extraction process.

railway softening   See "excess lime-soda softening."

rain forest   See "tropical rain forest."

Rainlogger   Stormwater sampling unit by American Sigma, Inc.

Rake-O-Matic   Hydraulically operated, reciprocating rake bar screen formerly offered by BIF.

Ram®   Waste compactor by S&G Enterprises, Inc.

RAMP   Rural Abandoned Mine Program.

Rampactor   Compaction device by SRS Industrial Engineering.

Ram-Rod   Dewatering screw press by Ketema, Inc.

RAMS   Regional Air Monitoring System.

Ramtube®   Reverse pulse self-cleaning baghouse by Farr Co.

Ranney®   Well screen and caisson intake products by Layne Christensen, Ranney Division.

Ranney Intake   Surface water intake system utilizing a passive screen/caisson arrangement by Layne Christensen, Ranney Division.

RAP   Remedial action plan.

RaPID Assay™   Reagent kit for field soil analysis by Strategic Diagnostics, Inc.

Rapid Decanter®   Solid bowl centrifuge by Krauss Maffei Corp.

Rapid Gravity Dewatering™   Inclined gravity filter for sludge dewatering by Wil-Flow, Inc.

rapid mix   A physical water treatment process that involves rapid and complete mixing of coagulants or conditioning chemicals.

rapid sand filter   Granular media filter in which water flows downward through a sand filter bed at rates typically ranging from 80 to 320 L/min/m^2 (2 to 8 gpm/sq ft) of surface area.

Rapidor   Pressure leaf filter by Liquid-Solids Separation Corp.

RAPS   Regional Air Pollution Study.

RAS   See "return activated sludge (RAS)."

rasp   A machine that grinds waste into a manageable size and controls odors.

RATA   See "relative accuracy test audit (RATA)."

rated capacity   The basis for calculating the period of time or volume of treated product delivered by a water softener, demineralizer, or filter between regenerations, backwashing, or servicing.

RatedAeration®   Circular steel-activated sludge wastewater treatment plant formerly offered by USFilter/Envirex.

rate-of-flow controller   A device that automatically controls the rate of flow of a fluid.

Ratio   Turbidimeter by Hach Co.

RatioFlo™   Flow-to-polymer ratio valve by USFilter/Stranco.

raw sewage   Untreated wastewater and its contents.

raw sludge   Undigested sludge recently removed from a sedimentation basin.

raw water   Untreated surface or groundwater.

Raymond   Sludge incinerator formerly offered by GL&V/Dorr-Oliver, Inc.

Raymond Process   An aquifer bioremediation technology where groundwater is recovered from the aquifer, treated, amended with nutrients and oxygen, and reinjected.

RaySolv   Product group of CSM Worldwide/RaySolv.

Raysorb   Activated carbon VOC control system by CSM Worldwide/RaySolv.

RBC   See "rotating biological contactor (RBC)."

RBCA   See "Risk Based Corrective Action (RBCA)."

RCC   Resource Conservation Co.

RCO   See "regenerative catalytic oxidizer (RCO)."

RCP   Reinforced concrete pipe.

RCRA   See "Resource Conservation and Recovery Act (RCRA)."

RCRIS   Resource Conservation and Recovery Information System

RDF   See "refuse derived fuel (RDF)."

rDNA   Recombinant DNA.

RDV   Reference dose values.

Reacher   Reciprocating rake bar screen by Schloss Engineered Equipment.

reactant   Any substance taking part in a chemical reaction.

reaction rate   The rate at which a chemical reaction occurs.

reactivation   The process of removing adsorbed organics and restoring the adsorptive characteristics of an adsorbent, usually by thermal or chemical means.

Reactivator®   Solids contact clarifier by Graver Co.

reactive airways dysfunction syndrome (RADS)   An asthma-like condition resulting in hyper-responsiveness to inhaled materials that may develop after long-term exposure to hydrogen sulfide.

reactive wall   A permeable vertical wall constructed of a reactive mixture and installed below grade to treat groundwater that flows through it.

reactive waste   A solid waste that is normally unstable and readily undergoes violent change, generates toxic gases or fumes, or is capable of detonation or explosion.

reactivity   The tendency for a chemical substance to combine with other elements and compounds.

reactor   The container or tank in which a chemical or biological reaction is carried out.

Reactor-Thickener   Sludge-thickening device using mixers and dewatering screens by JDV Equipment Corp.

React-pH™   pH-stable-activated carbon by Calgon Carbon Corp.

reaeration   The absorption of oxygen into water under conditions of oxygen deficiency.

reagent   A chemical added to a system to bring about a chemical reaction.

reagent grade water   High purity water suitable for use in making reagents for use in analytical procedures.

reasonable maximum exposure   The maximum exposure reasonably expected to occur in a population.

reasonably available control measures (RACM)   A broadly defined term referring to technological and other measures for pollution control.

reasonably available control technology (RACT)   Control technology that is reasonably available, technologically and economically feasible, and usually applied to existing sources in nonattainment areas; in most cases it is less stringent than new source performance standards.

reboiler   An evaporator-condenser unit which produces secondary steam after condensation of primary steam. Used to isolate steam systems and avoid cross contamination of boiler chemicals.

recalcining   Recovery of lime from water or wastewater sludge, usually with a multiple hearth furnace.

recarbonation   The reintroduction of carbon dioxide into water, usually during or after lime-soda softening.

receiving water   Surface water body that receives effluent discharge from a wastewater treatment plant.

receptor   A molecule that can selectively bind to another substance in the initial stages of a toxic response.

receptor cell   A nerve ending or group of nerve endings specialized for the reception of stimuli.

recharge   The natural or artificial process of replenishing an aquifer.

recharge area   A land area in which water reaches the zone of saturation from surface infiltration.

recharge rate   The quantity of water per unit of time that replenishes or refills an aquifer.

reciprocating rake bar screen   An automatic bar screen with a single rake that is raised and lowered to clean a stationary bar rack.

reclaimed water   Wastewater that has been treated to a level that allows for its reuse for a beneficial purpose.

Recla-Mate®   Modular physical chemical treatment plant by USFilter/Microfloc.

reclamation   The process of improving or restoring the condition of land or other material to a better or more useful state.

Recla-Pac   Package biological treatment plant by USFilter/Microfloc.

RECOjet   Gas scrubber by USFilter/Gutling.

recombinant bacteria   A microorganism whose genetic makeup has been altered by deliberate introduction of new genetic elements, and whose offspring also contain these new genetic elements.

recombinant DNA   The new DNA formed by combining pieces of DNA from different organisms or cells.

recommended maximum contaminant level (RMCL)   The maximum level of a contaminant in drinking water at which no known or anticipated adverse effect on human health would occur, including an adequate margin of safety. Recommended levels are nonenforceable health goals. See also "maximum contaminant level (MCL)."

reconstructed source   Facility in which components are replaced to such an extent that the fixed capital cost of the new components exceed 50% of the capital cost of constructing a comparable brand new facility. New-source performance standards may be applied to sources

record of decision (ROD)   A public document that explains which cleanup alternative(s) will be used at National Priorities List sites.

recovery   In reverse osmosis processes, recovery indicates the amount/percentage of product water recovered from the feed stream.

recovery rate   Percentage of usable recycled materials that have been removed from the total amount of municipal solid waste generated in a specific area or by a specific business.

recreational waters   Any waterbody used for recreational activities such as swimming, boating, or fishing.

Rectangulaire   Package wastewater treatment unit formerly offered by USFilter/Envirex.

recycle ratio   The recycled flow rate divided by the influent flow rate in an activated sludge wastewater treatment system or other process system.

recycling   The process by which recovered materials are transformed into new products. See also "water recycling."

red bag waste   Medical or infectious wastes.

Red Fox   Sewage treatment systems for marine applications by Red Fox Environmental, Inc.

Red List   A list of 23 dangerous substances, designated by the U.K., whose discharge to the water should be minimized.

Red Rubber™   Bar screen toothed rake segments of cast urethane by Rubber Millers, Inc.

red tide   A reddish discoloration of water caused by an excessive growth of certain microbes, whose toxins can cause massive fish kills.

red water   Water whose reddish color usually results from the precipitation of iron salts or the presence of microbes that depend on iron or manganese.

Red-B-Gone®   Rust and iron stain remover by Pro Products, Corp.

redox potential   See "oxidation-reduction potential (ORP)."

reducing agent   Any substance that can give up electrons in a reaction.

reduction   A chemical reaction where an element or compound gains electrons causing a decrease in valence.

Redux™   Dissolved air flotation system by ABS Pumps, Inc.

REECO®   Product line of Dürr Environmental, Inc.

reed   Any of a variety of tall, slender grasses grown in wet areas.

reed bed   Tertiary wastewater treatment system where organics remaining in secondary effluent are used as nutrients in the growth of reeds. Also used for biosolids volume reduction.

REEF®   Fine pore floor mounted diffuser by Environmental Dynamics Inc.

Reel Auggie®   Sludge and compost mixer by Knight Manufacturing Corp.

ReelAer   Horizontal cage surface aerator formerly offered by Walker Process Equipment.

reentry interval   The period of time immediately following the application of a pesticide during which unprotected workers should not enter a field.

reference dose (RfD)   The exposure level of a carcinogenic contaminant thought to be without significant risk to humans when ingested daily over a specified time period.

reformulated gasoline (RFG)   Gasoline whose composition has been changed from conventional gasoline sold in 1990 to reduce air pollutants released from evaporation and exhaust emissions.

Refotex   Fine bubble diffuser by Refractron Technologies Corp.

Refractite   Ceramic filter membrane by Refractron Technologies Corp.

refractory   A highly heat resistant material used as a liner in a furnace or incinerator.

refractory organics   Organic substances that are difficult or unable to be metabolized in a biological system.

refrigerant   A substance which, by undergoing a phase change, lowers the temperature of its environment. Commercial refrigerants, which include CFCs and HFCs, are liquids whose latent heat of vaporization results in cooling.

refueling emissions   Emissions released during vehicle refueling.

refuse   All solid waste material discarded as useless.

refuse derived fuel (RDF)   Fuel produced from municipal solid waste through shredding, pyrolysis, or other methods.

refuse reclamation   Conversion of solid waste into useful products, e.g., composting organic wastes to make soil conditioners and separating aluminum and other metals for recycling.

Regal®   Gas chlorinator by Chlorinators, Inc.

Regenair®   Regenerative air blower by Gast Manufacturing, Inc.

regenerant   A chemical solution used to restore the exchange capacity of ion exchange resin.

regenerate The process of restoring exchange capacity of an ion exchange material.

regenerative catalytic oxidizer (RCO) An emissions control device that utilizes a catalyst to accomplish VOC oxidation.

Regenerative Environmental Equipment Company Product line acquired by Dürr Environmental, Inc.

regenerative thermal oxidizer (RTO) An emissions control device that utilizes heat to accomplish VOC oxidation.

Re-Gensorb™ VOC and HAP removal system by M&W Industries, Inc.

Regional Administrator The administrative head of each of the ten regions organized by the U.S. EPA.

Reg-U-Flo® Vortex valve flow control device by H.I.L. Technology, Inc.

regulated asbestos-containing material (RACM) Friable asbestos material or nonfriable ACM that will be or has been subjected to sanding, grinding, cutting, or abrading or has crumbled, been pulverized, or reduced to powder in the course of demolition or renovation operations.

regulated medical waste Any solid waste generated in the diagnosis, treatment, immunization, or research of human beings or animals, or in the production or testing of related biological products.

Rehydro-Floc™ Aluminum chlorohydrate flocculant solution by Reheis, Inc.

reject The waste stream containing impurities rejected in a treatment process, most commonly applied to reverse osmosis, electrodialysis, and ultrafiltration systems.

reject staging Reverse osmosis process configuration where the reject from one stage is used as feedwater on a subsequent stage to increase water recovery. Also called "brine staging."

rejection In reverse osmosis, the percentage of dissolved solids removed from the feedwater as it passes through the semipermeable membrane. Also called "salt rejection."

relative accuracy test audit (RATA) A comparison between an emissions stack tester's readings and readings obtained from a CEM.

relative humidity The total amount of water vapor present in the air, expressed as a percentage of the maximum amount that the air could hold at a given temperature.

release Any occurrence where a regulated substance discharges, spills, pumps, pours, emits, injects, dumps, disposes, or escapes into the air, soil, or water.

Reliant™ Conventional package and modular filtration product line by Roberts Filter Group.

REM Registered Environmental Manager.

rem Roentgen equivalent man. A measure of the effective radiation dose absorbed by human tissue.

remedial action The actual construction or implementation phase of a Superfund site cleanup that follows remedial design.

remedial investigation and feasibility study (RI/FS) An evaluation of the risks associated with a hazardous waste site that includes the process of selection of an appropriate remedy.

remedial response   Long-term action that stops or substantially reduces a release or threat of a release of hazardous substances that is serious but not an immediate threat to public health.

remediation   The treatment of waste to make it less toxic and/or less mobile, or to contain a site to minimize further release.

Remedi-Cat™   Catalytic oxidizer by Global Technologies.

REMS   RCRA Enforcement Management System.

rendering plant   A plant that converts grease and livestock carcasses into fats, oils, and other products.

renewable resource   A resource that theoretically cannot be totally consumed due to its ability to reproduce or regenerate.

Renneburg   Sludge dryer manufacturing division of Heyl & Patterson, Inc.

RenovAir™   Biofiltration system for VOC reduction by Envirogen.

Reo-Pure   Reverse osmosis system by Great Lakes International, Inc.

reportable quantity (RQ)   Quantity of a hazardous substance that triggers reports under CERCLA. If a substance exceeds its RQ, the release must be reported to the National Response Center, the SERC, and community emergency coordinators for areas likely to be affected.

Reporter™   Multiprobe for measuring water conditions by Hydrolab Corp.

repowering   Rebuilding and replacing major components of a power plant instead of building a new one.

representative sample   A portion of material or water that is as nearly identical in content and consistency as possible to that in the larger body of material or water being sampled.

reserve capacity   Extra treatment capacity built into solid waste and wastewater treatment plants and interceptor sewers to accommodate flow increases due to future population growth.

reservoir   An artificial or natural pond, lake, basin, or tank which is used to store or control water.

residence time   The period of time that a volume of liquid remains in a tank or system.

residual   Amount of a pollutant remaining in the environment after a natural or technological process has taken place, including the sludge remaining after initial wastewater treatment and the particulates remaining in air after it passes through a scrubbing or other treatment process.

residual chlorine   See "chlorine residual."

residual disinfectant concentration   The concentration of a disinfectant after a stated contact time. See also "C × T."

residual risk   The extent of health risk from air pollutants remaining after application of the Maximum Achievable Control Technology (MACT).

residue   Solid or semisolid material remaining after processing, evaporation, or incineration.

resin   A material having ion exchange properties used in ion exchange systems.

resin beads   Spherical beads with ion exchange properties used in ion exchange systems.

Resinator®   Ion exchange system by Aqualogic Inc.

resistance   (1) In water conditioning, the opposition offered by water to the flow of electricity through which it may be used to estimate the mineral content; the reciprocal of electrical conductance. (2) For plants and animals, the inborn or acquired ability to withstand poor environmental conditions or attacks by chemicals or disease.

resistivity   A measure of resistance to the flow of electricity, used as an accurate measure of a water's ionic purity.

Resi-Tech   Division of Waterlink/Aero-Mod Systems.

Resolv-R2®   Solvent recovery and waste disposal system by PBR Industries.

Resource Conservation and Recovery Act (RCRA)   A 1976 U.S. law, amended in 1984, to regulate management and disposal of solid and hazardous wastes.

resource recovery   The process of obtaining matter or energy from materials formerly discarded.

respiration   Intake of oxygen and discharge of carbon dioxide as a result of biological oxidation.

respirator   Personal respiratory protection device certified to meet minimum government performance standards.

respirometer   An instrument used to study the character and extent of respiration.

restoration   Measures taken to return a site to pre-violation conditions.

Retec®   Heavy metal recovery systems offered by USFilter/Industrial Wastewater Systems.

retentate   The portion of the feed solution rejected by the membrane in a pressure-driven membrane process.

retention pond   A basin used for wastewater treatment and/or storage.

retention time   The length of time water or wastewater will be retained in a unit treatment process or facility.

Re-Therm®   Thermal VOC oxidation unit by Dürr Environmental, Inc.

Retox™   Regenerative thermal oxidizer by Adwest Technologies, Inc.

retrofit   Addition of a pollution control device on an existing facility without making major changes to the generating plant.

Retroliner   Forms for filter underdrain rehabilitation by Roberts Filter Group.

return activated sludge (RAS)   Settled activated sludge that is returned to mix with raw or primary settled wastewater.

return sludge   See "return activated sludge (RAS)."

reuse   Using a product or component of municipal solid waste in its original form more than once.

reverse deionization   A deionization system having an anion exchange resin ahead of a cation exchange resin.

reverse osmosis (RO)   A method of separating water from dissolved salts by passing feedwater through a semipermeable membrane at a pressure greater than the osmotic pressure caused by the dissolved salts.

reversible effect   An effect which is not permanent; especially adverse effects which diminish when exposure to a toxic chemical is ceased.

Revolver™   Rotary adsorption system for VOC abatement and solvent recovery by Vara International.

Rex   Water/wastewater equipment product line offered by USFilter/Envirex.

Rex Chainbelt   Former name of USFilter/Envirex.

Reynolds number   A nondimensional number that measures the state of turbulence in a fluid system. It is calculated as the ratio of inertia effects to viscous effect.

RF/AS   Roughing filter/activated sludge.

RfD   See "reference dose (RFD)."

RFG   See "reformulated gasoline (RFG)."

RFP   Request for proposal.

RFQ   Request for quotation.

RHA   Rivers and Harbors Act.

rheology   The study of the deformation and flow of substances.

rhizosphere   The zone of intermingled roots and soil.

RHRS   Revised Hazard Ranking System

RI/FS   See "remedial investigation and feasibility study (RI/FS)."

***RibbonFlow*™**   Baffled clearwell by CBI Walker, Inc.

ribonucleic acid (RNA)   One of two types of long-chain molecules containing hereditary material vital to reproduction.

Rich Tech   Equipment product line by USFilter/Aerator Products.

Richards of Rockford   Former equipment manufacturer acquired by Aqua-Aerobic Systems, Inc.

RIFS   Remedial investigation and feasibility study.

rift   A shallow, rapidly flowing and usually rocky area of water in a stream.

RIGA   Former equipment manufacturer.

***Riga-Sorb*™**   Activated carbon adsorbers by Farr Co.

right of free capture   The concept that groundwater belongs to the person who owns the land above it and that he is free to capture and use as much as desired.

rill   A small channel eroded into the soil surface by runoff which can be easily smoothed or obliterated by normal tillage.

Rim-Flo   Peripheral feed, circular clarifier by USFilter/Envirex.

Ringelman test   A method of quantifying the opacity of an air pollution emission by comparing it to a set of standard disks having increasing degrees of discoloration from light grey (number 1) through black (number 5).

***Ringlace*®**   Attached growth biomedia by Ringlace Products, Inc.

RingSparjer   Air injection diffuser by Walker Process Equipment.

rinse   The portion of an ion exchange regeneration cycle in which fresh water is used to remove spent and excess regenerant from the resin column.

***Rio Linda*®**   Chlorine dioxide generator by Vulcan Performance Chemicals.

rip rap   Broken stone or rocks placed compactly or irregularly on dams, levees, dikes, or similar embankments for protection against the action of waves or currents.

riparian habitat   Areas adjacent to rivers and streams with a high density, diversity, and productivity of plant and animal species relative to nearby uplands.

riparian rights   A landowner's rights to the water on or bordering his property, including the right to prevent diversion or misuse upstream.

Rippl diagram   A graph which plots cumulative flow versus time, and is used to design storage reservoirs and equalization basins.

rise rate   See "overflow rate (OFR)."

rising film evaporator   An evaporator using vertical heat transfer surfaces where liquor on one side of the surface is boiled by steam condensing on the other side, causing vapors to rise carrying the liquid upward as a film.

risk   A measure of the probability that damage to life, health, property, and/or the environment will occur as a result of a given hazard.

risk assessment   Qualitative and quantitative evaluation of the risk posed to human health and/or the environment by the actual or potential presence and/or use of specific pollutants.

Risk Based Corrective Action (RBCA)   A process using the principles of exposure assessment, toxicity, and mobility to make cost-effective corrective actions on sites while protecting human health and the environment.

risk based targeting   The direction of resources to those areas identified as having the highest potential or actual adverse effects on human health and/or the environment.

risk factor   Characteristic (e.g., race, sex, age, obesity) or variable (e.g., smoking, occupational exposure level) associated with increased probability of a toxic effect.

risk management   The process of evaluating and selecting alternative regulatory and nonregulatory responses to risk which requires the consideration of legal, economic, and behavioral factors.

risk specific dose   The dose associated with a specified risk level.

river basin   The land area drained by a river and its tributaries.

RJ Environmental   Former name of USFilter/RJ Environmental.

RKL®   Pinch valve by Moyno Industrial Products.

RLL   Rapid and large leakage

RM-10®   Clay-based flocculating agent for wastewater by Colloid Environmental Technologies Co.

R-MAP   Regional Management Assessment Program.

RMCL   Recommended maximum contaminant level. This term is being discontinued in favor of "MCLG."

RMP   "Roberts Manhattan Process."

RMP   Risk management plan.

RMPR   Risk Management Program Rule.

RMS™   Rinsewater maintenance system by USFilter/Industrial Wastewater Systems.

RNA   See "ribonucleic acid (RNA)."

RO   See "reverse osmosis (RO)."

Robo™   Bar screen by Vulcan Industries, Inc.

Robo Rover™   Traversing bar screen by Vulcan Industries, Inc.

Robo Stat™   Stationary bar screen by Vulcan Industries, Inc.

RoClean™   Membrane cleaners by Avista Technologies.

rodding   A method of cleaning tubes or sewers using long rods which are able to remove or dislodge debris.

rodenticide   A chemical or agent used to destroy rats or other rodent pests, or to prevent them from damaging food or crops.

RODTOX   Rapid oxygen and toxicity tester by SAMI.

Roebelt   Belt filter press by Roediger Pittsburgh, Inc.

Roedos   Mixing system for dry and liquid polymers by Roediger Pittsburgh, Inc.

Roefilt   Sieve drum concentrator by Roediger Pittsburgh, Inc.

Roeflex   Fine bubble diaphragm diffuser by Roediger Pittsburgh, Inc.

Roemix   Lime post treatment mixing system for dewatered sludge by Roediger Pittsburgh, Inc.

roentgen   Unit of exposure dose of X-ray or gamma radiation.

Roepress   Belt filter press by Roediger Pittsburgh, Inc.

Roevac   Vacuum sewage system by Roediger Pittsburgh, Inc.

Ro-Flo®   Sliding vane compressor by A-C Compressor Corp.

ROG   Reactive organic gas.

Rogun   Reverse osmosis membrane cleaning solution by BetzDearborn-Argo District.

RollAer   Aerobic digestion aeration equipment by Walker Process Equipment.

Roll-Dry   Internally fed rotary fine screen by Schlueter Co.

Rolling Grit   Aerated tank grit washing and removal unit by Walker Process Equipment.

Romembra   Reverse osmosis membrane elements by Toray Industries, Inc. and Ropur AG (Europe).

Romicon®   Hollow fiber membrane filtration products by Koch Membrane Systems, Inc.

Romi-Kon™   Oil/water emulsion separator by Koch Membrane Systems, Inc.

RomiPure™   Hollow fiber ultrafiltration products by Koch Membrane Systems, Inc.

Roots   Centrifugal compressor and blower product line by Dresser Industries/Roots Division.

Roplex   Live bottom feeder for solids and storage piles by Hindon Corp.

Roptic®   Filter cake sensor by Rosenmund.

RoQuest™   Reverse osmosis coagulants and flocculants by Avista Technologies.

Rosep™   Reverse osmosis systems by Graver Co.

Rossmark   Former name of USFilter/Rossmark.

Rotadisc®   Sludge dryer by Atlas-Stord, Inc.

Rotafilt®   Wastewater fabric filter by Hans Huber GmbH.

Rotafine   Rotary fine screen by Waste-Tech, Inc.

Rotamat®   Screening equipment product line by Lakeside Equipment Corp. (U.S.) and Hans Huber GmbH (Europe).

rotameter   A variable area liquid flow meter.

Rotamix   Digester mixing system combining chopper pumps and mixing nozzles by Vaughan Co., Inc.

Rotapak   Screw type screenings compactor by Longwood Engineering Co., Ltd.

Rota-Rake   Circular sludge collector by Graver Co.

Rotarc   Arc-type bar screen by John Meunier, Inc.

rotary collector   Rotating mechanisms used in circular clarifiers to collect and remove settled solids.

rotary distributor   Rotating pipe that evenly distributes wastewater on the surface of a trickling filter.

rotary drum screen   Cylindrical screen used to remove floatable and suspended solids.

rotary drum thickener   Rotating cylindrical screen used to thicken sludge.

rotary kiln incinerator   An incinerator consisting of a slowly rotating, long horizontal cylinder in which material is fed at one end and tumbled by the kiln to promote drying as it is conveyed to the other end.

Rotasieve   Externally fed rotary fine screen by Waste-Tech, Inc.

rotating biological contactor (RBC)   A fixed film biological treatment device where biological organisms are grown on circular discs and mounted on a horizontal shaft that slowly rotates through wastewater.

rotavirus   An enteric virus commonly found in domestic wastewater which causes diarrhea.

Rotex   Rotating grit removal system by Simon-Hartley, Ltd.

Rotho®   Hose pump by USFilter/Wallace & Tiernan.

rotifer   A very small aerobic, multicellular animal that feeds on organic matter in wastewater.

Rotoair Disc   Submerged rotating biological contactor by Pro-Equipment, Inc.

Rotobelt   In-channel fine screen by Dontech, Inc.

Roto-Brush   Rotary screen brush cleaning device by Dontech, Inc.

Roto-Channel   Combination bar screen and compacting conveyor by Dontech, Inc.

RotoClean   Screenings washer by Parkson Corp.

RotoClear   Microscreen formerly offered by Walker Process Equipment.

Rotoco®   Continuous duty granular media filter by Baker Process.

RotoDip   Manually controlled slotted pipe skimmer by Walker Process Equipment.

Rotodip®   Volumetric feeder for liquids or slurries by BIF.

Rotodisintegrator   Debris grinder by USFilter/Zimpro.

Roto-Drum   Internally fed rotary fine screen and thickener by Dontech, Inc.

Roto-Guard®   Horizontal drum screen/thickener by Parkson Corp.

Roto-Kone®   Gravity belt thickener drainage elements by Komline-Sanderson Engineering Corp.

Rotoline   Rotary distributor for trickling filter formerly offered by USFilter/Envirex.

Rotomite   Sludge handling dredger by SRS Crisafulli, Inc.

Rotopac®   Screw type screenings compactor by John Meunier, Inc.

Rotopass   Externally fed rotary fine screen by Passavant-Roediger GmbH.

Roto-Press   Combination rotary fine screen and dewatering press by Dontech, Inc.

Roto-Press   Screenings compactor by Roto-Sieve AB.

RotoPress®   Screenings compactor by Parkson Corp.

Rotopress®   Shaftless screw compactor by Andritz-Ruthner, Inc.

rotor   See "brush aerator."

Rotordisk®   Rotating biological contactor by CMS Group, Inc.

Rotorobic   Package rotating biological contactor system by Waterlink Separations, Inc.

Rotoscoop®   Self-cleaning volumetric feeder by Wyssmont Co., Inc.

Roto-Scour   Sand filter underdrain system by Graver Co.

Rotoscreen   In-channel bar screen by Waterlink Separations, Inc.

RotoSeal   Rotary distributor for trickling filters by Walker Process Equipment.

Roto-Sep™   Primary wastewater treatment system by Dontech, Inc.

Rotoshear   Internally fed rotary fine screen by Waterlink Separations, Inc.

Roto-Sieve®   Internally fed rotary fine screen by Roto-Sieve AB.

Roto-Skim   Rotary pipe skimmer by USFilter/Envirex.

Rotosludge®   Drum-type rotary sludge thickener by Waterlink Separations, Inc.

RotoSorb   VOC/HAP concentrator by Waterlink/Barnebey Sutcliffe.

Rotospir®   Shaftless screw conveyor by Andritz-Ruthner, Inc.

Rotostep   In-channel bar screen by Waterlink Separations, Inc.

Rotostrainer   Externally fed rotary fine screen by Waterlink Separations, Inc.

Rotosweep   Filter media surface agitator by Roberts Filter Group.

Rototherm®   Agitated thin film evaporator by Artisan Industries, Inc.

Roto-Thickener™   Rotary drum sludge thickener by Dontech, Inc.

Roto-Trak   Sludge plows or chicanes used with gravity sludge dewatering by Komline-Sanderson Engineering Corp.

Roto-Trols   Pressure operated pump controller by Healy-Ruff Co.

Rotox   Submersible aeration and mixing system by USFilter/Davis Process.

rough fish   Fish not prized for eating, such as gar and suckers, which are usually more tolerant of changing environmental conditions than game species.

roughing filter   A high-rate filter designed to receive high hydraulic or organic loading rates as a first, or intermediate, treatment step.

route of exposure   The avenue by which a chemical comes into contact with an organism, and which may include inhalation, ingestion, dermal contact, or injection.

ROV   Remotely operated vehicle.

Rover   Rust removing compounds by Hach Co.

ROWPU   Reverse osmosis water purification unit.

Roxidizer®   VOC and air toxics control system by Tellkamp Systems, Inc.

Royce Equipment   Screening and equipment manufacturer acquired by USFilter/Rex & Link-Belt Products.

ROZ3   Catalyst-impregnated activated carbon by Norit Americas Inc.

rpm   Revolutions per minute.

RQ   See "reportable quantity (RQ)."

RSD   Risk-specific dose.

RSDS   Vacuum-assisted rapid sludge dewatering system by the U.S. Environmental Products, Inc.

RSI   See "Ryznar Stability Index (RSI)."

RSPA   Research and Special Programs Administration of U.S. Department of Transportation.

RTECS   Registry of Toxic Effects of Chemical Substances.

RTI Evaporator   Wastewater evaporator product line acquired by USFilter/Dewatering Systems.

RTO   See "regenerative thermal oxidizer (RTO)."

RTR   Reinforced thermosetting resin.

rubbish   Combustible and noncombustible solid waste from residential and commercial sources.

run   The time period or continuous course during which a unit operates or a test occurs.

rundown screen   See "static screen."

running losses   Evaporation of motor vehicle fuels from a fuel tank while a vehicle is in use.

runoff   Rainwater, leachate, or other liquid that drains over land and reaches a drain, sewer, or body of water.

RUP   Restricted use pesticide.

rupture disk   A diaphragm designed to burst at a predetermined pressure differential.

RWH   Restricted hazardous waste.

Ryznar Stability Index (RSI)   A scale used to evaluate the corrosion or scaling potential of water.

S

S&A   Sampling and analysis.

S&L   Smith & Loveless, Inc.

SAB   See "Science Advisory Board (SAB)."

SAB Reactor   Package wastewater treatment plant by Biosab, Inc.

SAC   Starved air combustion.

SAC™   Sludge age control system by United Industries, Inc.

sacrificial anode   A sacrificial piece of metal, usually zinc or magnesium, electrically connected to a more noble metal in an electrolyte. The anode goes into solution at an accelerated rate to protect the more noble metal from corrosion.

SAE   Society of Automotive Engineers.

safe   Condition of exposure under which there is a practical certainty that no harm will result to exposed individuals.

Safe Drinking Water Act (SDWA)   A U.S. Act assuring that public water supplies are free of contaminants which may cause health risks; it prevents endangerment of underground sources of drinking water.

safe water   Water that does not contain harmful bacteria, toxic materials, or chemicals and is considered safe for drinking even though it may have taste, odor, color, and certain mineral problems.

safe yield   The annual amount of water that can be taken from a source or supply over a period of years without depleting that source beyond its ability to be replenished naturally in "wet years."

Safgard   Rotary fine screen products by Schlueter Co.

sag line   See "inverted siphon."

salination   See "salinization."

saline   Containing or resembling sodium chloride or similar salts.

saline water   See "salt water."

salinity (1) The concentration of dissolved salts in water. (2) The total dissolved solids in water after all carbonates and organic matter have been oxidized.

salinization The accumulation of salts in a soil to the extent that plant growth is inhibited, usually occurring as a result of excessive irrigation in an arid area. Also spelled "salination."

Salmonella An aerobic bacteria that is pathogenic in humans and chiefly associated with food poisoning.

Salmonellosis A common type of food poisoning characterized by a sudden onset of gastroenteritis caused by eating food contaminated with Salmonella bacteria.

salt A class of ionic compounds formed by the combination of an acid and a base, of which sodium chloride is one of the most common examples.

salt flux The amount of dissolved substances that are able to pass through a reverse osmosis membrane.

salt marsh A coastal marsh periodically flooded with salt water.

salt pan An accumulation or layer of salts in the soil that may be toxic to agricultural crops.

salt rejection In reverse osmosis, the ratio of salts removed to the original salt concentration.

salt splitting The conversion of salts to their corresponding acids or bases, usually by means of an ion exchange system.

salt water Water containing a dissolved salt concentration greater than 10,000 mg/L.

salt water intrusion The intrusion of salt water into a body of fresh surface water or groundwater.

salting out A procedure in which salt is added to a solution to cause an organic compound to precipitate so that it can be physically removed.

Salt-Master Water softener brine reclamation system by Culligan International Corp.

salvage The utilization of waste materials.

SAM™ Status alert modem to monitor disinfectant dosing system by Strantrol, Inc.

SAM™ Surge-anoxic mix wastewater treatment technology by Fluidyne Corp.

sampler A device used with or without flow measurement to obtain a portion of water or waste for analytical purposes.

sampling well See "monitoring well."

Sanborn Technologies Product line of Waterlink Biological Systems.

sanctions Actions taken by the federal government for failure to plan or implement a State Improvement Plan (SIP), and which may include withholding of highway funds and a ban on construction of new sources of potential pollution.

sand Any rock fragment between 1/16 mm and 2 mm in diameter.

Sand Dollar Sludge harvesting machine for sludge drying beds by Cherrington Corp.

sand drying bed See "sludge drying bed."

sand filter See "granular media filtration."

Sandfloat Combination dissolved air floatation and sand filter treatment system by Krofta Engineering Corp.

SandPIPER® Diaphragm pump by Warren Rupp, Inc.

Sandsep®   Screw-type grit classifier by Spirac.

Sandwash   Hydrocyclone by Axsia Serck Baker, Inc.

SANE   Sulfur and nitrogen emissions.

Sanilec®   Sodium hypochlorite generating systems by Exceltec International Corp.

Sanilo™   Water treatment product line by USFilter Corp.

Sani-Sieve   Gravity fed static screen by Dontech, Inc.

sanitary connection   A single family residential connection or single commercial or industrial connection to a public water supply system.

sanitary landfill   See "landfill (LF)."

sanitary sewer   Collection system of underground piping used to remove sanitary wastewater.

sanitary sewer overflow (SSO)   Overloaded operating condition of a sanitary sewer that results from inflow/infiltration.

sanitary wastewater   Domestic wastewater without storm and surface runoff that originates from sanitary conveniences.

San-I-Tech™   Grease interceptor by Scienco/FAST Systems.

Sanitron™   Ultraviolet water purifier by Atlantic Ultraviolet Corp.

SanTech   Product line of Waterlink Biological Systems.

Sanuril®   Hypochlorite tablet disinfection system by Exceltec International Corp.

SAP   Scientific Advisory Panel.

Saphyr®   A sludge treatment process by USFilter.

saponify   The conversion of a fat or grease into a soap by reaction with an alkali.

saprophytic bacteria   Bacteria that feed on dead or nonliving organic matter.

SAR   (1) See "sodium absorption ratio (SAR)." (2) See "supplied air respirator (SAR)."

SARA   See "Superfund Amendments and Reauthorization Act."

SASS   Source assessment sampling system.

Satellite   Electrically driven rotary distributor for fixed film reactor by Simon-Hartley, Ltd.

Sation®   Water treatment product line by USFilter Corp.

saturated steam   Vapor in equilibrium with water at the boiling temperature and containing no liquid.

saturated zone   See "zone of saturation."

saturation   The maximum concentration of a phase or material that can be contained within another phase or another material.

Saturation Index   See "Langelier Saturation Index (LSI)."

SAV   Submerged aquatic vegetation.

SAV715   Stainless steel sludge collector chain by Hitachi Maxco, Ltd.

SavagePlate™   Molded filter underdrain by Tetra Process Technologies.

savanna   A tropical or subtropical grassland characterized by scattered trees. Also spelled "savannah."

Save-All   Clarifier designed for paper mill fiber recovery by Walker Process Equipment.

save-all   Separation device used in a paper mill to reclaim fibers and fillers from white water.

SBA   Strong-base anion exchanger.

SBC   Submerged biological contactor.

SBOD   Soluble BOD. See "biochemical oxygen demand (BOD)."

SBR   See "sequencing batch reactor (SBR)."

SBS   See "Sick Building Syndrome (SBS)."

SC™   Package spray-type deaerating heater by Graver Co.

SCADA   Supervisory control and data acquisition. The hardware and software systems that gather and control data from a remote site.

SCADA-Flo™   Open channel transmitter by Marsh-McBirney, Inc.

scale   A mineral deposit or precipitate that forms on the interior surface of containers or water lines as a result of a heating or other physical or chemical change.

Scalper™   Inclined conveyor screening belt by Derrick Corp.

scanning electron microscope (SEM)   A microscope with a magnification range from 20X to 200,000X at a resolution of 100 Å, where illumination is provided by a beam of electrons which scan the specimen surface.

ScanRDI™   Microbial detection system by Chemunex, Inc.

scarp   A steep, almost perpendicular slope.

Scavenger   Robotic scrubber and vacuum sludge removal system by Aqua Products, Inc.

scavenging   (1) The unauthorized and/or uncontrolled removal of materials at any point in a solid waste management system. (2) The removal of a substance by converting it to another form or adsorbing it onto another compound.

SCBA   Self-contained breathing apparatus.

SCC   Source classification code.

SCD   See "streaming current detector (SCD)."

Scentoscreen   Portable gas chromatograph for VOC analysis by Sentex Systems, Inc.

SCFM   Standard cubic feet per minute.

Schistosoma   A flatworm or blood fluke that is highly parasitic to snails during one phase of its life and to humans during another.

schistosomiasis   A waterborne disease of tropical and subtropical regions transmitted to humans who wade or bathe in water infested by *Schistosoma*, with freshwater snails acting as intermediate hosts.

schmutzdecke   A biologically active layer that forms on the top of slow sand filters to aid in the removal of suspended solids.

Schoop Process   Process for coating steel that uses a blast of air to spray a mist of molten metal onto the surface to be protected.

SchreiberFlex®   Fine bubble diffuser by Schreiber Corp.

Schumacher Filters   Former name of USFilter/Schumacher.

Science Advisory Board (SAB)   A group of external scientists who advise the EPA on science and policy.

scientific method   An orderly method of obtaining, organizing, and applying new knowledge.

Scion®   Short cycle ion exchange system by USFilter/Rockford.

Scoop-A-Fish   Traveling water screen fish collection trough by Norair Engineering Corp.

ScorGuard®   Organic cooling tower water treatment additive by Ecolab Inc.

Scorpion®   Ultraviolet light disinfection system by Capital Controls Co.

scouring velocity   The minimum velocity required to carry away material accumulations in a conduit or pipeline.

Scour-Pak®   Granular media gravity depth filter by Graver Co.

SCOVOx™   Catalyst/adsorber technology for VOC emissions destruction by Goal Line Environmental Technologies.

SCR   (1) See "selective catalytic reduction (SCR)." (2) Silicon controlled rectifier.

scrap   Materials discarded from manufacturing operations that may be suitable for reprocessing.

screening   (1) A treatment process using a device with uniform openings to retain coarse solids. (2) A preliminary test method used to separate according to common characteristics.

screenings   The material removed by a screening device.

screenings press   A mechanical press used to compact and/or dewater material removed from mechanical screening equipment.

Screenings Washer Monster™   Washer/grinder for wastewater screenings by JWC Environmental.

Screezer   Combination screening and dewatering device by Waste-Tech, Inc.

screw conveyor   A conveyor utilizing a helical screw rotating within a trough to convey material.

screw pump   A low lift, high capacity pump that raises water by means of a slowly rotating inclined shaft fitted with a helical blade which revolves in a trough or pipe. Also known as "Archimedes' screw pump."

Screwpeller™   Centrifugal screw impeller used in a surface aerator by Aeration Industries, Inc.

scroll centrifuge   See "solid bowl centrifuge."

scrubber   A device used to removal particulates or pollutant gases from combustion or chemical process exhaust streams.

scrubbing   The removal of impurities from an air or gas stream by entraining the pollutants in a water spray.

ScruPac™   Screw-typed screenings compactor by Vulcan Industries, Inc.

Scru-Peller®   Sludge pump by Yeomans Chicago Corp.

SCSA   Soil Conservation Society of America.

SCUBA™   Filter underdrain by USFilter/General Filter.

SCUBA™   Self-contained gate and valve actuator by Rodney Hunt Co.

scum   Floatable materials found on the surface of primary and secondary settling tanks consisting of food wastes, grease, fats, paper, foam, and similar matter.

scum breaker   A device installed in a sludge digester to break up scum.

scum collector   A mechanical device for removing scum from the surface of a settling tank.

Scum Sucker™   Telescopic pipe for scum removal by United Industries, Inc.

scum trough   A trough used in a primary sedimentation basin to remove scum and convey it from the basin.

Scumbuster™   Pump used to chop solids in a digester scum blanket by Vaughan Co., Inc.

SCWO   See "supercritical water oxidation (SCWO)."

SDI   Strategic Diagnostics, Inc.

SDI   See "Silt Density Index (SDI)."

SDWA   See "Safe Drinking Water Act (SDWA)."

sea   A large body of salt water that is wholly or partly enclosed by land.

Sea Cell   In-situ sodium hypochlorite generator by Baker Hughes Process Systems.

Sea Devil   Floating oil skimmer by Vikoma International Ltd.

sea lettuce   Common seaweed that can grow in nuisance concentrations in the presence of excess nutrients.

sea level   The average surface level of the sea, uninfluenced by tidal movement or waves, used as a reference for elevation.

Sea Screen   Seawater filtration system by Baker Hughes Process Systems.

Sealtrode   Sealed pump controller by Yeomans Chicago Corp.

Seaskimmer   Floating oil skimmer by Vikoma International Ltd.

seawall   A wall built to protect a coastline from erosion, and caused by wave action.

seawater   General term for sea or ocean water, with a typical total dissolved solids concentration of 35,000 mg/L.

Secchi disk   A small disk, divided into black and white quadrants, that is lowered into water to visually observe water clarity and estimate the depth of the euphotic zone.

Secchi disk depth   The water depth at which a Secchi disk is no longer visible. In a lake, this depth is approximately equal to the euphotic zone.

Secodyne   Product group of Polydyne, Inc.

second order reaction   A reaction in which the rate of change is proportional to the square of the concentration of one of the reactants or to the product of the concentrations of two different reactants.

second stage biochemical oxygen demand   See "nitrogenous oxygen demand (NOD)."

secondary clarifier   A clarifier following a secondary treatment process designed for gravity removal of suspended matter.

secondary contaminant   A contaminant that affects drinking water taste, odor, or aesthetics.

secondary drinking water regulations   Nonenforceable regulations applying to public water systems and specifying the maximum contamination levels that, in the judgment of the EPA, are required to protect the public welfare, and which apply to any contaminants that may adversely affect the odor o.

secondary effluent   Treated wastewater leaving a secondary treatment facility, usually having a BOD_5 and suspended solids of less than 30 mg/L.

secondary emissions   Emissions that occur as a result of the construction or operation of a facility but do not come from the facility.

secondary materials   Materials that have been manufactured and used at least once and are to be used again.

secondary maximum contaminant levels (SMCLs)   Guidelines that address taste, odor, color, and other aesthetic aspects of drinking water but do not present health risks.

secondary pollutant   A pollutant formed in the environment as the result of a chemical reaction of two or more primary pollutants or naturally occurring elements.

secondary sludge   The sludge from the secondary clarifier in a wastewater treatment plant.

secondary standards   National ambient air quality standards designed to protect welfare, including effects on soils, water, crops, vegetation, manmade materials, animals, wildlife, weather, visibility, and climate.

secondary treatment   The treatment of wastewater through biological oxidation after primary treatment.

second-stage BOD   See "nitrogenous biochemical oxygen demand (NBOD)."

secure landfill   Landfill which segregates and isolates hazardous materials from contacting each other, the groundwater, or the atmosphere.

secure maximum contaminant level   Maximum permissible level of a contaminant in water delivered to the free flowing outlet of the ultimate user, or of contamination resulting from corrosion of piping and plumbing caused by water quality.

Sedifloat   Water and wastewater treatment unit by Krofta Engineering Corp.

Sediflotor®   Dissolved air flotation unit by Infilco Degremont, Inc.

sediment   The solid material that settles from a liquid.

sediment yield   The quantity of sediment arriving at a specific location.

sedimentation   The removal of settleable suspended solids from water or wastewater by gravity in a quiescent basin or clarifier.

sedimentation basin   A quiescent tank used to remove suspended solids by gravity settling. Also called clarifiers or settling tanks, they are usually equipped with a motor driven rake mechanism to collect settled sludge and move it to a central discharge point.

SEE   Schloss Engineered Equipment.

seed   (1) Crystalline particles added to a supersaturated solution to induce precipitation. (2) Well-digested sludge used to seed a sludge digester.

seepage   Percolation of water through the soil from unlined canals, ditches, laterals, watercourses, or water storage facilities.

seepage pit   A covered excavation that receives septic tank effluent and permits its effluent to seep through the bottom and sides of the excavation.

seepage spring   A spring occurring where the water table breaks the ground surface. Also called a "gravity spring."

Seghers Pelletech®   Indirect sludge dryer and pelletizing unit by Wheelabrator Water Technologies, Inc.

Seghodryer   Indirect contact sludge dryer by Seghers Better Technology USA.

selective catalytic reduction (SCR)   Flue gas treatment process for the removal of NOX by reduction with ammonia to form elemental nitrogen and water.

selective pesticide   A chemical designed to affect only certain types of pests, leaving other plants and animals unharmed.

Selectofilter   Revolving drum screen strainer by USFilter/Diffused Air Products Group.

Selector Plug Flow™   Biological wastewater treatment process by USFilter/Industrial Wastewater Systems.

Selectostrainer   In-line strainer formerly offered by USFilter/Headworks Products.

Selex®   Graded density cartridge filters by Osmonics, Inc.

self-sustaining   The point at which a process provides sufficient energy to operate without the need for supplementary fuel.

SelRO®   Membrane filtration systems by LCI Corp.

SEM   See "scanning electron microscope (SEM)."

semi-confined aquifer   An aquifer partially confined by soil layers of low permeability through which recharge and discharge can still occur.

semipermeable   A membrane that does not have measurable pores, but through which smaller molecules can pass.

senescence   The aging process sometimes used to describe lakes or other bodies of water in advanced stages of eutrophication.

senescent lake   A very old lake, practically full of sediment and rooted water plants, which will eventually become a marsh.

sensible heat   Heat measurable by temperature alone.

Sension™   Water analyzer meter by Hach Co.

sensitivity   The ability of a unit or instrument to respond to a small difference in values.

Sentinel™   Filter backwash control system by Roberts Filter Group.

Sentre-Fier   Rotary fine screen by Dontech, Inc.

Sentry   Groundwater gasoline recovery system by Douglas Engineering.

Sentry-II®   VOC sampler by BIOS International Corp.

SEPA®   Reverse osmosis membranes by Osmonics, Inc.

Separator-Plus™   Parallel plate separator by USFilter/Davco.

SepraEight®   Deep bed condensate polisher by Graver Co.

sepralators   Membrane elements.

septage   The settled solids produced in individual on-site wastewater treatment systems including septic tanks and cesspools.

septic   Condition characterized by bacterial decomposition under anaerobic conditions.

septic system   An onsite system designed to treat and dispose of domestic wastewater typically consisting of a tank that receives waste from a residence or business and a system of tile lines or a pit for disposal of the sludge that remains after decomposition of the soluble organic matter.

septic tank   A domestic wastewater treatment device principally used for individual residences that combines sedimentation, sludge digestion, and sludge storage in a single or dual compartmented tank. See also "septic system."

septicity   The condition that results from biological degradation of organic matter in wastewater under anaerobic conditions, usually producing hydrogen sulfide or other odorous compounds.

Septra™   Pleated backwash filters by Pall Corp.

septum   A permeable material used to support filter medium.

sequencing batch reactor (SBR)   Treatment process characterized by the interruption of flow to the reactor during the sedimentation and decanting phase of treatment.

Sequest-All   Sequestering agent used to control iron scaling and corrosion by Sper Chemical Corp.

sequestering agent   A chemical compound such as EDTA that binds with other compounds or ions so they cannot be involved in chemical reactions. See also "chelating agent."

sequestration   The formation of a stable, water soluble complex with an ion in solution to prevent precipitation or scaling. See also "chelation."

Sequox™   Biological wastewater treatment process for nutrient removal by Waterlink/Aero-Mod Systems.

Seral®   Laboratory water treatment product line by USFilter/Lowell.

Serck Baker   Former name of Axsia Serck Baker, Inc.

Ser-Ductor®   Air-free agitation system by Serfilco, Ltd.

Serfilco®   Wastewater treatment equipment products by Serfilco, Ltd.

seroepidemiology   The measuring of serum antibodies to specific pathogens in a population.

serogroup   A group of closely related organisms having one or more common antigens.

Serpentix®   Convoluted, self-cleaning belt conveyor by Serpentix Conveyor Corp.

service connector   The pipe that carries tap water from a public water main to a building.

service factor (SF)   A multiplier that when applied to the rated power indicates the permissible power loading that may be carried under the conditions specified.

service line sample   A one-liter sample of water collected according to federal regulations that has been standing for at least 6 hours in a service pipeline.

service pipe   The pipeline extending from the water main to the building served or to the consumer's system.

SES   Secondary emissions standard.

Sessil®   Polyethylene strip media for trickling filters by NSW Corp.

SETLdek   Clarifier tube settlers by Brentwood Industries, Inc.

setpoint   An input value to be maintained by a control device.

settleability   The tendency of suspended solids to settle.

settleable solids   That portion of suspended solids which are of a sufficient size and weight to settle to the bottom of an Imhoff cone in 1 hour.

settled sludge volume (SSV)   Volume of settled sludge measured at predetermined time increments for use in process control calculations.

settling chamber   A series of screens placed in the way of flue gases to slow the stream of air, thus helping gravity to pull particles into a collection device.

settling tank   A quiescent tank used to remove suspended solids by gravity settling. Also called clarifiers or sedimentation basins, they are usually equipped with a motor driven rake mechanism to collect settled sludge and move it to a central discharge point.

settling tubes   See "tube settlers."

settling velocity   The rate at which a particle settles through air or water.

7Q10   An abbreviation for a design stream flow rate describing the lowest stream flow for 7 consecutive days that would be expected to occur once in 10 years. See also "xQy."

sewage   See "wastewater."

sewage fungus   Common term for filamentous mass of fungi and bacteria resulting from high organic loadings that can clog or reduce efficiency of waste treatment equipment.

sewer   Collection system of underground piping used to remove wastewater.

Sewer Chewer™   Comminuter/sludge grinder by Yeomans Chicago Corp.

sewer gas   A gas mixture produced by anaerobic decomposition of organic matter usually containing high percentages of methane and hydrogen sulfide.

sewerage   The entire system of wastewater collection, treatment, and disposal.

sewershed   Land area that drains into a sewer.

SF   See "service factor (SF)."

SFA   Spectral flame analyzer.

SFT™   Sediment flushing tank by John Meunier, Inc.

ShallowTray   Aeration system for removal of volatile organics by Northeast Environmental Products, Inc.

Shann-No-Corr   Zinc metaphosphate corrosion inhibitor and sequestering agent by Shannon Chemical Corp.

Sharples®   Division of Alfa Laval Separation, Inc.

sharps   Hypodermic needles, pipettes, scalpel blades, blood vials, needles with attached tubing, broken or unbroken glass, and culture dishes used in animal or human patient care or treatment, or in medical, research, or industrial laboratories.

Sharpshooter   Polymer feed and control system by Norchem Industries.

Shartles   Product line by Thermal Black Clawson.

Shearfuser   Cast iron diffuser for anaerobic digestion by USFilter/Diffused Air Products Group.

sheet flow   Overland stormwater flow in a thin sheet of uniform thickness.

shell-and-tube heat exchanger   A tubular heat exchanger housed within the shell of a pressure vessel.

sherardizing   A process for protecting iron from corrosion by means of a corrosion resistant layer of zinc on the iron surface.

Shigella   A bacterium associated with dysentery that is transmitted through consumption of water or food contaminated with fecal matter.

shigellosis   A gastrointestinal disorder usually caused by food or waterborne organisms of the genus *Shigella*.

shock load   A sudden hydraulic or organic load to a treatment plant.

shore   (1) The land bordering a body of water. (2) To brace or give support.

short circuiting   Uneven flow through a vessel that results from density currents or inadequate mixing which allows some currents to leave the vessel more quickly than others.

short term exposure limit (STEL)   The maximal allowable level of a material in workplace air, usually measured over a 15-minute period.

short ton   See "ton."

Shriver®   Plate and frame filter press by Baker Process.

shute   The horizontal wire in woven wire mesh, also called the "weft" wire.

SHWL   Seasonal high water level.

SI unit   The international system of units (Système International) largely based on the metric system used for measuring length, mass, volume, and radiation.

SIC   See "Standard Industrial Classification (SIC)."

sick building syndrome (SBS)   Condition in which at least 20% of a building's occupants display symptoms of illness for more than 2 weeks and the source of the illness cannot be positively identified.

side hill screen   See "static screen."

side water depth (SWD)   The depth of water measured along a vertical interior wall of a basin or tank.

Side Winder   Environmental screens by Cook Screen Technologies, Inc.

SideCar™   RBC aeration system by Jones MacCrea, Inc.

siderite   A mineral of ferrous carbonate.

sidewall   The wall at the side of a structure.

Sidewall Separator   In-channel clarifier for an oxidation ditch by Lakeside Equipment Corp.

sieve analysis   A size distribution analysis of a filter sand sample using a series of standard sieve screens.

sieve size   The standard sieve size through which a sample of sand will pass.

sievert   Unit of radiation equal to the amount that produces the same damage to humans as 1 roentgen of high voltage X-rays.

SightWell   Circular clarifier with hydraulic suction type sludge removal system by Walker Process Equipment.

Sigma   Low speed surface aerator by Purestream, Inc.

Sigma Flight   Fiberglass sludge collector flight by USFilter/Envirex.

signal words   The words used on a pesticide label — "Danger, Warning, Caution" — to indicate level of toxicity.

significant deterioration   Pollution resulting from a new source in previously "clean" areas. See "prevention of significant deterioration (PSD)."

significant municipal facilities   Those publicly owned sewage treatment plants that discharge a million gallons per day or more and are therefore considered by states to have the potential to substantially affect the quality of receiving waters.

significant violations   Violations by point source dischargers of sufficient magnitude or duration to be a regulatory priority.

SIHI Pumps   Former name of Sterling Fluid Systems (USA).

SIHI-Halberg   Digester draft tube sludge mixer by SIHI Pumps, Inc.

Silent Pump   Portable sewage pumping station by Gorman-Rupp Co.

silica   A mineral composed of silicon and oxygen.

silicate   Any compound containing silicon, oxygen, and one or more metallic compounds.

siliceous   Composed of or containing silica or a silicate compound.

siliceous gel zeolite   A synthetic hydrated sodium aluminosilicate with ion exchange properties, once widely used in ion exchange water softeners. See also "gel zeolite."

silicosis   A lung disease caused by prolonged inhalation of silica dust which results in fibrosis or scarring of lung tissue.

Sil-Kleer®   Filter aid by Silbrico Corp.

silo   A tall, cylindrical storage vessel for dry solids.

Silo Pac   Chemical feed system by USFilter/Wallace & Tiernan.

Siloda™   High rate sludge composting process by USFilter/Krüger (North America) and OTV.

silt   Individual mineral particles ranging in size between fine sand and clay.

silt density index (SDI)   A measure of the fouling tendency of water based on the timed flow of a liquid through a membrane filter at a constant pressure.

silting   The deposition of silt or sediment in a waterbody.

silvaculture   Forest management for the cultivation and harvest of timber.

Silver Band   Granular media pressure filter by Baker Hughes Process Systems.

Silver Series™   Membrane diffuser by Sanitaire Corp.

Silverback™   Aqueous cleaner recovery system using ceramic membrane filters by USFilter/Industrial Wastewater Systems.

silviculture   Management of forest land for timber.

Simcar®   Turbine aerator by United Industries, Inc.

Simplex   Low speed surface mechanical aerators by USFilter/Asdor.

Simplex™   Single module air scrubber by USFilter/Davis Process.

Sim-Pre™   Biological nutrient removal process for wastewater treatment by USFilter/Envirex.

Simrake   Rotary bar screen by Simon-Hartley, Ltd.

Simspray   Rectangular distributor for fixed film reactor by Simon-Hartley, Ltd.

Simultech™   Biological nutrient removal system by Schreiber Corp.

Simul-Wash™   Simultaneous air/water filter backwash system by Tonka Equipment Co.

Sinclair   Internally fed rotary fine screen by Bielomatik London, Ltd.

sink   Place in the environment where a compound or material collects.

sinking   Controlling oil spills by using an agent to trap the oil and sink it to the bottom of the body of water where the agent and the oil are biodegraded.

sinter   To bond and partly fuse masses of metal particles together through the use of heat below the melting point, without actually liquefying.

SIP   See "State Implementation Plan (SIP)."

siphon   A closed conduit, a portion of which lies above the hydraulic grade line, resulting in a pressure less than atmospheric and requiring a vacuum within the conduit to start flow.

SITE   See "Superfund Innovative Technology Evaluation (SITE)."

site assessment program   A means of evaluating hazardous waste sites through preliminary assessments and site inspections to develop a Hazard Ranking System score.

Sitepro™   Remediation site control system by Geotech ORS Environmental Equipment.

siting   The process of choosing a location for a facility.

SIU   Significant industrial user.

Siverseries™   Packaged desalination system by Matrix Desalination, Inc.

SK®   Variable area flowmeter by McCrometer, Inc.

skid mounted   Equipment or equipment packages mounted on a horizontal structure or platform to facilitate handling and/or installation.

Skim-Kleen®   Oil removal belt skimmer by Tenco Hydro, Inc.

skimming   The process of removing or diverting water and/or floating matter from the surface of a liquid.

Skim-Pak™   Floating, self-adjusting weir skimmer by Douglas Engineering.

Skim-Pak™   Floating skimmer by ABS Pumps, Inc.

skip   Bar screen cleaning rake.

SKRAM   Acoustic fish behavioral control device by USFilter/Rex & Link-Belt Products.

SL™   Granular media filter underdrain by F.B. Leopold Co., Inc.

slag   Waste residues produced by metal smelting and coal gasification.

slake   The process of mixing lime with water to accomplish a chemical combination.

slaked lime   See "hydrated lime."

slash and burn   An agricultural practice where vegetation is cut, allowed to dry, and burned prior to soil cultivation and planting.

Slide Gate   Screenings press by Andritz-Ruthner, Inc. (Western Hemisphere) and USFilter/Contra-Shear.

slime   (1) Substances of a viscous organic nature, usually formed from a microbiological growth, which attach themselves to other objects, forming a coating. (2) The coating of biomass that accumulates in trickling filters or sand filters and periodically sloughs away to be collected in clarifiers.

Slinger®   Sludge spreader by Knight Manufacturing Corp.

Slo-Mixer   Axial flow paddle flocculators by USFilter/Envirex.

slop oil   Separator skimmings and tramp oil generated during refinery startup, shutdown, or abnormal operation.

slough   A swamp, bog, or marsh, especially if part of an inlet or backwater.

sloughing   The disattachment of accumulated biological solids from trickling filter media.

slow sand filter   Sand filter characterized by low flow rates relying on the formation of a layer of solids on the top of the sand bed to accomplish most of the filtration.

sludge   Accumulated and concentrated solids generated within the wastewater treatment process that have not undergone a stabilization process.

sludge age   The average time microbial cell remains in an activated sludge system. It is equal to the mass of cells divided by the rate of cell wasting from the system.

Sludge Age Controller™   Sludge age control system by United Industries, Inc.

sludge blanket   The accumulated sludge hydrodynamically suspended in a clarifier or other enclosed body of water.

sludge bulking   A phenomenon that occurs in activated sludge plants where a sludge does not readily settle or concentrate.

sludge cake   Dewatered residue from a filter press, centrifuge, or other sludge dewatering device.

sludge collector   Mechanisms used in clarifiers to collect and remove settled solids from the tank bottom.

sludge conditioning   See "conditioning."

Sludge Detention Optimizer™   Sludge thickening/conditioning process by Dontech, Inc.

sludge dewatering   The removal of a portion of the water contained in sludge by means of a filter press, centrifuge, or other mechanism.

sludge digestion   See "digestion."

sludge dryer   A device utilizing heat for the removal of a large portion of water within sludge.

sludge drying bed   A partitioned area consisting of sand or other porous material upon which sludge is dewatered by drainage and evaporation. Also known as "drying bed" or "sand drying bed."

Sludge Expert   Automatic belt press control and management system by Alpine Technology, Inc.

Sludge Gun®   Sludge level detector by Markland Specialty Engineering, Ltd.

Sludge Guzzler   Hydraulically driven sludge pump by Guzzler Manufacturing, Inc.

Sludge Mate™   Container filter by Flo Trend Systems, Inc.

sludge stabilization   A treatment process to convert sludge to a stable product for ultimate disposal or use and reduce pathogens to produce a less odorous product.

Sludge Sucker™   Clarifier sludge removal device by USFilter/General Filter.

sludge volume index (SVI)   The volume in milliliters occupied by 1 gram of settled sludge after settling for 30 minutes in a graduated cylinder.

Sludgebuster™   Sewage shredder by ZMI/Portec Chemical Processing.

SludgeCleaner®   Sludge and scum screen and compactor by Parkson Corp.

SludgeMaster®   Sludge dryer by USFilter/Davis Process.

SludgeMaster®   Submersible air-powered pump by Warren Rupp, Inc.

SludgeMIZER   Sludge dryer by Fen-Tech Environmental, Inc.

Sludgepactor   Sludge screen and compactor by Waste-Tech, Inc.

SludgePress™   Biosolids dewatering products by Enviroquip, Inc.

Sludgifier   Lagoon dredge by VMI Inc.

Sludglite®   Sludge blanket level detector by Ecolotech Corp.

slug load   A sudden hydraulic or organic load to a treatment unit.

sluice gate   Manual or power-operated gate used to isolate a channel from flow.

slurry   A suspension of a relatively insoluble chemical in water, usually having a suspended solids concentration of 5000 mg/L or more.

Slurrycup™   Grit removal system by Eutek Systems, Inc.

Slurrystore®   Slurry storage system by A.O. Smith Engineered Storage Products.

small calorie (cal)   See "calorie."

small quantity generator (SQG)   A generator of between 100 and 1000 kg/month of hazardous wastes. Sometimes called a "squeegee."

small-quantity handlers (SQHs)   Universal waste handlers who accumulate up to 5000 kilograms of wastes.

Smart Skimmer   Oil removal system by Douglas Engineering.

SmartFilter™   Traveling bridge filter by Agency Environmental, Inc.

SmartRO™   Reverse osmosis system by Water and Power Technologies.

SMBS   Sodium metabisulfite.

SMCL   Secondary maximum contaminant level.

SMCLs   EPA-mandated maximum contaminant level in drinking water based on taste, odor, or aesthetics.

SMCRA   See "Surface Mining Control and Reclamation Act (SMCRA)."

smelter   A facility that melts or fuses ore, often with an accompanying chemical change, to separate its metal content.

SMOA   Superfund Memorandum of Agreement.

smog   A type of air pollution characterized by reduced visibility due to atmospheric particulates and elevated levels of photochemical oxidants.

Smogless™   Wastewater treatment and sludge drying and incineration plants by USFilter/Smogless.

smoke   The suspended matter in an exhaust emission which obscures the transmission of light.

smoke number (SN)   A dimensionless term quantifying smoke emissions.

Smooth-tex   Rectangular woven mesh for screening equipment by USFilter/Rex & Link-Belt Products.

SMSA   Standard metropolitan statistical area.

SMX   Belt filter press by Andritz-Ruthner, Inc.

SN   See "smoke number (SN)."

SNAAQS   Secondary National Ambient Air Quality Standards.

Snail   Grit dewatering system by Eutek Systems, Inc.

SNARL   Suggested no adverse response level.

SNC   Significant noncompliers.

Snowflake Packing   Plastic packing media for air stripping applications by Norton Co.

SO₂   Sulfur dioxide.

SOC   See "synthetic organic chemicals (SOC)."

SOCMA   Synthetic Organic Chemical Manufacturers Association.

SOCMI   Synthetic organic chemical manufacturing industry.

SOD   Sediment oxygen demand.

soda ash   See "sodium carbonate."

sodium absorption ratio (SAR)   An expression describing the relative activity of sodium ions in exchange reactions with soil.

sodium aluminate   An auxiliary coagulant used in water treatment. Chemical formula is $Na_2Al_2O_4$.

sodium bisulfite   A liquid dechlorinating agent. Chemical formula is $NaHSO_3$.

sodium carbonate   A compound often used in water softening operations, also called "soda ash." Chemical formula is Na_2CO_3.

sodium chloride   The chemical name for common salt. Chemical formula is $NaCl$.

sodium cycle exchange   Ion exchange water softening process in which sodium on the ion exchange resin is exchanged for hardness and other ions in water. Sodium chloride is commonly used as the regenerant.

sodium hexametaphosphate (NaHMP)   An alkaline metallic element which is water soluble and commonly used as a sequestering or dispersing agent.

sodium hydroxide   Caustic soda. Chemical formula is NaOH.

sodium hypochlorite   A liquid chlorine solution frequently used as a water or wastewater disinfectant. Chemical formula is $Na(OCl)_2$.

sodium metabisulfite   A crystalline form of sulfur dioxide used to remove chlorine. Chemical formula is $Na_2S_2O_5$.

sodium sulfite   An oxygen scavenger used in boiler and cooling water systems. Chemical formula is Na_2SO_3.

sodium-free water   Bottled water containing less than 5 milligrams of sodium per serving.

soft detergents   Cleaning agents that break down in nature. See also "linear alkyl sulfonate (LAS)."

soft water   Any water containing less than 17.1 mg/L (1 gpg) of calcium or magnesium expressed as calcium carbonate.

softening   Treatment process that involves the removal of calcium and magnesium ions from water.

SOG   Stripper off-gas.

soil   The combination of mineral and organic matter that supports plant life on the surface of the earth.

soil adsorption field   A sub-surface area containing a trench or bed with clean stones and a system of piping through which treated sewage may seep into the surrounding soil for further treatment and disposal.

soil conditioner   An organic material like humus or compost that helps soil absorb water, build a bacterial community, and take up mineral nutrients.

soil erodibility   A measure of the soil's susceptibility to raindrop impact, runoff, and other erosional processes.

soil flushing   In-situ remediation technique where large volumes of water or other solutions are injected into the soil or groundwater to flush hazardous contaminants to a recovery point.

soil gas   Gaseous elements and compounds in the small spaces between particles of the earth and soil. Such gases can be moved or driven out under pressure.

soil horizon   A zone or layer within a soil made distinctive by its color, texture, or mineral content.

soil sterilant   A chemical that temporarily or permanently prevents the growth of all plants and animals depending on the chemical.

soil vapor extraction (SVE)   Technique to remove VOCs and promote bioremediation of compounds in unsaturated soils.

Sokalan®   Antiscalant for desalination of seawater by BASF.

sol   Colloidal dispersion of solids in liquid.

solar constant   The rate that the sun's radiant energy is received per unit area on a horizontal surface at the top of the earth's atmosphere.

solar pond   Pond used to accomplish evaporation via direct solar heating.

solar still   A distillation device that utilizes solar energy.

solder   Metallic compound used to seal joints between pipes, and until recently, usually contained 50% lead. Use of lead solder containing more than 0.2% lead is now prohibited for pipes carrying drinking water.

sole-source aquifer   An aquifer that supplies 50% or more of the drinking water of an area.

solid bowl centrifuge   Continuous operation centrifuge consisting of a cylindrical, tapered bowl and an internal helical scroll, both revolving at slightly different speeds to separate solids from water by means of a centrifugal force. Also called "scroll centrifuge" or "decanter."

solid waste   Garbage, refuse, sludge, and other discarded material resulting from community activities or commercial or industrial operations.

solid waste disposal   The final placement of refuse that is not salvaged or recycled.

solid waste management   Supervised handling of waste materials from their source through recovery processes to disposal.

Solidex   Screw press system by Hosokawa Bepex Corp.

solidification   A process where materials are added to waste to produce a solid.

solidification and stabilization   Removal of wastewater from a waste or the chemical change that makes it less permeable and susceptible to transport by water.

solids balance   A mathematical representation of a treatment system that defines the amount of solids entering and exiting each unit treatment process.

solids contact clarifier   A clarifier in which liquid passes upward though a solids blanket and discharges at or near the surface.

solids retention time (SRT)   The mass of solids in a vessel (kg) divided by the solids removed (L/d).

Solidur   UHMW polyethylene components for chain and flight sludge collectors by Solidur Plastics Co.

Soliquator   Clarification unit by Orival, Inc.

Solo™   Controllerless downwell cleanup pumps by QED Environmental Systems, Inc.

Solu Comp®   Water measurement instrument by Rosemount Analytical, Inc.

solubility   The amount of a substance that can dissolve in a solution under a given set of conditions.

solubility product   The equilibrium constant that describes the reaction by which a precipitate dissolves in pure water to form its constituent ions.

soluble   Capable of being dissolved in a fluid.

solum   The upper layers of soil including the top soil (A-horizon) and intermediate soil (B-horizon).

solute   A substance dissolved in a fluid.

solution   A liquid that contains dissolved solute.

Solvay Interox   Product group of Solvay America.

Solvay Minerals   Product group of Solvay America.

solvent   Liquid capable of dissolving or dispersing one or more substances.

solvent extraction   The process of selectively extracting a liquid constituent from a wastewater using an organic solvent. Also called "liquid-liquid extraction (LLE)."

Solvo-Salvager®   Vacuum assisted distillation equipment by Westport Environmental Systems.

Som-A-System®   Sludge thickening process by Somat Corp.

sone   A subjectively determined unit of the loudness of a sound.

Sonix 100   Tank-mounted chlorinator by USFilter/Wallace & Tiernan.

Sonozaire™   Electronic odor control device by Howe-Baker Engineers, Inc.

soot   Carbon dust formed by incomplete combustion.

sorbent   A solid material used to concentrate dissolved solids.

Sorb-N-C™   Packaged stack gas sorbent by Church & Dwight Co. Inc.

Sorbond®   Sludge stabilizer by Colloid Environmental Technologies Co.

sorption   The concentration of dissolved solids through absorption or adsorption on a solid.

SOS   Stormwater overflow screens by John Meunier, Inc.

SOTE   Standard oxygen transfer efficiency.

sough   A ditch used to drain mine water.

SOUR   See "specific oxygen uptake rate (SOUR)."

sour environment   Environment containing significant amounts of hydrogen sulfide.

sour gas   Natural gas containing impurities such as hydrogen sulfide which give the gas an acrid odor.

sour water   Water containing hydrogen sulfide or other malodorous compounds.

source   The point of origin or part of a facility that generates or releases an excess of a substance.

Southern Oscillation   A circulation pattern in the atmosphere that brings low pressure to the eastern Pacific and high pressure to the Indian Ocean. See also "El Niño/Southern Oscillation (ENSO)."

SOx   See "sulfur oxides (SOX)."

Soxhlet Extraction Method   An extraction process using trichlorofluoroethane to determine the oil and grease content in a liquid.

SoyGold®   Biodegradable, nontoxic solvent by Ag Environmental Products LLC.

sparingly soluble compounds   Term used to describe compounds with solubility ranges from near zero to a few thousand milligrams per liter.

Sparjair   Package wastewater treatment plant by Walker Process Equipment.

Sparjer   Aeration products by Walker Process Equipment.

SparjLift   One- or two-level air injection pump by Walker Process Equipment.

sparkling water   Carbonated water for which the carbon dioxide content comes from the same source as the water.

SPCC   (1) Spill prevention control and countermeasures. (2) Spill prevention, containment, and countermeasure.

SPCCP   See "Spill Prevention Control and Countermeasures Plan (SPCCP)."

SPE   Secondary particulate emissions.

special waste   Items such as household hazardous waste, bulky wastes including refrigerators and pieces of furniture, and tires or used oil.

species A population of similar organisms that interbreed among themselves.

specific conductance The measure of electrical conductance of water or a water solution and the reciprocal of specific resistance. Usually stated in micromhos per cm.

specific gravity The ratio of the density of a substance to the density of water.

specific oxygen uptake rate (SOUR) An indicator of a sludge's odor-causing potential by aerating a sludge sample and measuring the rate of oxygen depletion.

specific resistance A measure of total ionized solids concentration determined by the resistance of a 1 cm cube of water to the passage of electricity under standard conditions.

specific ultraviolet absorbance (SUVA) The amount of ultraviolet light that can be absorbed per unit of total organic carbon which may indicate DBP formation potential.

SpectraGuard™ Liquid antiscalant and stabilizer for reverse osmosis systems by Professional Water Technologies, Inc.

spectrophotometer An instrument for measuring the amount of electromagnetic radiation absorbed by a sample as a function of wavelength.

Spectrum™ Aeration-mixing systems by Environmental Dynamics Inc.

Speedi-Berm™ Secondary containment system by Aero Tec Laboratories, Inc.

Spektrotherm® Low pressure, high intensity ultraviolet lamps by PCI-Wedeco Environmental Technologies, Inc.

spent caustic Waste product formed when petrochemicals and refined hydrocarbons are treated with caustic soda solutions to remove impurities.

spent regenerant Wastes from the regeneration of an ion exchange system.

Sphaerotilus A filamentous bacteria which commonly causes sludge bulking in activated sludge wastewater treatment plants.

Spher-Flo Single stage sewage pumps by Aurora Pump.

sphericity A measure of roundness and wholeness of filter media and ion exchange resin beads.

Spill Prevention Control and Countermeasures Plan (SPCCP) Plan covering the release of hazardous substances as defined in the Clean Water Act.

spill water Water released from an impoundment.

spillway The channel designed to direct overflowing water at the top of a dam.

Spira-Cel® Spiral wound, cross flow filtration membranes by Celgard LLC.

Spiracone® Conical tank upflow clarifier by USFilter/General Filter.

Spiractor® Lime softener by USFilter/Warren.

Spiraflo Peripheral feed clarifier by Lakeside Equipment Corp.

Spirafloc Peripheral feed clarifier with flocculation zone in outer raceway by Lakeside Equipment Corp.

Spiragester Combination digester and clarifier by Lakeside Equipment Corp.

Spiragrit Grit removal system by Lakeside Equipment Corp.

Spiral Flow® Water/solids separator by Alar Engineering Corp.

spiral heat exchanger A type of heat exchanger used in heavy fouling applications.

Spiral Scoop Dissolved air flotation skimming device by Krofta Engineering Corp.

Spiralflow Dewatering screen fabric by Geschmay GmbH.

Spiralift™   Screw pump by USFilter/Zimpro.

Spiralklean®   Screenings washer by Parkson Corp.

Spiraltek™   Dry sump filters by Osmonics, Inc.

Spira-Pac   Package digester clarifier combination by Lakeside Equipment Corp.

Spiratex™   Point-of-use filtration system by Osmonics, Inc.

Spirathickener   Peripheral feed gravity thickener by Lakeside Equipment Corp.

Spiratrex   Ultrafiltration membrane by Osmonics, Inc.

Spira-Twin Spiragester   Clarifier/digester unit for primary and secondary sedimentation of trickling filter effluents by Lakeside Equipment Corp.

Spiravac   Peripheral feed clarifier with vacuum assisted sludge removal by Lakeside Equipment Corp.

Spirolift®   Screw-type vertical conveyor by Spirac.

Spiropac®   Screw-type compactor by Spirac.

Spiropress®   Screw-type solids dewatering press by Spirac.

Spirosand®   Shaftless grit classifier by Andritz-Ruthner, Inc.

Spirovortex   Activated sludge treatment system including tertiary filtration by GL&V/Dorr-Oliver, Inc.

Split-ClarAtor™   Secondary clarifier by Waterlink/Aero-Mod Systems.

splitter box   A chamber that equally divides incoming flow into two or more streams.

spoil   Dirt or rock removed from its original location by strip-mining, dredging, or construction.

sponge ball cleaning   The use of flexible sponge balls added to a recirculating liquid to scour scale or other deposits that form on the inside of condenser or heat exchanger tubes.

spore   A reproductive cell or seed of a microbe, often dormant or environmentally resistant.

sprawl   Unplanned development of open land.

spray dryer   Sludge-drying device utilizing centrifugal force to atomize sludge into fine particles and spray them into the top of a drying chamber.

spray irrigation   The spreading of treated wastewater on agricultural land by spraying.

spray tower scrubber   A device that sprays alkaline water into a chamber where acid gases present to aid in the neutralizing of the gas.

Spray-Film®   Vapor compression distillation unit by Aqua-Chem, Inc.

Spraymaster®   Low-headroom packaged deaerator by Cleaver-Brooks.

spring   A natural flow of water from the ground.

spring turnover   See "turnover."

spring water   Bottled water collected from an underground formation from which the water flows naturally from the surface, or a bore hole that taps the spring and is located near where the spring emerges.

Sprint™   Submersible pump product line by Crane Pumps & Systems.

Sprout-Bauer   Former name of Andritz-Ruthner, Inc. screening equipment product line.

Spyder™   Fixed-grid sludge withdrawal system by Roberts Filter Group.

SQBE   Small quantity burner exemption.

SQG   See "small quantity generator."

SQHs   See "small-quantity handlers."

Squarex   Circular sludge collector mechanism with pivoted corner extensions for square basins by GL&V/Dorr-Oliver, Inc.

squeegee   See "small quantity generator (SQG)."

SR-7™   Ion exchange resins by Sybron Chemicals, Inc.

SRC   Solvent-refined coal.

SRF   See "State Revolving Fund (SRF)."

SRM   Standard reference method.

SRT   See "solids retention time (SRT)."

SS   See "suspended solids (SS)."

SSA   Sole source aquifer.

SSAC   Soil site assimilated capacity.

SSI   Screening Systems International.

SSMS   Spark source mass spectrometry.

SSO   See "sanitary sewer overflow (SSO)."

SSPC   Society for Protective Coatings. Formerly "Steel Structures Painting Council."

SSPMA   Sump and Sewage Pump Manufacturers Association.

SST   See "stainless steel (SST)."

SSV   See "settled sludge volume (SSV)."

Stabilaire   Package contact stabilization treatment plant formerly offered by USFilter/Envirex.

stability index   See "Langelier Saturation Index (LSI)."

stabilization pond   A large shallow basin used for wastewater treatment by natural processes involving the use of algae and bacteria to accomplish biological oxidation of organic matter.

StablCal™   Formazin turbidity standards by Hach Co.

stable air   A motionless mass of air that holds instead of dispersing pollutants.

stack   (1) A vertical conveyance method for the elevated discharge and dispersion of pollutants. (2) Common term for the key element of an electrodialysis unit consisting of multiple membrane cells with electrodes on both ends.

stack effect   Air, as in a chimney, that moves upward because it is warmer than the ambient atmosphere.

stack gas   See "flue gas."

StackMasterIMS™   Continuous emissions monitor by Molecular Analytics LLC.

stage   One of several units of a flash evaporator, each of which operates at a successively lower pressure.

stagnant   Motionless, not flowing in a current or stream.

Stahlermatic®   Rotating biological contacter wastewater treatment system by Stahler GmbH.

stainless steel (SST)   Corrosion resistant steel containing a minimum of 12% chromium as the principal alloying element.

stakeholder   Any organization, governmental entity, or individual that has a stake in or may be impacted by a given approach to environmental regulation, pollution prevention, energy conservation, etc.

STAKfilter   High volume vertical pressure screen filter by Everfilt Corp.

Stak-Tracker™   Continuous emissions monitoring system by GE Reuter-Stokes.

Standard Industrial Classification (SIC)   A U.S. government numbering system used to categorize industrial facilities.

Standard Methods   Common abbreviation for the reference book *Standard Methods for the Examination of Water and Wastewater*, widely used in water and wastewater testing and analysis, and published jointly by APHA, AWWA, and WEF.

standard plate count   See "heterotrophic plate count (HPC)."

standard seawater   A widely accepted "standard" total dissolved solids concentration of approximately 36,000 mg/L, considered to be typical of most seawaters.

standard solution   A solution whose strength or reacting value per unit volume is known.

standpipe   A vertical, cylindrical water storage tank of uniform diameter with a height greater than its diameter.

Stanley Compo-Cast   Former manufacturer of nonmetallic sludge collector wear shoes now made by Trusty Cook, Inc.

stapling   The entanglement of stringy or fibrous debris on a mesh or bar rack.

Star Filter®   Filter press by Star Systems, Inc.

STAR®   Anaerobic package plants by ADI Systems, Inc.

StarScreen   In-channel fine screen by Sernagiotto Technologies.

Sta-Sieve   Static fine screen by SWECO Engineering Corp.

Stata-Tube Mixer™   Motionless mixer by TAH Industries, Inc.

State Implementation Plan (SIP)   Plans to implement air quality standards required of each state under the Clean Air Act.

State Revolving Fund (SRF)   A U.S. EPA program awarding grants to states for use in community projects which comply with public health protection needs, with an emphasis on small drinking water systems.

static head   The vertical distance between a fluid's supply surface level and free discharge level.

Static Mixaerator™   Static mixing aerators by JDV Equipment Corp.

static mixer   Motionless device consisting of fixed baffles incorporated into channels or pipelines to create turbulence and mix additives added upstream.

static pile composting   A composting method where piles of municipal wastewater solids are aerated to eliminate the need for remixing.

static screen   Fine screen using a stationary, inclined screen deck that acts as a sieve to remove solids from liquids. Also known as a "rundown screen" or "sidehill screen" or "static screen."

static tube diffuser   A coarse bubble diffuser consisting of a vertically oriented cylinder with internal baffles to promote air/water mixing.

stationary source   A fixed-site producer of pollution, mainly power plants and other facilities using industrial combustion processes.

stator   The stationary member of an electric motor or generator.

Stato-Screen   Static screen by Vulcan Industries, Inc.

Stauffer Chemical   Former name of Rhone-Poulenc Basic Chemical Co.

steady state   An equilibrium condition that exists in a system.

Steady Stream   Turbidimeter by GLI International.

steam chest   The steam chamber adjacent to a heat exchanger tubesheet.

steam stripper   The process of removing volatile and semi-volatile contaminants from liquid where steam and liquid are passed countercurrently through a packed tower.

STEL   See "short term exposure limit (STEL)."

STEM   Scanning transmission-electron microscope.

Stengel Baffle   Inlet baffle for rectangular sludge collector by USFilter/Zimpro.

stenothermophiles   Bacteria that grow best at temperatures above 60°C.

step aeration   Variation of the activated sludge process where settled wastewater is introduced at several points in the aeration tank to equalize the F/M ratio.

Step Screen®   In-channel fine screen by Hans Huber GmbH.

Stepaire   Circular package step aeration wastewater treatment plant formerly offered by USFilter/Envirex.

sterile   Free from bacteria or other microorganisms.

sterilization   The destruction or removal of all living organisms within a system.

Ster-L-Ray™   Germicidal ultraviolet lamps by Atlantic Ultraviolet Corp.

SternPAC™   Polyaluminum coagulant by Sternson Ltd.

Stiff & Davis Index   Index used to determine the saturation point of calcium carbonate in seawater or other highly saline water.

STIG   Steam injected gas turbine.

still   Apparatus used in distillation.

stilling well   A tube or chamber used to dampen waves or surges in a large body of water usually used for purposes of water level measurement.

Sti-P³®   Steel tank standard for double wall underground tank with cathodic protection by Steel Tank Institute.

stock solution   A concentrated chemical solution, often used as a reagent.

stoichiometric   The ratio of chemical substances reacting in water that corresponds to their combining weights in the theoretical chemical reaction.

stoker   A mechanical device that feeds solid fuel to a furnace.

Stokes' Law   Law that defines the settling velocity of a particle based on its density and size.

stop log   A removable wooden, steel, or concrete bulkhead which fits into vertical grooves in a channel to stop water flow.

storage lagoon   A lagoon constructed with a sealed bottom used to store and collect solids.

Stord   Former name of Atlas-Stord, Inc.

Storm King™   Vortex-type separation system by H.I.L. Technology, Inc.

storm sewer   Collection system of underground piping used to remove water resulting from precipitation runoff.

Stormceptor®   Stormwater treatment device by Stormceptor Corp.

StormFilter™   Stormwater filtration and management system by Stormwater Management.

StormGate™   High flow bypass by Stormwater Management.

StormTreat™   Stormwater collection and treatment system by StormTreat Systems, Inc.

stormwater   Water resulting from precipitation runoff.

STP (1) Sewage treatment plant. Equivalent to "WWTP." (2) Standard temperature and pressure.

Straightline® Group of wastewater treatment products by USFilter/Envirex.

strainer A device that retains solids but allows liquids to flow through.

Strain-O-Matic Self-cleaning strainer by Hayward Industrial Products, Inc.

StrainPress® Screenings and trash removal compactor by Parkson Corp.

Stranco Former name of USFilter/Stranco.

Strantrol Chemical controller for disinfection systems by USFilter/Stranco.

Strata Clear Water flotation separator by Smith & Loveless, Inc.

StrataMix Fine bubble diffuser by Wilfley Weber, Inc.

Strata-Sand™ Continuously backwashed gravity sand filter by F.B. Leopold Company, Inc. (U.S.) and Simon-Hartley, Ltd. (U.K.).

Stratavap Evaporator system by Licon, Inc.

stratosphere The level of the atmosphere containing most of the earth's ozone that lies between the troposphere and the mesosphere.

stray current corrosion Electrochemical corrosion caused by stray currents leaking from an electrical installation.

Stream Saver™ Automated spill control system by ILC Dover, Inc.

streaming current The net ionic and colloidal surface charges of suspended solid particles in solution.

streaming current detector (SCD) Measuring device used to detect and monitor the net electrical charge of particles in solution after coagulants have been added.

Streamline Flow proportional sampler by American Sigma, Inc.

Streamline Rectangular chain and flight collector by Purestream, Inc.

streptococcus A genus of bacteria that includes some of the most common human pathogens.

stress corrosion cracking Formation of cracks caused by the action of a corrosive medium in combination with tensile stress.

stress relieving Heat treatment carried out to reduce internal stresses in steel.

Stress-Key Pre-cast concrete packaged sewage treatment plants by Marolf, Inc.

strip cropping Growing crops in a systematic arrangement of strips or bands that serve as barriers to wind and water erosion.

strip mining A method of mining where surface soil and strata are removed to gain access to the mineral deposits.

stripper Device used to remove volatile and semi-volatile contaminants from water.

Stripper® Multistaged diffused-bubble aeration system by Lowry Aeration Systems, Inc.

Stripperator Treatment unit for hydrocarbon-contaminated water by Ejector Systems, Inc.

strong acid An acid that approaches 100% ionization in dilute solutions.

strong acid cation exchanger A cation exchange resin with an exchange site capable of splitting neutral salts to form their corresponding free acids.

strong acid ion exchanger Ion exchange using a cationic resin that contains exchangeable functional groups derived from a strong acid.

strong base anion exchanger An anion exchange resin with an exchange site capable of splitting neutral salts to form their corresponding free bases.

strong base ion exchange   Ion exchange using an anionic resin that contains exchangeable functional groups.

strong base load factor z   The sum of alkalinity, sulfate, chloride, silica, and carbon dioxide expressed as calcium carbonate equivalents.

Stuart-Carter   Walking beam flocculator by JDV Equipment Corp.

SU   Standard unit.

subchronic   Of intermediate duration, usually used to describe studies or levels of exposure between 5 and 90 days.

subdrainage   The control and removal of excess groundwater, usually by means of pipe drains that intercept seepage and/or lower the water table.

subituminous coal   A grade of coal with a heat content higher than that of lignite but lower than bituminous.

sublimation   The process of changing a gas to a solid, or a solid to a gas, without going through the liquid phase.

submerged aquatic vegetation   Vegetation such as sea grasses that cannot withstand excessive drying and therefore live with their leaves at or below the water surface.

submerged MBR   An MBR configuration in which the membranes are immersed directly within the activated sludge reactor.

submerged tube evaporator   An evaporator where steam enters a tube bundle submerged in the fluid to be boiled.

submerged weir   A weir which when in use has a water level on the downstream side at an elevation equal to, or higher than, the weir crest. Also called a "drowned weir."

subnatant   Liquid remaining beneath the surface of floating solids.

Suboscreen®   In-channel rotary fine screen by Andritz-Ruthner, Inc. (Western Hemisphere) and USFilter/Contra-Shear.

Subrotor   Progressing cavity pump by Monoflo.

subsidence   The lowering of the natural land surface due to a reduction of the fluid pressure, removal of underlying supporting material, compaction due to wetting, or added loads on the land surface.

subsonic flow   Liquid flow at a speed that is less than the speed of sound in the fluid.

substrate   The organic matter or nutrients used as food substances during biological wastewater treatment.

subtropics   The regions bordering the tropical zone.

Sucoflow®   Aeration diffuser by Huber+Suhner AG.

suction head   The head between the centerline of a pump and the water level on the pump's suction side.

Sulf Control®   Sulfide inhibitor for prevention of hydrogen sulfide formation by NuTech Environmental Corp.

sulfamic acid   Acid often used as a cleaning agent. Chemical formula is HSO_3NH_2.

sulfate   The divalent, negative SO_4 anion or an ester containing the anion.

sulfate-reducing bacteria   Bacteria capable of reducing sulfate or other forms of oxidized sulfur to hydrogen sulfide gas.

SulfaTreat®-OC   Hydrogen sulfide removal process by SulfaTreat Co.

Sulfaver   Reagent chemical used to determine phosphates concentration in water by Hach Co.

Sulfex®   Sulfide precipitation process for heavy metal removal by USFilter/Industrial Wastewater Systems.

sulfide   The divalent, negative S anion or a salt containing the anion.

sulfonator   Device used to inject and meter sulfur dioxide to dechlorinate water.

sulfonic acid   An organic acid used to provide some cation exchange resin with its ion exchange capability. Chemical formula is SO_3H.

SulfOx™   Biological oxidant for elimination of odors and toxicity from wastewater and sludges by Sybron Chemicals, Inc.

Sulftech®   Reducing agent by General Chemical Corp.

sulfur   A flammable, nonmetallic element with many commercial uses. Chemical symbol is S.

sulfur oxides (SOX)   Air contaminants resulting from the combustion of fuels containing sulfur in the presence of oxygen.

sulfuric acid   A toxic corrosive acid capable of dissolving most metals. Chemical formula is H_2SO_4.

Sulfur-Rite™   A hydrogen sulfide and light mercapton scavenger system by USFilter/Gas Technologies.

sullage   Wastewater that drains from a home, farmyard, refuse heap, or street.

Sulzer   Screenings grinder by GL&V/Dorr-Oliver, Inc.

Sumigate®   Rubber dam for combined sewer overflow applications by Rodney Hunt Co.

sump   A pit or reservoir that serves to collect water or wastewater for subsequent removal from the system.

sump pump   A pump used to remove water or wastewater from a sump or wet well.

Sumpaire™   Self-aspirating jet aerator by ABS Pumps, Inc.

Sump-Gard®   Vertical centrifugal pump by Vanton Pump & Equipment Corp.

Suparator®   Oil/water separator by Lemacon Techniek B.V.

super austenitic stainless steel   Stainless steel alloyed with more than 4% molybdenum.

Super Blend™   Cellulose acetate reverse osmosis membranes by TriSep Corp.

Super Detox™   Encapsulation process to chemically stabilize furnace dust and other heavy metal residues by Conversion Systems, Inc.

Super Dome™   Ceramic dome diffuser by Ferro Corp.

Super Shredder®   In-line macerator by Franklin Miller, Inc.

Super Sieve Screen   Sieve screen by Sizetec, Inc.

Superblock II®   Filter underdrain with water and air/water backwash capabilities by F.B. Leopold Co., Inc.

Super-Cel®   Diatomaceous earth filter media by Celite Corp.

Supercell   Dissolved air flotation unit by Krofta Engineering Corp.

Superceptor®   Pump station for oily and greasy wastewater by Thermaco, Inc.

superchlorination   Chlorination with excess dosages which insures disinfection with short contact times and which produces free or combined residuals so large as to require dechlorination.

supercool To cool a substance in its liquid state at a temperature below its freezing point.

supercritical water oxidation (SCWO) High temperature/pressure wastewater treatment process where organic material is oxidized at temperatures above a fluid's critical point.

Super-D-Canter® Solid bowl decanter centrifuge by Alfa Laval Separation, Inc.

Superdraw Supernatant withdrawal unit by Walker Process Equipment.

Superfloc® High charge cationic flocculant by Cytec Industries, Inc.

Superfund A U.S. federal law authorizing identification and remediation of unsupervised hazardous waste sites. See "CERCLA."

Superfund Amendments and Reauthorization Act A 1986 U.S. law passed to reauthorize and expand CERCLA and require public disclosure of chemical release information and development of an emergency response plan.

Superfund Innovative Technology Evaluation (SITE) EPA-supported research, development, and demonstration projects designed to develop new remediation technologies.

Superfuser Fine bubble diffuser by USF/Envirex.

superheat The sensible heat in a gas above the amount needed to maintain the gas phase.

superheated steam Steam with additional heat added after vaporization, increasing its temperature and energy.

Supermal Pearlitic malleable iron chain material by Jeffrey Chain Corp.

supernatant Liquid above the settled sludge layer in a sedimentation basin.

Superpulsator® Solids contact clarifier utilizing inclined plates and intermittent pulsing to expand the sludge blanket by Infilco Degremont, Inc.

SuperSand™ Continuous sand filter by Waterlink Separations, Inc.

supersaturation A solution containing more of a dissolved substance than is usually possible at equilibrium.

SuperScraper™ Bottom sludge scraper by Waterlink Separations, Inc.

SuperSettler™ Inclined plate clarifier by Waterlink Separations, Inc.

SuperSkimmer™ Surface scum skimmer by Waterlink Separations, Inc.

Superslant™ Inclined plates for clarification by Filtronics, Inc.

Supersorbon Solvent recovery plant by Dedert Corp.

Superstill Former name of VaPure vapor compression still by Paul Mueller Co.

Superthickener Large diameter center pier gravity thickener by GL&V/Dorr-Oliver, Inc.

Super-Trasher Vertical sludge pump by Vanton Pump & Equipment Corp.

supplied air respirator (SAR) Respirator using an airline or tank to provide air to the facepiece.

support gravel Layers of graded gravel between underdrain openings and filter media to prevent media from leaking into the underdrain.

Supracell Dissolved air flotation unit by Krofta Engineering Corp.

Suprex™ Mixed bed-type condensate polisher by Graver Co.

surcharge (1) The height of wastewater in a sewer manhole above the crown of the sewer when the sewer is flowing completely full. (2) Loads on a system

beyond those normally anticipated. (3) An extra monetary charge imposed —
especially on flows into a wastewater collection system — when set quantity
or quality limits are exceeded.

SURF® Two-stage contact clarifier/filter by USFilter/General Filter.

surface aerator Mechanical aeration device consisting of a partially submerged
impeller attached to a motor and mounted on floats or a fixed structure.

surface condenser A condenser, usually of the shell-and-tube design, that provides a suitable heat transfer surface area for condensing to occur where
cooling water and process fluid remain separated.

surface impoundment A natural topographic depression, man-made excavation,
or diked area formed primarily of earthen materials to hold an accumulation
of liquid wastes or waste-containing free liquids.

surface loading rate A criteria used for design of sedimentation tanks expressed
as flow per day per unit of basin surface area. Also called "overflow rate."

surface mining See "strip mining."

Surface Mining Control and Reclamation Act (SMCRA) A 1977 U.S. law that
set performance standards for environmental protection to be met at most
surface mining operations for coal.

surface runoff Precipitation, snow melt, or irrigation in excess of what can infil-
trate the soil surface and be stored in small surface depressions.

Surface Scatter On-line turbidimeter by Hach Co.

surface tension The force acting on a liquid surface that results in a minimum
liquid surface area. Produced by the unbalanced inward pull exerted on the
layer of surface molecules by molecules below the liquid surface.

surface wash An auxiliary high pressure water spray system used to agitate and
wash the surface of granular media filters.

surface water Water from sources open to the atmosphere including lakes, reservoirs, rivers, and streams.

Surface Water Treatment Rule A 1989 EPA regulation that set maximum contaminant level goals for constituents including *Giardia lamblia* and viruses
for public water systems using surface water sources or groundwater under
the direct influence of surface water.

surface-active agent See "surfactant."

Surfact Process to upgrade activated sludge system with air driven rotating biological contactors by USFilter/Envirex.

surfactant A surface-active agent such as a detergent which, when mixed with
water, generally increases its cleaning ability, solubility, and penetration, while
reducing its surface tension.

Surfaer Slow speed surface aerator by USFilter/Aerator Products.

Surfpac™ Vertical-type PVC trickling filter media by the former American Surfpac Corp.

surge-flow irrigation A surface irrigation technique that involves intermittent
application of irrigation water to a field for 50% of an irrigation cycle and
diversion to another area for the remainder of the cycle.

suspect material Building material suspected of containing asbestos.

suspended growth process Biological wastewater treatment process where the microbes and substrate are maintained in suspension within the liquid.

suspended loads Sediment particles maintained in the water column by turbulence and carried with the flow of water.

suspended solids (SS) Solids captured by filtration through a glass wool mat or 0.45 micron filter membrane.

suspension A system in which very small particles are uniformly dispersed in a liquid or gaseous medium.

suspensoid Colloidal dispersion of solids in liquid.

sustainable development Growth that does not result in ecological damage.

Sutorbilt® Air blower by Gardner Denver Blower Division.

SUVA See "specific ultraviolet absorbance (SUVA)."

SVE See "soil vapor extraction (SVE)."

SVI See "sludge volume index (SVI)."

SVOC Semi-volatile organic compound.

SVP-Pure™ Single vessel process for chlorine dioxide technology by EKA Chemicals.

SW846 Test methods for evaluating solid waste.

swamp A type of wetland dominated by woody vegetation but without appreciable peat deposits. See also "wetlands."

swamp gas See "marsh gas."

Swan® Analytical instruments by Industrial Analytics, Corp.

SWD See "side water depth (SWD)."

SWDA Solid Waste Disposal Act.

sweep flocculation Coagulation/flocculation process using relatively large amounts of iron or metal salts to form voluminous floc particles used to trap smaller particles.

sweet environment Environment containing no, or negligible amounts of, hydrogen sulfide.

sweet water Brackish water that may be used for drinking even though it may not meet potable water standards.

sweeten To remove sulfur contaminants from petroleum products.

swelling Condition that exists as a result of intrusion of water into a particle.

Swing-Flex™ Check valve by Val-Matic Valve & Manufacturing Corp.

Swingfuser® Removable aeration header and drop pipe assembly by USFilter/Diffused Air Products Group.

Swingtherm® Regenerative catalytic oxidizer by Kvaerner Chemetics.

SwingUp Removable aeration header and drop pipe assembly by USFilter/Aerator Products.

Swingwirl Vortex flow meter by Endress+Hauser.

Swirl-Flo™ Solids/liquid separation process by H.I.L. Technology, Inc.

SwirlMix Complete mix package wastewater treatment plant by Walker Process Equipment.

Swiss Combi System Sludge drying and pelletization system by Wheelabrator Water Technologies, Inc.

SWRO Seawater reverse osmosis.

SWTR   See "Surface Water Treatment Rule."

SybrSorb™   Oil and oleophiphilic compound absorbent by Sybron Chemicals, Biochemical Division.

SymBio™   Wastewater nitrification/denitrification treatment process by Enviroquip, Inc.

symbiotic   A relationship between organisms of different species that benefits both members of the relationship such that neither could carry out certain activities alone.

syndet   Commonly used contraction for "synthetic detergent."

synergy   The combined action of two agents that results in a reaction greater than the sum of the individual agents acting alone.

synfuels   Liquid or gaseous fuels produced from coal, lignite, or other solid carbon sources.

synthetic detergent   Cleaning agents such as linear alkyl sulfonate that react with water hardness but whose products are soluble.

synthetic organic chemicals (SOC)   Manmade organic chemicals, some of which are volatile while others tend to stay dissolved in water instead of evaporating.

Syphonid   Recovery device for sinker contaminates by Science Application International Corp.

System-3   Oily water treatment system by Megator Corp.

systemic toxicity   Adverse effects caused by a substance that affects a body in a general rather than local manner.

T

T&O   Taste and odor.

T-2000®   Dry sorbent for use in baghouses by Solvay America.

tailings   Residue from the separation of useful values from ore.

tailpipe standards   Emissions limitations applicable to engine exhausts from mobile sources.

tailwater   The runoff of irrigation water from the lower end of an irrigated field.

Tait-Andritz   Former name of Andritz-Ruthner, Inc.

TAK®   Ultraviolet systems by PCI-Wedeco Environmental Technologies, Inc.

TAMS   Toxic air monitoring system.

tangential screen   See "static screen."

tank blanketing   See "gas blanket."

Tankleenor   Petroleum tank cleaner by Gorman-Rupp Co.

tannin   Colored compounds that form when plant matter degrades in water.

tapered aeration   Variation of the activated sludge process where the amount of air supplied in an aeration basin is tapered to match the demand exerted by the microbes.

tapered flocculation   A flocculation process utilizing multiple compartments and a gradually increasing velocity gradient.

tapeworm A parasitic flatworm that can live in the digestive tract or liver of vertebrates.

TAPPI Technical Association of the Pulp and Paper Industry.

tardigrade Minute water animals with segmented bodies and four pairs of unsegmented legs.

tare The empty weight of a vessel or container determined by deducting the weight of the contents from the total weight of the full load.

Targa Vapor compression water desalination unit by Mechanical Equipment Co., Inc.

Taskmaster® Screenings grinder by Franklin Miller, Inc.

taste and odors Two important characteristics of drinking water which are the targets of minimization or elimination in a water treatment plant.

Taulman/Weiss In-vessel composting system by USFilter/Davis Process.

taxis The involuntary movement of a free-moving cell or organism toward or away from an external stimulus such as light.

TBF See "traveling bridge filter (TBF)."

TBT See "top brine temperature."

TC See "total carbon (TC)."

TCA Trichloroethane.

TCD Thermo-compression distillation.

TCDD Tetrachlorodibenzoparadioxin. A dioxin byproduct.

TCDF Tetrachlorodibenzofuran.

TCE See "trichloroethylene (TCE)."

TCF Totally chlorine free.

TCF® Horizontal tube cartridge filter by Ropur AG.

TCLP See "toxicity characteristic leaching procedure (TCLP)."

TCR See "Total Coliform Rule (TCR)."

TCRI Toxic Chemical Release Inventory.

TD Toxic dose.

TDH See "total dynamic head (TDH)."

T-Disc™ Disc-type oil skimmer by Vikoma International Ltd.

TDS See "total dissolved solids (TDS)."

Teacup™ Grit removal system by Eutek Systems, Inc.

TEC Tonka Equipment Co.

Technasand™ Continuous backwashing sand filter by WesTech Engineering Inc.

technology-based limitations Industry-specific effluent limitations applied to a discharge when it will not cause a violation of water quality standards at low stream flows.

TechXtract™ Chemical decontamination process by EET, Inc.

TecTank Bolted steel storage tank by A.O. Smith Engineered Storage Products.

tectonic Relating to changes in the structure and forces produced in the earth's crust.

Tecweigh® Volumetric feeder by Tecnetics Industries, Inc.

TEFC See "totally enclosed fan cooled (TEFC)."

TEG Tetraethylene glycol.

Tekkem   Lime slaker by RDP Technologies, Inc.

Tekleen™   Filter screen by Automatic Filters, Inc.

TEL   See "tetraethyl lead (TEL)."

telemetry   The process of transmitting measured data by radio to a distant station.

TeleTote™   Open channel electromagnetic flow meter by Marsh-McBirney, Inc.

TEM   See "transmission electron microscope (TEM)."

TEMA   Tubular Equipment Manufacturers Association.

temperate zone   The middle latitude regions of the earth, located between latitude 23°26′ and 66°33′ north or south, which have moderate climates and distinct summer and winter seasons.

temperature inversion   See "inversion."

temporary hardness   Hardness associated with bicarbonates of calcium and magnesium which precipitate upon boiling.

temporary threshold shift (TTS)   A temporary reduction in hearing ability resulting from noise overexposure.

Ten States Standards   Common name for "Recommended Standards for Wastewater Facilities," a report of the Wastewater Committee of the Great Lakes-Upper Mississippi River Board of State Public Health and Environmental Engineers.

Tenkay®   Cartridge air filter by Farr Co.

tensile strength   The maximum tensile load per square unit of cross section that a material is able to withstand.

Tenten   Continuously backwashed gravity sand filter by F.B. Leopold Company, Inc. (U.S.) and Simon-Hartley, Ltd. (U.K.).

TEOM®   Ambient air particulate sampler by Rupprecht & Patashnick Co., Inc.

teratogenic   A chemical or agent with properties which cause disfiguring by affecting the genetic characteristics of an organism.

terminal headloss   The headloss at the end of a filter run cycle signifying that the filter bed is filled with solids.

terminal settling velocity   The maximum sedimentation rate of an unhindered suspended particle.

terracing   Dikes built along the contour of sloping farm land which hold runoff and sediment to reduce erosion.

Terra-Gator®   Pressurized waste sludge injection system by Ag-Chem Equipment Co.

TerraStor™   Temporary containment system by ModuTank, Inc.

tertiary effluent   The effluent discharged from a tertiary treatment process.

tertiary filtration   The use of a granular media filter to improve secondary wastewater effluent quality.

tertiary treatment   The use of physical, chemical, or biological means to improve secondary wastewater effluent quality.

TES Filter   Package, dual media gravity filter by USFilter/Davco.

tetrachloroethylene (PCE)   See "perchloroethylene (PCE, also PERC)."

tetraethyl lead (TEL)   An antiknock additive used in gasoline. The U.S. phased out its use in 1989 due to concerns over lead emissions.

***Tetratex*®**   Microporous PTFE membrane for filtration by Tetratec.

tetratogenesis   The introduction of nonhereditary birth defects in a developing fetus by exogenous factors such as physical or chemical agents acting in the womb to interfere with normal embryonic development.

Texas Star   Circular membrane diffuser by Aeration Research Co.

TF/AS   See "trickling filter-activated sludge (TF/AS)."

TF/SC   Trickling filter-solids contact.

***TFC*®**   Reverse osmosis membrane by Koch Membrane Systems, Inc.

***TFCL*®**   Thin film composite reverse osmosis membrane by Koch Membrane Systems, Inc.

***TFCS*™**   Reverse osmosis elements by Koch Membrane Systems, Inc.

***TFM*®**   Reverse osmosis membrane by Osmonics Desal.

TGNMO   Total gaseous nonmethane organics.

thalweg   The line of maximum depth in a river or stream.

THC   Total hydrocarbons.

theoretical oxygen demand (ThOD)   Determination of organic matter in a water or wastewater calculated on the basis of the chemical formula of the constituents.

therapeutic index   The ratio of the dose required to produce toxic or lethal effects to dose required to produce nonadverse or therapeutic response.

***ThermaGrid*™**   Regenerative thermal oxidizer by McGill Airclean Corp.

thermal oxidizer   An emissions control device that utilizes heat to accomplish VOC oxidation.

thermal pollution   The discharge of heated effluent into the environment resulting in an undesired increase in ambient temperature, above that resulting from natural solar radiation.

thermal treatment   Processes such as incineration or pyrolysis which use elevated temperatures to treat hazardous waste.

thermal value   See "heat value."

***ThermoBlender*™**   Sludge and quicklime blending unit by RDP Technologies, Inc.

thermocline   The middle layer in a stratified lake that results from varying water densities.

thermocompressor   A steam ejector that uses high pressure steam to increase the pressure of a lower pressure steam.

***ThermoFeeder*™**   Chemical feed unit by RDP Technologies, Inc.

ThermoFlo   Spray coolers by Aqua-Aerobic Systems, Inc.

thermophiles   Bacteria that grow best at temperatures between 45° and 60°C.

thermophilic digestion   Sludge digestion within a thermophilic range of approximately 45° to 60°C.

***Thermo-Sludge Dewatering*™**   Float-type sludge thickening process by Dontech, Inc.

***Thermox*®**   Flue gas analyzer by Ametek Inc., Process & Analytical Division.

thickener   A tank, vessel, or apparatus used to reduce the proportion of water in a slurry or sludge.

thickening   A procedure used to increase the solids content of sludge by removing a portion of the liquid.

***ThickTech*™**   Rotary drum thickener by Waterlink Separations, Inc.

thin film evaporator   Evaporator where liquid flows or is sprayed over heat transfer surfaces, usually tubes, in a thin, turbulent film.

Thioguard®   Odor and corrosion control product by Premier Chemicals.

thiols   See "mercaptans."

Thio-Red   Dissolved metal precipitant by Etus, Inc.

thixotrope   A colloid whose properties are affected by mechanical treatment.

thixotropy   The time-dependent property of some emulsions and sludges to change rheological and physical characteristics when left at rest.

THM   See "trihalomethane (THM)."

THMFP   See "trihalomethane formation potential (THMFP)."

ThOD   See "theoretical oxygen demand (ThOD)."

Thomas Conveyor   Former name of USFilter/Thomas Conveyor

3DP™   Three-belt dewatering filter by Baker Process.

threshold dose   The minimum dose of a substance to produce an effect.

threshold level   Time-weighted average pollutant concentration values, exposure beyond which is likely to adversely affect human health.

threshold limit value (TLV)   The maximal allowable workplace air level for a chemical.

threshold odor number (TON)   The number of dilutions of odor-free water required to eliminate the odor in a water sample.

Thru Clean   Back cleaned mechanical bar screen by USFilter/Headworks Products.

TIC   See "total inorganic carbon (TIC)."

tidal marsh   Low, flat marshlands traversed by channels and tidal hollows and subject to tidal inundation. Normally the only vegetation present is salt tolerant bushes and grasses.

tide   The periodic rising and falling of the sea that results from the gravitational attraction of the moon.

tide gate   A swinging gate in a sewer pipe that prevents seawater from entering the system during high tides.

Tideflex™   Air diffuser and flexible check valve by Red Valve Co., Inc.

TIE   Toxicity identification evaluation.

Tile   Ceramic tile filter underdrain by Roberts Filter Group.

tile field   A system used for subsurface discharge of treated wastewater effluent using open-jointed tile placed on gravel fill.

tilted plate separator   Oil separation device utilizing inclined plates to separate free nonemulsified oil and water based on their density difference.

time-lag   The time interval between the beginning of an event and the response that occurs as a result of the event.

time-weighted average (TWA)   In air sampling, the average air concentration of contaminants during a given period.

TIN   Total inorganic nitrogen.

tinajero   A home water filtration device where water is collected in a clay pot after filtration through a porous stone.

tines   The teeth or prongs of a bar screen cleaning rake.

tipping   The dumping of the contents of a waste truck at a waste disposal facility.

tipping fee   The fee charged to dispose of solid waste at a sanitary landfill.

Tipping Scum Weir   Pivoting scum weir by F.B. Leopold Co., Inc.

TISE   Take it somewhere else. See also "NIMBY."

***Titan ASD 200*™**   Reverse osmosis antiscalant/dispersant by Professional Water Technologies, Inc.

***Titan-90*™**   Multi-media automatic backwashing system by Serfilco, Ltd.

titer   The concentration of a substance in solution as determined by titration.

***Titeseal*®**   Copolymer fluid control gate by Plasti-Fab, Inc.

titration   A method of determining the concentration of a dissolved substance in terms of the smallest amount of a reagent required to bring about a given effect in reaction with a known volume of test solution.

Titraver   Chemicals used in analysis of water hardness by Hach Co.

TKN   See "total Kjeldahl nitrogen (TKN)."

***TLC*™**   Thin layer composite reverse osmosis membrane by Osmonics, Inc.

TLV   See "threshold limit value (TLV)."

TLV-C   TLV-Ceiling.

TLV-STEL   TLV-Short term exposure limit.

TLV-TWA   TLV-time weighted average.

TMDL   Total maximum daily load. An estimate of the capacity of a specific waterbody to assimilate pollution and still achieve designated uses.

TMP   See "transmembrane pressure (TMP)."

TMRC   Theoretical maximum residue contribution.

TMU   Multiple-tank unit cars used to carry chlorine ton cylinders.

***TMX*®**   Thermatrix, Inc.

TNCWS   See "transient, noncommunity water system (TNCWS)."

TNT   Trinitrotoluene.

TOA   Trace organic analysis.

TOC   See "total organic carbon (TOC)."

TOD   See "total oxygen demand (TOD)."

toilet   A small room with a bowl-shaped fixture for urination and defecation. Also can refer specifically to the fixture.

Tolhurst   Laboratory basket centrifuge by Baker Process/Ketema.

ton   A unit of weight equal to 2000 pounds or 907.2 kilograms. Also called "short ton."

TON   See "threshold odor number (TON)."

ton container   A 1-ton chlorine storage container.

TonkaFlo   Pumps for RO/UF/DI applications by Osmonics, Inc.

tonnage   The amount of waste that a landfill accepts, usually expressed in tons per month.

tonne   See "metric ton."

Tonozone   Ozone generator by Praxair-Trailigaz Ozone Co.

top brine temperature   The maximum temperature of the fluid being evaporated in an evaporator system.

topography   The physical features of a surface area including relative elevations and the position of natural and manmade features.

topping cycle   Cogeneration system where electricity is produced by the combustion process, and then byproduct heat is scavenged for thermal processes.

Tornado®   Aspirating type surface aerator by Aeromix Systems, Inc.

Torpedo™   Clarifier foam and grease removal unit by United Industries, Inc.

Torpedo Filter   Floating microfilter by BTG, Inc.

Torpedo Pump   Reverse osmosis pump by Pumps Unlimited.

Torque-Flow   Grit handling vortex slurry pump by Envirotech Pumpsystems.

TorusDisc   Sludge dryer/cooler by Hosokawa Bepex Corp.

Torvex®   Catalytic oxidation system by Süd-Chemie Prototech Inc.

total carbon (TC)   A quantitative measure of both total inorganic and total organic carbon in water and wastewater, determined instrumentally by chemical oxidation to carbon dioxide and subsequent infrared detection in a carbon analyzer.

Total Coliform Rule (TCR)   An EPA rule regulating known pathogens in drinking water.

total dissolved solids (TDS)   The weight per unit volume of all volatile and nonvolatile solids dissolved in a water or wastewater after a sample has been filtered to remove colloidal and suspended solids.

total dynamic head (TDH)   The total energy a pump must impart to water to move it from one point to another, measured as the difference in height between the free water surface level on the discharge and suction sides of a pump.

total inorganic carbon (TIC)   The sum of all inorganic carbon species in a water or wastewater.

total Kjeldahl nitrogen (TKN)   The sum of the organic plus ammonia nitrogen in a water sample which is determined by digesting and distilling the sample and then measuring the ammonia concentration in the distillate.

total organic carbon (TOC)   Measurement of organic matter in a water or wastewater that can be oxidized in a high temperature furnace.

total oxygen demand (TOD)   Measurement of organic matter in a water or wastewater that can be converted to stable end products in a platinum-catalyzed combustion chamber.

total solids (TS)   The sum of dissolved and suspended solids in a water or wastewater. Matter remaining as residue upon evaporation at 103 to 105°C.

total suspended particulates (TSP)   The concentration of all airborne particulate matter usually expressed as micrograms of particulate per cubic meter of sampled air.

total suspended solids (TSS)   The measure of particulate matter suspended in a sample of water or wastewater. After filtering a sample of a known volume, the filter is dried and weighed to determine the residue retained.

total toxic organics (TTO)   The sum of the concentrations of all toxic organic compounds found in a sample.

total trihalomethane (TTHM)   The sum of the concentration of trihalomethane compounds rounded to two significant figures.

totally enclosed fan cooled (TEFC)   Designation for motor enclosure that is not airtight but doesn't allow free exchange of air between the inside and outside of the motor case. Exterior cooling is provided by an integral external fan.

totally enclosed nonventilated (TENV)   Designation for motor enclosure that is not airtight, but constructed so as to prevent free exchange of air between the inside and outside of the motor case.

TotalSep   Oil/water separator by Hydro-Flo Technologies, Inc.

TotalTreat   Package wastewater treatment system by USFilter/Industrial Wastewater Systems.

TotlSep   Oil/water separator by Hydro Flo Technologies.

Toveko®   Continuously operating sand filter by USFilter/Krüger.

Tow-Bro®   Suction-type sludge removal system for circular clarifiers by USFilter/Envirex.

Tower Press   Belt filter press by Roediger Pittsburgh, Inc.

Towerbrom®   Bromine-based biocide by Calgon Corp.

Towermaster   Pressurized sand filter for cooling tower applications by Axsia Serck Baker, Inc.

TOX   (1) Total organic halogen. (2) Tetradichloroxylene.

ToxAlarm™   On-line toxicity monitor by Anatel Corp.

TOXFP   Total organic halogen formation potential.

toxic   Capable of causing an adverse effect on biological tissue following physical contact or absorption.

toxic chemical   Any chemical listed in the EPA rules as "Toxic Chemicals Subject to Section 313 of the Emergency Planning and Community Right-to-Know Act of 1986."

toxic cloud   Airborne plume of gases, vapors, fumes, or aerosols containing toxic materials.

Toxic Release Inventory (TRI)   Database of toxic releases in the U.S. compiled from SARA Title III section 313 reports.

toxic substance   Any chemical or material that may present an unreasonable risk of injury to one's health or the environment.

Toxic Substances Control Act (TSCA)   A 1976 U.S. law authorizing the EPA to collect information on chemical risks.

toxic waste   A waste that can produce injury upon contact with or by accumulation in or on the body of a living organism.

toxicant   A substance that is toxic to another organism.

toxicity   The property of being poisonous or causing an adverse effect on a living organism.

toxicity characteristic leaching procedure (TCLP)   Method used to determine the amount of a hazardous substance which will leach from a solid when the solid is subjected to water.

toxicology   The study of adverse effects of chemical on living organisms.

Toxilog   Portable single gas detector by Biosystems, Inc.

toxin   A poisonous material that can cause damage to biological tissue following physical contact or absorption.

TPAD   Temperature-phased anaerobic digestion, which combines thermophilic and mesophilic anaerobic digestion.

TPC®   Potassium permanganate by Carus Chemical Co., Inc.

TPH   Total petroleum hydrocarbons.

TPY   Tons per year.

TQM   Total Quality Management.

Trac Pump   Sludge-dredging system by E.S.G. Manufacturing, LLC.

trace elements   (1) Any elements in a water or wastewater that are present in very low concentrations. (2) Elements present in minor amounts in the earth's crust.

trace organics   Organic matter present in water supplies in very low concentrations which originates from natural sources and the synthetic chemical industry.

Tractor Drive   Circular clarifier drive unit by Walker Process Equipment.

Trac-Vac®   Suction sludge removal mechanism by Baker Process.

TracWare™   Particle counting system by Chemtrac Systems, Inc.

TrailigazConcept™   Ozone generator by Praxair-Trailigaz Ozone Co.

transboundary pollutants   Air pollution that travels from one jurisdiction to another, often crossing state or international boundaries.

transducer   A device that receives energy from one system and retransmits it, often in another form, to another system.

Trans-Flo™   Rectangular tank secondary clarifier by USFilter/Envirex.

transient, noncommunity water system (TNCWS)   A noncommunity water system that does not serve 25 of the same nonresidents per day for more than 6 months per year.

transient water system (TWS)   A public water system that serves a nonresident population.

Transmax®   Medium/coarse bubble air diffuser by Enviroquip, Inc.

transmembrane pressure (TMP)   The average pressure across a membrane, measured as the hydraulic pressure differential from the feed side to the permeate side.

transmission electron microscope (TEM)   A microscope with a magnification range of 220X to 1,000,000X at a resolution of 2 Å, where illumination is provided by a beam of electrons which pass through the electron transparent specimen.

transmission lines   Pipelines that transport raw water from its source to a water treatment, and then to the distribution grid system.

transmissivity   The rate that water flows though an aquifer under a hydraulic gradient.

transmutation   Changing one element into another by changing the number of protons in the atom nucleus.

Trans-Pak®   Solid waste compacting/baling unit by Harris Waste Management Group, Inc.

transpiration   The loss of water from the leaves and stems of plants.

transpiration ratio   The ratio of the weight of water lost to atmosphere through transpiration and the weight of dry plant material produced.

transuranic wastes   Radioactive wastes containing isotopes above uranium in the periodic table which are byproducts of fuel assembly, weapons fabrication, and reprocessing.

Transvap   Mobile wastewater reduction/recovery system by Licon, Inc.

Trasar®   Control and diagnostic technology by Nalco Chemical Co.

trash   Combustible waste including up to 10% plastic or rubber scraps from commercial and industrial sources.

Trash Hog®   Self-priming, solids handling pumps by ITT A-C Pump.

trash rack   A coarse screening device using a parallel set of stationary bars typically spaced at 38 mm (1.5 inch) to 150 mm (6 inches).

trash rake   A mechanical screening device used to remove rough or debris or trash from a trash rack.

trash-to-energy   Burning trash to produce energy.

Travalift™   Traversing sludge collecting and pumping mechanism formerly offered by USFilter/Envirex.

Travcyl™   Precision metering pump by EncyNova International, Inc.

traveling band screen   See "traveling water screen (TWS)."

traveling bridge clarifier   A rectangular clarifier with the sludge removal mechanism supported by a traversing bridge.

traveling bridge filter (TBF)   A granular media filter with multiple compartments which can be individually cleaned by a moveable, bridge-mounted backwashing device without taking the entire filter out of service.

traveling screen   See "traveling water screen (TWS)."

traveling water screen (TWS)   Automatically cleaned screening devices employing chain-mounted wire mesh panels to remove floating or suspended solids from a channel of water.

Traypak™   Deaerator trays by Graver Co.

TRC   Total residual chlorine.

TRE   Toxicity reduction evaluation.

treatability study   A study in which a waste is subjected to a treatment process to determine whether it is amenable to treatment and/or to determine the treatment efficiency or optimal process conditions for treatment.

treated wastewater   Wastewater that has been subjected to one or more physical, chemical, and biological processes to reduce its pollution of health hazards.

treatment, storage, and disposal (TSD)   A description of a facility where hazardous waste is treated, stored, and/or disposed.

Trebler   Automatic sampler by Lakeside Equipment Corp.

Trey Deaerator™   Three-stage deaerator by USFilter/Rockford.

TRI   See "Toxic Release Inventory (TRI)."

trial burn   An incinerator test to demonstrate compliance of the unit with RCRA operating standards.

Triangle Brand®   Copper sulfate product by Phelps Dodge Refining Co.

Triboflow   Continuous particulate emissions monitor by Auburn Systems LLC.

Tricanter®   Centrifuge by Krauss Maffei Corp.

Tricellorator   Three compartment dissolved air flotation unit by Pollution Control Systems, Inc.

Tricep™   Granular resin filtration process for the removal of iron and copper oxides with oil by Graver Co.

trichloramine   A compound formed by the chlorination of water containing ammonia which may be the source of taste and odors.

trichloroethylene (TCE)   A chlorinated hydrocarbon used as an industrial cleaner or solvent which may cause organ damage or tumors in humans.

trickle irrigation   Method in which water drips to the soil from perforated tubes or emitters.

trickling filter   An aerobic, fixed film wastewater treatment process where organic matter present in a wastewater is degraded as it is distributed over a biological filter bed.

trickling filter-activated sludge (TF/AS)   Process combining trickling filter wastewater treatment followed by an activated sludge process to satisfy some unique treatment requirement.

Tricon   Concrete water treatment plant with buoyant media flocculator/clarifier by USFilter/Microfloc.

Tridair   Dissolved air flotation unit by Engineering Specialties, Inc.

Trident®   Modular water treatment plant with buoyant media flocculator/clarifier by USFilter/CPC.

trihalomethane (THM)   Disinfectant byproducts formed when chlorine reacts with organic compounds in water. These halogenated organics are named as derivatives of methane and include suspected carcinogens.

trihalomethane formation potential (THMFP)   An indirect measure of trihalomethanes determined by a laboratory test procedure that measures the amount of trihalomethane precursors in a sample.

Trimite™   Package water treatment plant with buoyant media flocculator/clarifier USFilter/Microfloc.

Tri-NOx®   NOx removal process by Tri-Mer Corp.

Triogen   UV disinfection systems by Ozonia North America.

Tri-Packs®   Plastic packing media by Jaeger Products, Inc.

triple point   The condition at which the solid, liquid, and gaseous states of a substance coexist simultaneously in equilibrium.

Triplex™   Three-stage air scrubber by USFilter/Davis Process.

Trisep™   Oil/solids/water separation filters by Graver Co.

Triton®   Surface mounted aerator/mixer by Aeration Industries, Inc.

Triton®   Wedgewire screen lateral underdrain system USFilter/Microfloc.

Tritor   Front cleaned bar screen and grit removal device by USFilter/Headworks Products.

Triturator   Screenings grinder by USFilter/Envirex.

TriZone™   Packaged water treatment/ozonation system by USFilter/Microfloc.

Troll   Submersible water and temperature level probe by In-Situ, Inc.

Tromax™   Trommel screen by Norkot Mfg. Co.

tromeling   The removal of small diameter solids using a trommel screen.

trommel screen   A cylindrical, rotating screen used to separate solid waste material according to size and density.

trophic level   Any of the distinct feeding levels in a food chain.

tropical   Term used to describe the region, climate, or vegetation characteristic of the tropical zone.

tropical rain forest   The densely forested equatorial regions of the world that experience very high annual rainfall.

tropical zone   The region of the earth located between the Tropic of Cancer at latitude 23°26′ north and the Tropic of Capricorn at latitude 23°26′ south, which has a warm/hot year-round climate.

troposphere   The lowest level of the atmosphere that extends to a height of between 9 and 16 km above the earth's surface.

***TroubleShooter*™**   Portable containment unit for hazardous waste spills by ThermaFab, Inc.

TRPH   Total recoverable petroleum hydrocarbons.

true color   The color in water caused by the presence of humic or fulvic acids which result from the decomposition of organic matter.

Tru-Grit   Grit washing and separation system by Waterlink Separations, Inc.

Tru-Gritter   Grit removal system by Vulcan Industries, Inc.

trunnion   A pivot or pin mounted on bearings used to rotate or tilt something.

Tru-Test   Automatic liquid sampler formerly offered by USFilter/Headworks Products.

TS   See "total solids (TS)."

TSCA   See "Toxic Substances Control Act (TSCA)."

TSD   See "treatment, storage, and disposal (TSD)."

TSDF   Treatment, storage, and disposal facility.

TSE   Treated sewage effluent.

TSP   See "total suspended particulates (TSP)."

TSS   See "total suspended solids (TSS)."

TSV   Telescoping sludge valve.

TTHM   See "total trihalomethane (TTHM)."

TTHM>0   Instantaneous total trihalomethane concentration.

TTHMFP   Total trihalomethane formation potential.

TTO   See "total toxic organics (TTO)."

TTS   See "temporary threshold shift (TTS)."

TU   Turbidity unit. See "nephelometric turbidity unit (NTU)."

Tub Scrubber   Single stage dry chemical air scrubber by Purafil, Inc.

tube settlers   A series of parallel inclined tubes used to increase the settling efficiency of sedimentation basins.

tubercles   Knob-like mounds of corrosion on pipe surfaces.

tuberculation   Development or formation of small mounds of corrosion products on the inside of iron pipe which roughen the pipe and increase its resistance to water flow.

tubesheet   Flat plate used to secure the ends of tubes in an evaporator, heat exchanger, or boiler.

TUc   Chronic toxicity unit.

Tuff-Span   Modular tank covers by Enduro Composite Systems.

***Tulsion*®**   High temperature ion exchange resin by Thermax, Ltd.

tundra   An ecosystem dominated by lichens, mosses, grasses, and woody plants found at high latitudes (arctic tundra) and high altitudes (alpine tundra). Arctic tundra is underlain by permafrost and is usually saturated.

***Tunnel Reactor*®**   In-vessel composting system by the former Waste Solutions.

turbid   Cloudy or opaque condition of water caused by unsettled particles or sediment.

turbidimeter   Instrument used to measure water turbidity by detecting the intensity of light scattered at angles from a beam of light projected through a water sample.

turbidity   A qualitative measurement of water clarity resulting from suspended matter that scatters or otherwise interferes with the passage of light through the water.

Turbo®   Mechanical surface aerator by Aeration Industries, Inc.

Turbo™   Reverse osmosis booster pump by Pump Engineering, Inc.

TurboBlade™   Low speed mixer with variable pitch impeller blades by Baker Process.

TurboClean™   Spiral wound membrane element by TriSep Corp.

Turbodrain   Gravity belt thickener by Gebr. Bellmer GmbH.

TURBO-Dryer®   Sludge dryer by Wyssmont Co., Inc.

Turbofill™   Random packed air stripper media by Diversified Remediation Controls, Inc.

TurboFlow™   Soil remediation exhausters and blowers by Invincible AirFlow Systems.

Turbo-Scour   Continuously cleaned package sand filters by Smith & Loveless, Inc.

Turboshredder   Cutter assembly for grinder pump by Homa Pump Technology.

TurboStripper™   Air stripping technology by Diversified Remediation Controls, Inc.

Turbotron®   Rotary lobe blower by Gardner Denver Blower Division.

Turbo-Vac®   Central vacuum system for dust and airborne particulates by PBR Industries.

Turbozone®   Catalytic odor oxidizer by RGF O3 Systems, Inc.

Turbulator   Rapid mixing unit by Walker Process Equipment.

turbulence   (1) The fluid property characterized by irregular variation in the speed and direction of movement of individual particles or elements of the flow. (2) A state of flow of water in which the water is agitated by cross currents and eddies, as opposed to laminar, streamlined, or laminar flow.

turbulent flow   A flow situation in which the fluid moves in a random manner with a Reynolds number usually greater than 4000.

turgid   Swollen or distended by a buildup of fluid.

turndown   A ratio expressing the maximum-to-minimum capacity of a process or device.

turnover   Seasonal (spring and fall) change that occurs in a lake's thermal gradients, resulting in the circulation of biological and chemical materials.

TVA   Tennessee Valley Authority.

TVC   Thermal vapor compression. See "vapor compression evaporation (VC)."

TVR   Thermal vapor recompression. See "vapor compression evaporation (VC)."

TWA   See "time-weighted average."

Twister   Fine bubble diffuser by MixAir Technologies, Inc.

Twister   Low speed surface aerator by Aeromix Systems, Inc.

TWL   Top water level.

two-tray clarifier   Space-saving wastewater clarifier arrangement where one longitudinal clarifier basin is located above another and both are operated in parallel or series.

TWPS   Tactical water purification system.

TWS   (1) See "traveling water screen (TWS)." (2) See "Transient water system."

TxPro™   Solids transmitter by BTG, Inc.

TxPro-2™   Suspended solids/turbidity transmitter by Zellweger Analytics, Inc.

Tync Aeration   Former name of Aeration Research Company.

type I settling   See "discrete particle settling."

type II settling   See "flocculant settling."

type III settling   See "hindered settling."

type IV settling   See "compression settling."

typhoid   Highly infectious disease of the gastrointestinal tract caused by waterborne bacteria.

Typhoon   High speed surface aerator by Aeromix Systems, Inc.

Tysul   Hydrogen peroxide oxidant by E.I. DuPont De Nemours, Inc.

TZ   Treatment zone.

U

U.S. Filter   Former name of USFilter Corp.

UAQI   Uniform Air Quality Index.

UARG   Utility Air Regulatory Group

UBC   Uniform Building Code.

UBlock™   Air/water filter underdrain by Tetra Process Technologies.

UCC   See "ultra clean coal (UCC)."

UCCI   Urea-formaldehyde foam insulation.

UCL   Upper control limit.

UDMH   Unsymmetrical dimethyl hydrazine.

UEL   Upper explosive limit.

UF   See "ultrafiltration (UF)."

UFL   Upper flammability limit.

UFRV   See "unit filter run volume (UFRV)."

UFW   See "unaccounted-for water (UFW)."

UHMW   Ultrahigh-molecular-weight.

UHR Filter   Dual media sand filter by Idreco USA, Ltd.

UIC   See "Underground Injection Control (UIC)."

UL®   Underwriters Laboratories, Inc.

ullage   The amount that a container or tank lacks of being full.

Ultima   Nonmetallic sludge collector flight by Budd Co.

ultimate BOD (BODu)   The amount of oxygen required to completely satisfy the carbonaceous and nitrogenous biochemical oxygen demand.

ultimate strength   The stress calculated on the maximum value of the force and the original area of the cross section which causes fracture of the material.

Ultimer®   Acrylamide polymer for sedimentation and sludge conditioning by Nalco Chemical Co.

Ultipleat®   High flow filtration element by Pall Corp.

Ultipor III®   Coreless filter elements by Pall Corp.

ultra clean coal (UCC)   Coal that is washed, ground into fine particles, and chemically treated to remove sulfur, ash, silicone, and other substances.

Ultrabar™   Ultrafiltration barrier process by F.B. Leopold Co., Inc.

Ultracept®   Oil water separators by Jay R. Smith Mfg. Co.

UltraChem™   Membrane for oily waste applications by Membrex, Inc.

Ultradome®   Domed, aluminum tank cover by Ultraflote Corp.

ultrafiltration (UF)   A low pressure (200-700 kPa, 20-100 psi) membrane filtration process which separates solutes in the 20 to 1000 angstrom (up to 0.1 micron) size range.

UltraFlex®   LPDE containment liner by SLT North America, Inc.

Ultraflow   Programmable open-channel flow measurement device by Monitek Technologies, Inc.

Ultra-Guard   Ultraviolet disinfection system by UV Systems Technology, Inc.

Ultramix™   High efficiency mixer motor by Lightnin.

ultrapure water   Water with a specific resistance higher than 1 megohm-cm.

UltraScrub   Wet scrubber by Tri-Mer Corp.

Ultrasep   Ion exchange membrane for pressure driven processes by Ionics, Inc.

ultrasonic   Acoustic waves with frequencies above 20 kHz, not audible to the human ear.

ultrasonic flowmeter   A measurement device that transmits and receives sound waves through the flow to determine liquid flow rate.

UltraStrip   Air stripper by GeoPure Continental.

Ultra-Sweep™   Spiral blade clarifier by Hi-Tech Environmental, Inc.

Ultratest   Polyelectrolyte used to enhance liquid/solid separation by Ashland Chemical, Drew Industrial.

Ultra-Urban™   Stormwater runoff filter screen by AbTech Industries.

ultraviolet light (UV)   Light rays beyond the violet region in the visible spectrum, invisible to the human eye.

Ultrex   Tubular membrane diffuser by Aeration Research Company.

Ultrion SVC®   Low molecular weight organic coagulant by Nalco Chemical Co.

Ultrox®   Organic destruction process using ultraviolet light and oxidants by USFilter/Industrial Wastewater Systems.

unaccounted-for water (UFW)   The fraction of water fed into a water distribution system which is not registered by the customers' meters.

unburned lime   Another term for "calcium carbonate."

unconfined aquifer   An aquifer containing water that is not under pressure.

underdrain   Flow collection and backwash water distribution system used to support the filter bed in most granular media filters. Also called "filter bottom."

underflow   The concentrated solids removed from the bottom of a tank or basin.

Underground Injection Control (UIC)   The program under the Safe Drinking Water Act that regulates the use of wells to pump fluids into the ground.

underground storage tank (UST)   A tank used to contain accumulations of regulated substances with 10% or more of its volume underground.

UNEP   United Nations Environment Program.

Uni-Dose™   Electronic metering pump by Liquid Metronics, Inc.

Uniflap   Cast urethane flap valve by Ashbrook Corp.

Uniflow   Sloped bottom settling tanks equipped with chain and flight sludge collector by USFilter/Envirex.

uniformity coefficient   Method of characterizing filter sand where the uniformity coefficient is equal to the sieve size in millimeters that will pass 60% of the sand divided by that size passing 10%.

UniMag™   Closed pipe flow meter by Isco, Inc.

UniMix   Vertical or horizontal flocculation unit by Walker Process Equipment.

Uni-Pac   Water treatment equipment products by Cochrane Inc.

Unipak™   Induced gas flotation system by Unicel, Inc.

UniPro™   Reverse osmosis pump system by David Brown Union Pump Co.

Unipure   Heavy metals package waste removal system by Unipure Corp.

Uni-Scour   Dual media filter with self-contained backwash storage by Smith & Loveless, Inc.

Unisweep™   Filter surface wash system by Roberts Filter Group.

Unisystem®   Activated sludge wastewater treatment system by Aeration Industries, Inc.

unit filter run volume (UFRV)   The amount of liquid filtered between backwashes expressed as the product of the filter run length and the filtration rate.

Unitank®   Continuous flow wastewater treatment system by Seghers Better Technology USA.

Unitech   Evaporator and crystallizer product line by Graver Co.

United States Filter   USFilter.

Unitherm™   Thermal VOC oxidation unit by Dürr Environmental, Inc.

Unitube   Suction-type sludge removal system for circular clarifiers by USFilter/Envirex.

Uni-Vayor   Sludge removal conveyor by the former Bowser-Briggs Filtration Co.

Univer   Chemicals used to determine water hardness by Hach Co.

Universal®   Sand filter underdrain by F.B. Leopold Co., Inc.

Universal RAI®   Rotary positive blowers by Dresser Industries/Roots Division.

Universal Screen   Reciprocating rake bar screen by Meurer Industries, Inc.

universal treatment standards (UTS)   Rule addressing sampling methods of metal constituents in hazardous wastes for compliance with treatment standards.

Universal Venturi Tube™   Flow measurement device by BIF.

universal wastes   Wastes that include used, nonlead acid batteries, recalled pesticides, and mecury-containing thermostats.

UniversaLevel   Continuous liquid level transmitter by Ametek Drexelbrook Engineering Co.

Unizone   Ozone generators by Praxair-Trailigaz Ozone Co.

unleaded gasoline   Gasoline containing very little or no "tetraethyl lead" anti-knock additives.

Unox®   A pure oxygen, activated sludge wastewater treatment process by Lotepro Corp (Western Hemisphere) and Linde-KCA-Dresden GmbH.

unreasonable risk   Under the Federal Insecticide, Fungicide, and Rodenticide Act, that is any unreasonable risk to man or the environment, taking into account the medical, economic, social, and environmental costs and benefits of any pesticide.

unsaturated zone   The area above the water table where soil pores are not fully saturated, although some water may be present.

unstable   Condition describing an element or compound that reacts spontaneously to form other elements or compounds.

Upcore™   Upflow countercurrent regeneration process and resins by Dow Chemical Co.

UPE®   Universal Process Equipment Co.

upflow clarifier   Clarifier where flocculated water flows upward through a sludge blanket to obtain floc removal by contact with flocculated solids in the blanket.

upflow filter   Granular media filter characterized by an upward flow of liquid through the filter bed.

UPS   Uninterruptible power supply.

upset   An unexpected disturbance of a process or operation.

uptake   The absorption or ingestion of an element or compound in an organism or substance.

UPW   Ultrapure water.

urban runoff   Stormwater from city streets and adjacent domestic or commercial properties which carries pollutants of various kinds into the sewer systems and receiving waters.

urea   A soluble nitrogen compound that is a component of mammalian urine.

urease   An enzyme that converts urea in urine into ammonia.

USAID   U.S. Agency for International Development.

USC   Unified soil classification.

USCM   U.S. Conference of Mayors.

USDA   U.S. Department of Agriculture.

USDW   Underground sources of drinking water.

USEPA   See "Environmental Protection Agency (EPA)."

user fee   Fee collected from only those persons who use a particular service, as compared to one collected from the public in general.

USF   USFilter.

USFS   United States Forest Service.

USGS   U.S. Geological Survey.

USP   U.S. Pharmacopeia.

USPHS   U.S. Public Health Service.

USP-purified water   Water meeting USP-quality requirements that has been purified by distillation, ion exchange, or other suitable processes from water complying with the U.S. EPA's drinking water regulations and contains no added substances.

UST   See "underground storage tank (UST)."

UTS   See "universal treatment standards (UTS)."

U-tube manometer   See "manometer."

UV   See "ultraviolet light (UV)."

UV$_{254}$ A measure of water's ultraviolet absorption at a 254 nanometer wavelength which is considered representative of the humic content of natural organic matter and a surrogate for disinfection byproduct precursors.

UV3000™   Package ultraviolet water and wastewater disinfection system by Trojan Technologies, Inc.

UV4000™   Ultraviolet wastewater disinfection system by Trojan Technologies, Inc.

UV8000™   Medium and low pressure ultraviolet water disinfection system by Trojan *Techno*logies, Inc.

Uvaster®   UV disinfection system by USFilter.

UVOX   Ultraviolet oxidation.

UVT™   Universal venturi tube flow measurement by BIF.

V

V*Sep®   Membrane filtration process using vibration to prevent fouling by New Logic International.

V. cholerae   See "*Vibrio cholerae.*"

Vacflush®   Aeration cleaning system by USFilter/Jet Tech.

Vacpac™   Specialized waste container by PacTec, Inc.

Vacuator   Vacuum-operated system for removal of floating solids and scum by GL&V/Dorr-Oliver, Inc.

VacuLift™   Vacuum sludge removal system by Waterlink Separations, Inc.

Vacu-Treat   Rotary vacuum filter by USFilter/Envirex.

vacuum   A space from which air and gas have been removed and the pressure is less than normal atmospheric pressure.

vacuum breaker   A backflow prevention device that automatically vents a water line to the atmosphere when subjected to partial vacuum.

vacuum deaerator   Device operating under vacuum to remove dissolved gases from a liquid.

vacuum filter   A dewatering device utilizing a cloth-covered cylindrical drum slowly rotating in a tank of sludge while subject to an internal vacuum. Sludge is drawn to the filter cloth while water passes through the sludge cake.

vacuum truck   A tank truck used to remove wastewater or other liquid wastes in which the air inside the tank is pumped out and the liquid to be removed rushes in to fill the empty space.

vadose zone   Designation of the layer of the ground below the surface but above the water table.

value engineering   The process of subjecting an engineering design to a peer review by a group of experts who challenge the design concepts in an attempt to optimize the initial design and project cost.

ValueMAX™   Reverse osmosis system by USFilter.

valve   A device used to regulate flow of fluids through a piping system.

Valve PAC™   Valve positioning controller by F.B. Leopold Co., Inc.

van der Waals' Force   An attractive force apparent in colloidal particles.

Vanguard®   Air stripper by Delta Cooling Towers, Inc.

Vanguard®   Self-cleaning bar screen by Nuove Energie s.r.l.

VAPCCIs   See "Vatavuk Air Pollution Control Indexes (VAPCCI)."

vapor   The gaseous phase of a material in the solid or liquid state at standard temperature and pressure.

Vapor Combustor™   Vapor extraction unit by QED Environmental Systems, Inc.

vapor compression evaporation (VC)   Evaporative system where vapor boiled off in the evaporator is mechanically compressed and reused as the heating medium.

Vapor Guard™   Fabric membrane odor control tank cover by ILC Dover, Inc.

Vapor Pacs   Replaceable carbon canister systems for VOC emission control by Calgon Carbon Corp.

vapor plumes   Flue gases visible because they contain water droplets.

vapor pressure   The pressure at which equilibrium is established between the liquid and gas phases of a substance.

vapor recovery system   A system to gather all vapors and gases discharged from a storage tank and process them to prevent their emission to the atmosphere.

vapor trail   See "contrail."

vaporization   The process where a substance changes from a liquid or solid to the gaseous state.

VaporMate™   Volatile organic carbon treatment system by North East Environmental Products, Inc.

VaPure®   Vapor compression still by Paul Mueller Co.

VAR   Vector attraction reduction.

variable declining-rate filtration   Filter operation where the rate of flow through the filter declines and the level of liquid above the filter rises throughout the filter run.

Variair   Diffuser by USFilter/Diffused Air Products Group.

variance   Government permission for a delay or exception in the application of a given law, ordinance, or regulation.

Vari-Ator   Variable energy controller for floating aerator by USFilter/Aerator Products.

Vari-Cant™   Jet aeration system by USFilter/Jet Tech.

Variflo®   Wastewater distributor nozzle for high rate trickling filter by USFilter/Microfloc.

Vari-Pond®   Centrifuge pond control system by Wesfalia Separator, Inc.

VariPort™   Anaerobic digester gas mixing system by USFilter/Envirex.

VariSieve™   Solids classifier by Krebs Engineers.

VariTank™   Bolted steel tank by ModuTank, Inc.

Varivoid™   High rate granular media filter by Graver Co.

Vatavuk Air Pollution Control Indexes (VAPCCI)   Set of U.S. EPA indexes used to escalate air pollution equipment control costs.

VATIP Voluntary Advanced Technology Incentives Program. An EPA program intended to encourage environmental performance beyond what is required by the Cluster Rule.

vault An above- or below-ground-reinforced concrete structure for storing radioactive or other hazardous waste materials.

vault privy A holding tank with a seat or seats to provide for excretion of human wastes directly into the tank.

V-Auto™ In-line strainer by Andritz-Ruthner, Inc. (Western Hemisphere) and Andritz Sprout-Bauer S.A. (Eastern Hemisphere).

VC See "vapor compression evaporation (VC)."

VCE Vapor compression evaporation.

VCM Vinyl chloride monomer.

V-Cone® Differential pressure flow design flow meter by McCrometer, Inc.

VDB Vacuum drying bed by U.S. Environmental Products, Inc.

VDS Vertical drum screen by Waste-Tech, Inc.

VE Visual emissions.

vector An insect or other carrier capable of transmitting a pathogen from one organism to another.

vector-borne transmission The transmission of an infectious agent from an insect or other vector to a human.

Vee-wire® Wedge-shaped wire by USFilter/Johnson Screens.

vegetative controls Nonpoint source pollution control practices that involve vegetative cover to reduce erosion and minimize loss of pollutants.

vehicle miles travelled (VMT) A measure of the extent of motor vehicle operation; the total number of vehicle miles travelled within a specific geographic area over a given period of time.

Vekton Nylon sludge collector sprockets by Norton Performance Plastics.

Vektor™ Mixer by Lightnin.

veliger A free-swimming, larval mollusk.

VeliGON® Molluscicide used to kill zebra mussels by Calgon Corp.

velocity cap The horizontal cap on a vertical, offshore water intake which results in a horizontal inflow, thus reducing fish entrainment.

velocity gradient (G value) A measurement of the degree of mixing imparted to water or wastewater during flocculation.

velocity head The kinetic energy in a hydraulic system.

velometer A device used to measure air velocity.

ventilation The act of admitting fresh air into a space in order to replace stale or contaminated air; achieved by blowing air into the space.

VentOXAL® Oxygenator system by Air Liquide America.

Vent-Scrub™ Powder-activated carbon adsorption unit by USFilter/Westates.

VentSorb Disposable granular activated carbon filters by Calgon Carbon Corp.

Ventura™ VOC oxidizer by Megtec Systems, Inc.

venturi effect An increase in the velocity of a fluid as it passes through a constriction in a pipe or channel.

venturi meter   A meter used to measure flows in closed conduits by registering the difference in velocity heads between the entrance and outlet of a contracted throat.

venturi scrubber   A wet scrubber used to remove particulates from gaseous emissions.

VEO   Visible emission observation.

VerderFlex   Hosepump by Verder, Inc.

vermiculture   A stabilization and conversion process using earthworms to consume organic matter in sludge.

Versipad   Oil absorption device for light to medium oils by Mother Environmental Systems, Inc.

VerTech™   Wet oxidation process by Air Products & Chemicals, Inc.

VertiCel™   Biological wastewater treatment process by USFilter/Envirex.

Verti-Flo   Vertical flow, rectangular tank clarifier by USFilter/Envirex.

Verti-Jet™   Vertical pressure leaf filter by USFilter Corp.

Vertimatic   Upflow sand filter by USFilter Corp.

Vertimill™   Lime slaker by Svedala Industries, Inc.

Verti-Press   High pressure filter press by Filtra-Systems Hydromation.

Verti-Press   Inclined screw-type solids/screenings press by Dontech, Inc.

Vertiscreen   Vertical stationary screen by Thermal Black Clawson.

very low sodium water   Bottled water containing 35 mg or less of sodium per serving.

Vestrip™   Vacuum extraction system by Ejector Systems, Inc.

VFD   Variable frequency drive.

Vibra-Matic™   Vibrating pressure leaf filter cleaning system by USFilter Corp.

Vibra-Screen   Vibrating fine screen by Triton Technologies.

Vibrasieve®   Vibrating fine screen by Andritz-Ruthner, Inc.

vibrating screen   A mechanical screening device whose screening surface is shaken or vibrated to enhance solid/liquid separation. A variation of this screen is used as a solids classifier.

Vibrio cholerae   The bacterium responsible for causing cholera. Also called "*V. cholerae.*"

Viggers Valve   Automatic sludge blow-off valve by Walker Process Equipment.

VIGR™   Electostatic precipitator by Wheelabrator Air Pollution Control, Inc.

Vincent Press   Horizontal dewatering screw press by Vincent Corp.

Vinyl Core   PVC biological filter media formerly offered by B.F. Goodrich Co.

viral   Pertaining to or caused by a virus.

VIRALT   A mathematical model used to calculate the concentration of viruses at the water table and well after water has been transported through a subsurface media.

Virchem®   Corrosion control product by Carus Chemical Co., Inc.

virtual water   Water embedded in water-intensive commodities such as cereals.

virulent   Extremely infectious or damaging to organisms.

virus   Smallest biological structure capable of reproduction which infect their host, producing disease.

viscera   The internal organs of an animal.

Viscomatic   Lime slaker by Infilco Degremont, Inc.

viscosity   The property of a fluid that offers resistance to flow due to the existence of internal friction.

Vistex   Grit removal system by Vulcan Industries, Inc.

Vitec™   Antiscalants and dispersants by Avista Technologies.

Vitox®   Oxygen injection system for wastewater treatment plants by BOC Gases.

vitrified clay   A kiln-fired clay product used to make pipe and brick.

VLA   Very large array. See "array."

VMT   See "vehicle miles travelled (VMT)."

V-notch weir   A weir with a triangular, v-shaped notch.

VO Nozzle   Variable orifice spray nozzle by USFilter/Rex & Link-Belt Products.

VOC   See "volatile organic compounds (VOC)."

VOC Wagon™   Self-contained thermal oxidizer by NAO Inc.

VOCarb   Activated carbon by USFilter/Westates.

void volume   The volume of the spaces between granular material such as filter media or ion exchange resin, expressed as a percentage of the total volume occupied by the material.

volatile   A substance that evaporates or vaporizes at a relatively low temperature.

volatile liquids   Liquids which easily vaporize or evaporate at room temperature.

volatile organic compounds (VOC)   Highly evaporative organic compounds often found in paints, solvents, and similar products.

volatile suspended solids (VSS)   Organic content of suspended solids in a water or wastewater. Determined after heating a sample to 600°C.

Volcano™   Continuous downflow sandfilter by Lighthouse Separation Systems, Inc.

Volclay®   Sodium bentonite soil sealant liner systems by Colloid Environmental Technologies Co.

Vollmar   Stormwater holding tank tipping bucket by Waterlink Separations, Inc.

Voluette   Ampules of standard reagents solutions for analytical laboratory use by Hach Co.

volume reduction   Processing waste materials to decrease the amount of space they occupy, usually by compacting or shredding, incineration, or composting.

Volumeter™   Flow monitoring instrument by Marsh-McBirney, Inc.

volumetric feeder   Dry chemical feeder that supplies a constant preset or proportional chemical delivery by volume and will not recognize a change in material density.

volumetric tank test   A test to determine the physical integrity of a storage tank where the volume of fluid in the tank is measured directly or calculated from product-level changes, and a marked drop in volume indicates a leak.

Vortair®   Low speed surface aerator by Infilco Degremont, Inc.

Vortechs™   Stormwater treatment system by Vortechnics, Inc.

vortex   A flow with a whirling, rotary motion forming a cavity in the center toward which particles are drawn.

Vortex™   Circular tank grit removal system with a turbine type rotor by Infilco Degremont, Inc.

vortex flow regulator   A funnel-shaped flow control device used to generate a uniform flow rate from a storage basin, tank, or cistern.

vortex grit removal   Grit removal system relying on a mechanically induced vortex to capture grit solids in the center hopper of a circular tank.

Vorti-Mix®   Submerged turbine aerator by Infilco Degremont, Inc.

Vostrip™   Air stripping tower by EnviroSystems Supply.

Voyager™   Portable GC by PerkinElmer Instruments.

VPS™   Vapor phase odor control system by NuTech Environmental Corp.

VRTO   Valveless regenerative thermal oxidizer. See "regenerative thermal oxidizer (RTO)."

VSD   Virtually safe dose.

VSR   Volatile solids reduction.

VSS   See "volatile suspended solids (VSS)."

VTC™   Vertical tube coalescing oil/water separator by AFL Industries, Inc.

VTE   Vertical tube evaporator.

VTSH®   Vertical turbine solids handling pump by Fairbanks Morse Pump Corp.

VTX   Vertical vortex pump by Yeomans Chicago Corp.

vulnerability analysis   Assessment of elements in the community susceptible to damage should a release of hazardous materials occur.

vulnerable zone   An area over which the airborne concentration of a chemical accidentally released could reach the level of concern.

VVC   Vacuum vapor compression.

W

W&T   USFilter/Wallace & Tiernan.

W/O   Water-in-oil emulsion.

WAC   Weak-acid cation exchanger.

WAC®   Polyaluminum chloride flocculant by Water Treatment Solutions division of Elf Atochem North America, Inc.

wadi   A dry stream or river bed which carries flash flood water.

Wagner   Sand filter underdrains by Infilco Degremont, Inc.

Walhalla Process   Sulfate and heavy metals precipitation and removal process by Graver Co.

walking beam flocculator   A mechanical flocculation device whose mixing paddles are attached to a horizontal beam that rocks back and forth to produce a reciprocating motion.

Wallace & Tiernan   Former name of USFilter/Wallace & Tiernan.

walnut shell filter   A filtration device that uses ground walnut shells as granular filter media to remove hydrocarbons and other suspended solids from water.

WAO   See "wet air oxidation (WAO)."

warp   The vertical wire in woven wire mesh.

WAS   See "waste activated sludge (WAS)."

Wash Press   Screenings washer/press by Lakeside Equipment Corp.

Washpactor   Sewage screenings washing and compacting unit by Waste-Tech, Inc.

wash-water trough   The trough located above filter media used to sluice away backwash water.

waste-activated sludge (WAS)   Excess activated sludge discharged from an activated sludge treatment process.

waste characterization   Identification of chemical and microbiological constituents of a waste material.

waste feed   The continuous or intermittent flow of wastes into an incinerator.

waste heat evaporator   An evaporator that uses the heat of a gas turbine, diesel engine jacket water, or exhaust gas.

waste load allocation   The maximum load of pollutants each waste discharger is allowed to release into a particular waterway.

Waste Management, Inc.   Former name of WMX Technologies, Inc.

waste minimization   Pollution control plan that includes both pollution source reduction and environmentally sound recycling.

waste oil emulsion   A thick, viscous waste that results from a water-in-oil emulsion.

waste oils   Lubricating oils that have completed their intended use cycle and must be treated and reused or receive proper disposal.

waste reduction   Using source reduction, recycling, or composting to prevent or reduce waste generation.

waste stabilization pond   A pond receiving raw or partially treated wastewater in which stabilization occurs.

Wastebuster II™   Biological wastewater treatment plant by HRI-Biotek.

Wastemaster™   Wastewater headworks treatment unit by WesTech Engineering Inc.

Wastewarrior   Ultrafiltration wastewater treatment system by Hyde Marine, Inc.

wastewater   Liquid or waterborne wastes polluted or fouled from household, commercial, or industrial operations, along with any surface water, stormwater, or groundwater infiltration.

water   A colorless, transparent liquid essential for all life which freezes as ice at 0°C (32°F) and vaporizes forming steam at 100°C (212°F). Water occurs naturally on earth as rivers, lakes, and oceans, and falls from the clouds as rain. Water's chemical formula is H_2O and it is a nonsymmetrically formed molecule of two hydrogen atoms which form an angle of 104.5 degrees with the oxygen atom lying between them.

water banking   The process of storing excess water as groundwater during wet years for later extraction and use during dry cycles.

Water Blaze   Wastewater evaporator by Landa, Inc.

Water Boy®   Package water treatment plant by USFilter/Microfloc.

Water Buffalo™   Reverse osmosis unit by Mechanical Equipment Co., Inc.

Water Champ   Chemical induction unit by USFilter/Stranco.

water closet (WC)   A common term for "toilet."

water demand   The water requirments for a particular purpose.

water dilution volume (WDV)   The volume of water required to dilute radioactive waste to a concentration that meets drinking water standards.

Water Eater   Wastewater evaporator by Equipment Manufacturing Corp.

water effect ratio (WER)   A test procedure that determines the capacity of site water to mitigate metal toxicity by comparing the toxic endpoint of site water to the toxic endpoint of laboratory water.

Water Factory 21   An Orange County, California treatment plant designed to produce high quality water from municipal wastewater for injection into aquifers, creating a coastal barrier and preventing sea water intrusion of underground water supplies.

water for injection (WFI)   Water purified to USP standards by distillation, ion exchange, or other suitable process, containing no additives, and intended for use as a solvent for preparation of parenteral solutions.

Water Free™   Grit and screenings gate by Tetra Process Technologies.

water hammer   Condition that may cause damage or rupture to a piping system resulting from a rapid increase in pressure that occurs in a closed piping system when the liquid velocity is suddenly changed.

water hyacinth   See "hyacinth."

Water Maze®   Wastewater clarifier by Landa, Inc.

water meter   A device installed in a pipe that measures and registers the quantity of water passing through it.

water pollution   The presence in water of enough harmful or objectionable material to damage the water's quality.

Water Pollution Control Corp.   Former name of Sanitaire Corp.

water purveyor   An agency or person that supplies water.

water reclamation   The restoration of wastewater to a state that will allow its beneficial reuse.

water recycling   Reclamation of effluent generated by a given user for on-site reuse by the same user.

water reuse   The beneficial use of reclaimed water for purposes such as irrigation, cooling, or washing.

water scarce   The condition existing in a country with annual, internal renewable freshwater resources of less than 1000 cubic meters per capita per year.

water splitting   An electrically driven membrane process that separates ionic species in solution through the use of a bipolar membrane that dissociates water into hydrogen and hydroxyl ions. Also referred to as "bipolar membrane electrodialysis."

water stress   The condition existing in a country with total freshwater resources of 1000 to 1600 cubic meters per capita per year with major problems in drought years.

water supplier   One who owns or operates a public water system.

water supply system   The collection, treatment, storage, and distribution of potable water from source to consumer.

water table   The upper limit of the portion of the ground saturated with water.

Water Tech   Former equipment manufacturer acquired by Roberts Filter Group.

Water Tiger™   Rotary and static fine screens by Komline-Sanderson Engineering Corp.

water vapor   The gaseous form of water.

water-based disease   A disease caused by hosts that either live in water or require water for part of their lifecycle, and are passed to humans when they are ingested or come in contact with the skin.

Waterbetter®   Filter cartridges by Harmsco Filtration Products.

waterborne   Of, or related to, something that can be carried by water.

waterborne disease   A disease transmitted through a drinking water supply.

waterbox   The chamber at the inlet end of a condenser tubesheet.

waterflood   A method of secondary oil recovery in which water is injected into an oil reservoir to force additional oil out of the formation into a producing well.

WaterKare   Dissolved air flotation product line by Kemco Systems, Inc.

Waterloo System   Groundwater sampling and monitoring system by Solinst Canada Ltd.

watermaker   A common term for a packaged, vapor compression thermal desalination unit.

watershed   The land areas that drain into a surface waterbody.

watershed approach   A coordinated framework for environmental management that addresses the highest priority problems within hydrologically defined geographic areas, taking into consideration both ground and surface water flow.

water-soluble   The description of a material that dissolves in water.

WaterSweet™   Granular media used for removal of hydrogen sulfide from drinking water by SultaTreat Co.

watertube boiler   Boiler in which the flame and hot combustion gases flow across the outside of the tubes and water is circulated within the tubes.

water-washed disease   A disease that results from inadequate sanitation or contact with contaminated water and can be prevented by washing with clean water.

Waterweb®   Mesh used in air scrubbers by Misonix, Inc.

waterwheel   A wheel driven by water flowing into buckets or falling on vanes on the periphery of the wheel.

waterworks   The system of reservoirs, treatment, and pumping by which a water supply is obtained and distributed to a community.

Waukesha Pump   Former name of Watson Marlow, Inc.

Wave Oxidation®   Fluctuating aerobic and anaerobic biological wastewater treatment system by Parkson Corp.

WaveOx®   Wastewater treatment process for biological nutrient removal by Parkson Corp.

WBA   Weak-base anion exchanger.

WC   See "water closet (WC)."

WDV   See "water dilution volume (WDV)."

weak acid   A poorly ionized acid that dissociates very little and produces few hydrogen ions in aqueous solution.

weak acid cation exchanger   Cation exchange products with functional groups incapable of splitting neutral salts to form corresponding free acids.

weak base anion exchanger   Anion exchange products with functional groups incapable of splitting neutral salts to form corresponding free bases.

Wedge-Flow™   Wedge wire filtration products by LEEM Filtration Products, Inc.

WedgePress   Belt filter press by the former Gray Engineering Co.

Wedgewater Filter Bed™   Gravity sludge dewatering filter bed by Gravity Flow Systems, Inc.

Wedgewater Sieve   Static bar screen by Gravity Flow Systems, Inc.

wedgewire   General term to describe trapezoidal or v-shaped wire.

WEF   Water Environment Federation.

weft   The horizontal wire in woven wire mesh, also called the "shute" wire.

weighted composite sample   A composite sample where the sample amounts are based on flow rates at the time of collection, i.e., the higher the flow, the larger the portion of sample taken at that time.

weir   A baffle over which water flows.

weir loading   The rate of flow out of a basin stated in terms of the volume of liquid passing over a stated length of weir per unit of time.

weir overflow rate   A measurement of the volume of water flowing over each unit length of weir per day.

Weiss   In-vessel composting system by USFilter/Davis Process.

well   A bored, drilled, or driven shaft or hole whose depth is greater than the largest surface dimension.

well monitoring   Measurement by on-site instruments or laboratory methods of the quality of water in a well.

well plug   A watertight and gastight seal installed in a bore hole or well to prevent movement of fluids.

well screen   A slotted or perforated well casing that allows passage of water while preventing solids from entering the well.

well water   Water from a hole bored, drilled, or otherwise constructed in the ground to tap an aquifer.

Well Wizard®   Groundwater monitor and sampling device by QED Environmental Systems, Inc.

Welles Products   Equipment product line acquired by USFilter/Aerator Products.

wellhead protection area   A protected surface and subsurface zone surrounding a well or wellfield supplying a public water system to keep contaminants from reaching the well water.

Wemco®   Product line of Baker Hughes Process Systems.

Wemco Pump   Product line of Envirotech Pumpsystems.

WER   See "water effect ratio (WER)."

WESP   See "wet electrostatic precipitator (WESP)."

Westates Carbon   Carbon products and services by USFilter/Westates.

Westchar™   Activated carbon by Osmonics, Inc.

Westchlor®   Polyaluminum coagulants by Westwood Chemical Corp.

Western Filter Company   Former manufacturer acquired by Osmonics, Inc.

WestRO™   Reverse osmosis and membrane products by Osmonics, Inc.

WET   See "whole effluent toxicity (WET)."

WET®   Product line of Waterlink Inc.

wet air oxidation (WAO) Process where sludge and compressed air are pumped into a pressurized reactor and heated to oxidize the volatile solids without vaporizing the liquid.

wet bulb temperature The temperature reading taken from a thermometer with a wetted wick surrounding its bulb.

wet electrostatic precipitator (WESP) An electrostatic precipitator whose collecting electrodes are rinsed with water to remove particulates.

wet scrubber An air pollution control device used to remove particulates and fumes from air by entraining the pollutants in a water spray.

wet steam Steam containing water droplets.

wet weather flow The flow in a combined sewer during snowmelt and rain events.

wet well A chamber in which water or wastewater is collected and to which the suction of a pump is connected.

Wetec™ Wet aeration installation system by Aeration Technologies, Inc.

wetlands Surface areas including swamps, marshes, and bogs which are inundated or saturated by groundwater frequently enough to support a prevalence of vegetation adapted for life in saturated soil conditions.

wetlands treatment A wastewater treatment system using the aquatic root system of cattails, reeds, and similar plants to treat wastewater applied either above or below the soil surface.

wetness index The precipitation for a given year expressed as a ratio of the mean annual precipitation.

wetted perimeter The length of wetted contact area between a stream of flowing water and the channel that contains it.

WFI See "water for injection (WFI)."

Wheeler Sand filter underdrain by Roberts Filter Group.

Whirl-Flo® Solids handling vortex pump by ITT A-C Pump.

Whirl-Wet-AG Airborne dust collector by Tri-Mer Corp.

Whispair® Rotary blower and gas pump product line by Dresser Industries/Roots Division.

white goods Large household appliances such as stoves, refrigerators, or washing machines which are discarded as wastes.

white water Filtrate from a paper- or board-forming machine, usually recycled for density control.

Whitewater Dissolved air flotation system by Water Resources Group, Inc.

WhiteWater™ Low profile air strippers by QED Environmental Systems, Inc.

WHM® Bulk chemical handling products by USFilter/Zimpro.

WHO World Health Organization.

whole effluent toxicity (WET) The aggregate toxic effect of an effluent measured directly by a toxicity test.

WHPP Wellhead protection program.

Wiese-Flo® Self-cleaning filter screen by USFilter/Headworks Products.

wildlife refuge An area designated for the protection of wild animals, within which hunting and fishing are either prohibited or strictly controlled.

Willett Pump Electro/hydraulic ram pump by USFilter/Dewatering Systems.

William Green   Former screening equipment manufacturer acquired by Brackett Geiger.

Willowtech®   Sludge mixer/blender by Ashbrook Corp.

windage emission   Emission of vapors resulting from wind blowing through a free-vented tank and educting some of the saturated vapors.

windbreak   A row of trees planted on level land as protection against winds and erosion.

Windjammer™   Brush aerator by United Industries, Inc.

windrow composting   A composting method where municipal wastewater solids are arranged in long triangular-shaped piles, called windrows, which are mechanically turned and remixed periodically.

Windrow-Mate   Compost odor control system by NuTech Environmental Corp.

Winklepress®   Belt filter press by Ashbrook Corp. (U.S.), and Gebr. Bellmer GmbH (Europe).

Winkler titration   A standard iodometric titration method of determining the dissolved oxygen level of a water or wastewater.

wire-to-wire efficiency   The efficiency of a pump and motor together.

witching   See "dowsing."

WLM   See "working level month (WLM)."

work   A force acting over a distance measured in joules or foot-pounds.

work exchanger   A mechanical energy recovery device where the energy of one fluid stream is transferred to another fluid.

working level month (WLM)   A unit of measure used to determine cumulative exposure to radon.

working load   An allowable recommended tensile load for chains used on conveyors, screens, or other applications of low relative speeds.

worm   A shank having at least one complete thread around the pitch surface.

worm gear   A gear with teeth cut on an angle to be driven by a worm; used to connect nonparallel, nonintersecting shafts.

WPCF   Water Pollution Control Federation. Former name of "Water Environment Federation (WEF)."

WPCP   Water pollution control plant.

WQA   Water Quality Act of 1987.

WQS   Water Quality Standard.

WRDA   Water Resources Development Act.

WRI   World Resources Institute.

Wring-Dry   Internally fed rotary fine screen by Schlueter Co.

WSA™   Former name of BG-WSA, Inc.

WSF   Water soluble fraction.

WSRA   Wild and Scenic Rivers Act.

WSTA   Water Science and Technology Association.

WSTB   Water Sciences and Technology Board.

WTP   Water treatment plant.

WTS   Former name of product group integrated into USFilter/Industrial Wastewater Systems.

WW   Wastewater.

WWEMA   Water and Wastewater Equipment Manufacturers Association.
WWF   World Wildlife Fund.
WWTF   Wastewater treatment facility.
WWTP   Wastewater treatment plant.
Wyss®   Fine bubble diffuser by Parkson Corp.

X

X-20™   Thin film composite reverse osmosis membrane by TriSep Corp.
XCELL™   Cell-less filter by USFilter/Davco.
xenobiotic   A description for any chemical compound that is foreign, and usually harmful, to the body of a living organism or biosystem.
Xentra   Continuous emissions analyzer by Servomex Co.
xeric   An organism that requires very little moisture to sustain life.
xeriscape   A landscape design used in arid climates that utilizes native and drought-tolerant plants in conjunction with water conservation techniques.
X-Flo™   Crossflow-type plastic trickling filter media by the former American Surfpac Corp.
XLP™   Rotary positive blower by Dresser Industries/Roots Division.
XL-Plus®   Solid bowl centrifuge by Baker Process.
XOtherm™   Spray cooling system by Environmental Dynamics Inc.
XP   See "explosion proof (XP)."
X-Pruf™   Explosion proof portable submersible pump by Crane Pumps & Systems.
xQy   A design stream flow condition that describes the lowest flow that will occur for (x) consecutive days not more than once every (y) years. See also "7Q10."
XR-5®   Containment geomembrane lining by Seaman Corp.
X-ray fluorescence   A technique used in the analysis of water-formed deposits and heavy metals or corrosion products in water.
Xtractor™   Sludge removal process by USFilter/Davco.
xylene   The common name for dimethylbenzene, a volatile organic compound used as a solvent and insecticide. Chemical formula is $C_6H_4(CH_3)_2$.

Y

yard waste   Grass clippings, leaves, and miscellaneous vegetative matter.
yellow cake   One of the final precipates in uranium extraction.
Yellow Jacket™   Floating hydrocarbon detection device by QED Environmental Systems, Inc.
yellow water   Water colored as a result of the presence of iron, also called "red water."

yellow-boy   Iron oxide flocculent usually observed as orange-yellow deposits in surface streams with excess iron content.

yield   The amount of a material produced by a process or chemical reaction.

yield point   The stress at which a substance undergoes significant change without an increase in applied stress.

Z

Z Chlor™   Chlorine and sulfite measurement and control unit by Bailey-Fischer & Porter.

ZD   See "zero discharge (ZD)."

zebra mussel   Freshwater mollusk that can foul water intake systems, treatment plants, and piping by attaching itself to a solid structure, eventually restricting flow.

ZeeWeed®   Membrane bioreactor technology by Zenon Environmental, Inc.

Zenobox   Membrane filtration system by Zenon Environmental, Inc.

Zenofloc   Polymer control system for belt filter press by Zenon Environmental, Inc.

ZenoGem®   Membrane bioreactor treatment process for tertiary treated wastewater by Zenon Environmental, Inc.

Zeo-Karb®   Sulfonated coal cation exchange process by USFilter/Warren.

Zeol™   Rotor concentrator for VOC abatement by Munters Zeol Corp.

zeolite   Minerals or synthetic resins that have ion exchange capabilities.

zeolite softening   Water softening process using a zeolite resin bed to accomplish ion exchange.

Zeo-Rex®   Oxidizing filter for iron and manganese removal by USFilter/Warren.

Zephyr®   Cartridge-type air filter by Farr Co.

Zephyr™   Induced air flotation system by Aeromix Systems, Inc.

zero discharge (ZD)   A facility that discharges no material to the environment.

zero liquid discharge (ZLD)   A facility that discharges no liquid effluent to the environment.

zero ODP   Zero ozone depletion potential.

zero order reaction   A reaction in which the rate of change is independent of the concentration of the reactant.

zero population growth (ZPG)   The situation where the birthrate is constant and equal to the death rate.

zero soft water   Water having less than 1.0 grain per gallon as calcium carbonate.

Zerofuel   Fluidized bed sludge incinerator by Seghers Better Technology USA.

zeta potential (ZP)   The voltage differential between the surface of the diffuse layer surrounding a colloidal particle and the bulk liquid beyond.

Zetag®   Cationic polyelectrolytes by Ciba Specialty Chemicals.

Zeta-Pak   Cartridge filter by Alsop Engineering Co.

Zickert   Product line of Waterlink, Inc.

ZID   See "zone of initial dilution (ZID)."

Zimmerman Process   Wet air oxidation process by USFilter/Zimpro.

Zimpress   Plate and frame sludge press by USFilter/Zimpro.

Zimpro Environmental   Former name of USFilter/Zimpro.

Zimpro Process   Thermal sludge conditioning process by USFilter/Zimpro.

ZLD   See "zero liquid discharge (ZLD)."

Z-list   OSHA's tables of toxic and hazardous air contaminants.

Z-metal   Pearlitic malleable iron chain material by USFilter/Envirex.

ZOC   See "zone of contribution."

ZOI   See "zone of incorporation."

zone of aeration   The comparatively dry soil or rock located between the ground surface and the top of the water table.

zone of contribution (ZOC)   The area surrounding a pumping well that includes all areas that supply water to recharge the well.

zone of incorporation (ZOI)   The depth to which soil on a landfarm is plowed or tilled to receive wastes.

zone of initial dilution (ZID)   The part of a lake or river where a discharge from an outfall first mixes with the receiving waters.

zone of saturation   The portion of the earth's crust below the water table where the pores are filled with water at greater than atmospheric pressure.

zooglea   A gelatinous matrix developed by growing bacteria associated with trickling filter beds and activated sludge floc.

zooplankton   Small aquatic animals that possess little or no means of propulsion.

ZP   See "zeta potential (ZP)."

ZPG   See "zero population growth (ZPG)."

ZRL   Zero risk level.

MANUFACTURERS DIRECTORY

A.B. Marketech, Inc.
9708 N. Range Line Rd.
Mequon, WI 53092
Phone: 262-255-7448
Fax: 262-241-3443

A.O. Smith Engineered Products
2101 S. 21st St.
Parsons, KS 67357
Phone: 316-421-0200
Fax: 316-421-9122
www.tectank.com

A-C Compressor Corp.
401 E. South Island St.
Appleton, WI 54915
Phone: 920-738-3080
Fax: 920-738-3141
www.accompressor.com

Abanaki Corp.
17387 Munn Rd.
Chagrin Falls, OH 44022
Phone: 440-543-7400
Fax: 440-543-7404
www.abanaki.com

ABB Air Preheater, Inc.
P.O. Box 372
Wellsville, NY 14895
Phone: 716-593-2700
Fax: 716-593-2721
www.abb.com

ABB Raymond
650 Warrenville Rd.
Lisle, IL 60532
Phone: 630-971-2500
Fax: 630-971-1076

ABS Pumps, Inc.
140 Pond View Dr.
Meriden, CT 06450
Phone: 203-238-2700
Fax: 203-238-0738
www.abspumpsusa.com

AbTech Industries
4110 N. Scottsdale Rd., Ste. 235
Scottsdale, AZ 85251
Phone: 602-874-4000
Fax: 602-970-1665

Acrison, Inc.
20 Empire Blvd.
Moonachie, NJ 07074
Phone: 201-440-8300
Fax: 201-440-4939
www.acrison.com

ACS Industries, Inc.
14211 Industry Rd.
Houston, TX 77053
Phone: 713-434-0934
Fax: 713-433-6201

ADI Systems, Inc.
1133 Regent St., Ste. 300
Fredericton, NB E3B 3Z2
Phone: 506-452-9000
Fax: 506-459-3954
www.adi.ca

Advanced Microbial Systems
P.O. Box 540
Shakopee, MN 55379
Phone: 612-445-4251
Fax: 612-445-7233
www.amsmicrobes.com

Advanced Polymer Systems
123 Saginaw Dr.
Redwood City, CA 94063
Phone: 650-366-2626
Fax: 650-365-6490
www.advancedpolymer.com

Advanced Sensor Devices, Inc.
430 Ferguson Dr.
Mountian View, CA 94043
Phone: 650-960-3007
Fax: 650-960-0127

Advanced Separation Technologies
5315 Great Oak Dr.
Lakeland, FL 33815
Phone: 941-687-4460
Fax: 941-687-9362
www.advsep.com

Advanced Structures, Inc.
200 Industrial Parkway
Escondido, OH 44024
Phone: 440-286-4116
Fax: 800-942-7659
www.pentairwater.com

Adwest Technologies, Inc.
1175 N. Van Horne Way
Anaheim, CA 92806-2506
Phone: 714-632-9801
Fax: 714-632-9812
www.adwestusa.com

Aeration Industries, Inc.
P.O. Box 59144
Minneapolis, MN 55459-0144
Phone: 952-448-6789
Fax: 952-448-7293
www.aireo2.com

Aeration Tehnologies, Inc.
P.O. Box 488
North Andover, MA 01845
Phone: 978-475-6385
Fax: 978-475-6387
www.aertec.com

Aero Tec Laboratories, Inc.
Spear Rd. Industrial Park
Ramsey, NJ 07446-1251
Phone: 201-825-1400
Fax: 201-825-1962
www.atlinc.com

Aeromix Systems, Inc.
2611 N. Second St.
Minneapolis, MN 55411-1634
Phone: 612-521-8519
Fax: 612-521-1455
www.aeromix.com

Aeropulse, Inc.
1746 Winchester Rd.
Bensalem, PA 19020
Phone: 215-245-7554
Fax: 215-245-7849
www.aerpulse.com

AFL Industries, Inc.
3661 West Blue Heron Blvd.
Riviera Beach, FL 33404
Phone: 561-844-5200
Fax: 561-844-5246

Ag Environmental Products LLC
9804 Pflumm Rd.
Lenexa, KS 66215
Phone: 913-599-6911
Fax: 913-599-2121
www.soygold.com

Ag-Chem Equipment Co.
5720 Smetana Dr.
Minnetonka, MN 55343-9688
Phone: 612-933-9006
Fax: 612-933-7432
www.agchem.com

Agar Corp.
1600 Townhurst
Houston, TX 77043
Phone: 713-464-4451
Fax: 713-464-7741

Agency Environmental, Inc.
8362 Veterans Highway, Ste. 101
Millersville, MD 21108
Phone: 410-729-8490
Fax: 410-729-8492

Air Liquide America
5230 S. East Ave.
Countryside, IL 60525
Phone: 708-482-8400
Fax: 708-579-7702

Air Products & Chemicals, Inc.
7201 Hamilton Blvd.
Allentown, PA 18195
Phone: 610-481-4911
Fax: 610-481-5084

Air-O-Lator Corp.
8100 Paseo
Kansas City, MO 64131
Phone: 816-363-4242
Fax: 816-363-2322
www.airolator.com

Airmaster Aerator
P.O. Box 156
Wisner, LA 71378
Phone: 318-724-7626
Fax: 318-724-6528
www.airmasteraerator.com

AirSep Corp.
260 Creekside Dr.
Buffalo, NY 14228-2075
Phone: 716-691-0202
Fax: 716-691-0707
www.airsep.com

Airvac, Inc.
P.O. Box 528
Rochester, IN 46975
Phone: 219-223-3980
Fax: 219-223-5566
www.airvac.com

Alar Engineering Corp.
9651 W. 196th St.
Mokena, IL 60448
Phone: 708-479-6100
Fax: 708-479-9059

Albright & Wilson Americas
P.O. Box 4439
Glen Allen, VA 23058-4439
Phone: 804-968-6300
Fax: 804-968-6385
www.albright-wilson.com

Alfa Laval Separation, Inc.
955 Mearns Rd.
Warminster, PA 18974-0556
Phone: 215-443-4000
Fax: 215-443-4155
www.alfalaval.com

Alpine Technology, Inc.
1250 Capitol of Texas Hwy, 2-300
Austin, TX 78746
Phone: 512-329-2809
Fax: 512-328-4792

Alsop Engineering Co.
P.O. Box 3449
Kingston, NY 12402
Phone: 914-338-0466
Fax: 914-339-1063

Alzeta Corp.
2343 Calle del Mundo
Santa Clara, CA 95054
Phone: 408-727-8282
Fax: 408-727-9740

Ambient Technologies Inc.
2999 N.E. 191 St., Ste. 407
Aventura, FL 33180
Phone: 305-937-0610
Fax: 305-937-2137
www.ide-tech.com

AMCEC, Inc.
2525 Cabot Dr., Ste. 205
Lisle, IL 60532
Phone: 630-577-0400
Fax: 630-577-0401
www.amcec.com

American Bio Tech, Inc.
3223 Harbor Dr.
St. Augustine, FL 32095
Phone: 904-825-1500
Fax: 904-825-1524

American International Chemical, Inc.
17 Strathmore Rd.
Natic, MA 01760
Phone: 508-655-5805
Fax: 508-655-0927

American Minerals, Inc.
901 E. Eighth Ave., Ste. 200
King of Prussia, PA 19406
Phone: 610-337-8030
Fax: 610-337-8033
www.ceramics.com/enviro-blend

American Products
1049 Southeast Holbrook Ct.
Port St. Lucie, FL 34952
Phone: 561-398-9881
Fax: 561-398-9840
www.americanproducts1.com

American Sigma, Inc.
P.O. Box 820
Medina, NY 14103
Phone: 716-798-5580
Fax: 716-798-5599
www.americansigma.com

AMETEK Drexelbrook
205 Keith Valley Rd.
Horsham, PA 19044
Phone: 215-674-1234
Fax: 215-674-2731
www.drexelbrook.com

Ametek Inc., Process & Analytical
  Division
150 Freeport Rd.
Pittsburgh, PA 15238
Phone: 412-828-9040
Fax: 412-826-0399
www.thermox.com

Ametek, PMT Products
820 Pennsylvania Blvd.
Feasterville, PA 19053-7886
Phone: 215-355-6900
Fax: 215-355-2937
www.ametekusg.com

AMETEK Rotron Biofiltration
75 North St.
Saugerties, NY 12477
Phone: 914-246-3711
Fax: 914-246-3802
www.rotrontmd.com

Amwell, Inc.
1740 Molitor Rd.
Aurora, IL 60505
Phone: 630-898-6900
Fax: 630-898-1647

Analytical Technology, Inc.
P.O. Box 879
Oaks, PA 19456-9969
Phone: 610-917-0991
Fax: 610-917-0992
www.analyticaltechnology.com

Anatel Corp.
2200 Central Ave.
Boulder, CO 80301
Phone: 303-442-5533
Fax: 303-447-8365
www.anatel.com

Andersen Instruments, Inc.
500 Technology Ct.
Smyrna, GA 30082-5210
Phone: 770-319-9999
Fax: 770-319-0336
www.anderseninstruments.com

Andritz Sprout-Bauer S.A.
10 Ave. de Concyr
F-45071 Orleans Cedex 2, France
Phone: 33-3851-5738
Fax: 33-3863-1565

Andritz-Ruthner, Inc.
1010 Commercial Blvd. S.
Arlington, TX 76017
Phone: 817-465-5611
Fax: 817-468-3961
www.andritz.com

Anthratech Western, Inc.
7260 12th St. SE
Calgary, Alberta, T2H 2S5
Phone: 403-255-7377
Fax: 403-255-3129

Applied Biochemists, Inc.
W175 N11163 Stonewood Dr.
Germantown, WI 53022
Phone: 262-255-4449
Fax: 262-255-4268
www.appliedbiochemists.com

Applied Process Technology, Inc.
35 Wellington
Conroe, TX 77034
Phone: 409-539-4099
Fax: 409-539-4089
www.centra-flo.com

Applied Regenerative Technologies Co.
45 Park Place S.
Morristown, NJ 07960
Phone: 201-984-8811
Fax: 201-326-1845

Applied Spectometry
W226 N555G Eastmound Dr.
Waukesha, WI 53186
Phone: 414-896-2650
Fax: N/A
www.chemscan.com

APV Crepaco, Inc.
9525 Bryn Mawr Ave.
Rosemont, IL 60018
Phone: 847-678-4300
Fax: 847-678-4407
www.apv.com

Aqua Ben Corp.
1390 N. Manzanita St.
Orange, CA 92867
Phone: 714-771-6040
Fax: 714-771-1465
www.aquaben.com

Aqua Care Systems
9542 Hardpan Rd.
Angola, NY 14006
Phone: 716-549-2500
Fax: 716-549-3950

Aqua Magnetics International, Inc.
915-B Harbor Lake Dr.
Safety Harbor, FL 34695
Phone: 727-447-2575
Fax: 727-726-8888

Aqua Products, Inc.
25 Rutgers Ave.
Cedar Grove, NJ 07009
Phone: 973-857-2700
Fax: 973-857-8981

Aqua-Aerobic Systems, Inc.
P.O. Box 2026
Rockford, IL 61130-0026
Phone: 815-654-2501
Fax: 815-654-2508
www.aqua-aerobic.com

Aqua-Chem, Inc.
P.O. Box 421
Milwaukee, WI 53201
Phone: 414-359-0600
Fax: 414-577-2723
www.aqua-chem.com

AquaClear Technologies Corp.
1550 Harbor Bou, Ste. 130
West Sacramento, CA 95691
Phone: 916-372-6826
Fax: 916-372-6827

Aqualogic Inc.
30 Devine St.
North Haven, CT 06473
Phone: 203-248-8959
Fax: 203-288-4308
www.aqualogic.com

Aqualytics, Inc.
7 Powderhorn Dr.
Warren, NJ 07059-5191
Phone: 732-563-2800
Fax: 732-563-2816
www.aqualytics.com

AquaPro, Inc.
1573 Fairway View Dr.
Birmingham, AL 35244
Phone: 205-988-5824

Aquarium Systems
8141 Tyler Blvd.
Mentor, OH 44060
Phone: 440-255-1997
Fax: 440-255-8994

Aquatrol Ferr-X Corp.
P.O. Box 531
Toms River, NJ 08754
Phone: 732-505-3100
Fax: 732-505-3038

AquaTurbo Systems
P.O. Box 189
Springdale, AR 72765
Phone: 501-927-1300
Fax: 501-927-0700

Arch Chemicals
1200 Lower River Rd.
Charleston, TN 37310
Phone: 423-780-2600
Fax: 877-321-2724
www.archwaterchemicals.com

Arlat, Inc.
6 Bram Ct.
Brampton, Ontario, Canada L6W 3R6
Phone: 905-457-1700
Fax: 905-457-1730
www.arlat.com

Artisan Industries, Inc.
73 Pond St.
Waltham, MA 02451
Phone: 781-893-6800
Fax: 781-647-0143
www.artisanind.com

Asahi America
P.O. Box 653
Malden, MA 02148-0005
Phone: 781-321-5409
Fax: 781-321-4421
www.asahi-america.com

Ashbrook Corp.
P.O. Box 16327
Houston, TX 77222
Phone: 281-449-0322
Fax: 281-449-1324

Ashland Chemical, Drew Industrial
One Drew Plaza
Boonton, NJ 07005
Phone: 973-236-7600
Fax: 973-263-4495
www.ashchem.com

Astraco Water Engineering
P.O. Box 52
8560 AB Balk,
Phone: 31-(0)514-608765
Fax: 31-(0)514-608766

ATA Technologies Corp.
1240 Valley Belt Rd.
Cleveland, OH 44131
Phone: 216-459-1930
Fax: 216-459-1958
www.ata-technologies.com

Atlantic Ultraviolet Corp.
375 Marcus Blvd.
Hauppauge, NY 11788
Phone: 516-273-0500
Fax: 516-273-0771
www.atlanticuv.com

Atlas Polar Co., Hydrorake Division
P.O. Box 160 Postal Station O
Toronto, Canada M4A 2N3
Phone: 416-751-7740
Fax: 416-751-6475
www.atlaspolar.com

Atlas-Stord, Inc.
309 Regional Rd. S.
Greensboro, NC 27409
Phone: 336-668-7727
Fax: 336-668-0537
www.atlas-stord.com

Auburn Systems LLC
8 Electronics Ave.
Danvers, MA 01923
Phone: 978-777-2460
Fax: 978-777-8820
www.auburnsys.com

Aurora Pump
800 Airport Rd.
Aurora, IL 60542-1494
Phone: 630-859-7000
Fax: 630-859-7060
www.pentairpump.com

Automatic Filters, Inc.
2672 S. La Cienega
Los Angeles, CA 90034
Phone: 310-839-2828
Fax: 310-839-6878
www.tekleen.com

Automation Products, Inc.
3030 Max Roy St.
Houston, TX 77008-9981
Phone: 713-869-0361
Fax: 713-869-7332
www.dynatrolusa.com

Avista Technologies
133 North Pacific St., Ste. E
San Marcos, CA 92069
Phone: 760-744-0536
Fax: 760-744-0619
www.avistatech.com

AWI
P.O. Box 900488
Sandy, UT 84090-0488
Phone: 801-576-1930
Fax: 801-523-8450

Axsia Serck Baker, Inc.
8601 Jameel Ste. 190
Houston, TX 77040
Phone: 713-934-8900
Fax: 713-934-8688
www.axsia.com

Azur Environmental
2232 Rutherford Rd.
Carlsbad, CA 92008-8883
Phone: 760-438-8282
Fax: 760-438-2980
www.azurenv.com

B.F. Goodrich Co.
9911 Brecksville
Brecksville, OH 44141
Phone: 216-447-5000
Fax: 216-447-5250

Badger Meter, Inc.
P.O. Box 581390
Tulsa, OK 74158
Phone: 918-836-8411
Fax: 918-832-9962

Bailey Polyjet
P.O. Box 8070
Fresno, CA 93747
Phone: 559-252-4491
Fax: 559-453-9030

Bailey-Fischer & Porter
125 E. County Line Rd.
Warminster, PA 18974
Phone: 215-674-6000
Fax: 215-674-7183
www.ebpa.com/bfp/

Baker Hughes Process Systems
14990 Yorktown Plaza
Houston, TX 77040
Phone: 713-934-4100
Fax: 713-934-4108
www.bakerhughes.com

Baker Process - Municipal Division
669 West 200 S.
Salt Lake City, UT 84101-1020
Phone: 801-526-2000
Fax: 801-526-2014
www.bakerhughes.com

Baker Process/Ketema
P.O. Box 1406
El Cajon, CA 92022
Phone: 619-449-0202
Fax: 619-449-0883

Baler Equipment Co.
P.O. Box 25150
Portland, OR 97225
Phone: 503-292-4118
Fax: 503-297-5991
www.baler-eqpt.com

Barnes Pumps, Inc.
P.O. Box 603
Piqua, OH 45356-0603
Phone: 419-774-1511
Fax: 419-774-1530

BASF
Carl Bosch Strasse 38
67056 Ludwigshafen, Germany
Phone: 49-621-60-42258
Fax: 49-621-60-72944

BASYS Technologies
1345 Northland Dr.
Mendota Heights, MN 55120
Phone: 612-452-3300
Fax: 612-683-8888
www.basystech.com

Bayer Corp.
100 Bayer Rd.
Pittsburgh, PA 15205
Phone: 412-777-2000
Fax: 412-777-4109
www.bayerus.com

Beaird Industries, Inc.
601 Benton Kelly St.
Shreveport, LA 71106-7198
Phone: 318-865-6351
Fax: 318-868-1701
www.beairdindustries.com

BEKO Condensate Systems Corp.
4140 Tuller Rd.
Dublin, OH 43017
Phone: 614-798-5275
Fax: 614-798-5276

Bethlehem Corp.
25th & Lennox St.
Easton, PA 18045
Phone: 610-258-7111
Fax: 610-258-8154

BetzDearborn, Inc.
4636 Somerton Rd.
Trevose, PA 19053-6783
Phone: 215-355-3300
Fax: 215-953-5524
www.betzdearborn.com

BetzDearborn-Argo District
185 Bosstick Blvd.
San Marcos, CA 92069
Phone: 760-727-2620
Fax: 760-727-3380
www.argoscientific.com

BG-WSA, Inc.
2969 South Chase Ave.
Milwaukee, WI 53207
Phone: 414-481-4133
Fax: 414-481-4134
www.bgwsa.com

BIF
16490 Chillicothe Rd.
Chagrin Falls, OH 44023
Phone: 440-543-5885
Fax: 440-543-9128
www.bifwater.com

BioChem Technology, Inc.
100 Ross Rd., Ste. 201
King of Prussia, PA 19406
Phone: 610-768-9360
Fax: 610-768-9363
www.biochemtech.com

BioLab, Inc.
Tenax Rd., Trafford Park
Manchester, M17 1WT England
Phone: 44-161-875-3875
Fax: 44-161-875-3175
www.wateradditives.com

BioLab Inc.
P.O. Box 1489
Decatur, GA 30031-1489
Phone: 404-378-1761
Fax: 404-371-0373
www.wateradditives.com

Bioprime, Ltd.
14 N. Main St.
White River Junction, VT 05001
Phone: 802-291-6123
Fax: 802-291-6124
www.bioprime.com

Biorem Technologies, Inc.
7496 Wellington Rd.
Guelph, Ontario, Canada N1H 6H9
Phone: 519-767-9100
Fax: 519-767-1824
www.bioremtechnologies.com

BIOS International Corp.
10 Park Pl.
Butler, NJ 07405
Phone: 973-492-8400
Fax: 973-492-8270
www.biosint.com

Bioscience, Inc.
1550 Valley Center Pkwy, Ste. 140
Bethlehem, PA 18017
Phone: 610-974-9697
Fax: 610-691-2170
www.bioscienceinc.com

Biosystems, Inc.
651 S. Main
Middletown, CT 06481
Phone: 860-344-1079
Fax: 860-344-1068
www.biosystems.com

Biothane Corp.
2500 Broadway, Drawer 5
Camden, NJ 08104
Phone: 856-541-3500
Fax: 856-541-3366
www.biothane.com

Biothermica International, Inc.
3333 Cavendish Blvd. Ste. 440
Montreal, Canada H4B QM5
Phone: 514-488-3881
Fax: 514-488-3125

Biotrol
10300 Valley View Rd., Ste. 107
Eden Prairie, MN 55344
Phone: 612-942-8032
Fax: 612-942-8526

Bioway America
9003H Lincoln Dr. W.
Marlton, NJ 08053
Phone: 609-268-6845
Fax: 609-268-3482
www.bioway.net

Blace Filtronics, Inc.
2310 E. 2nd St.
Vancouver, WA 98661
Phone: 360-750-7709
Fax: 360-750-7715

Blue-White Industries
14931 Chestnut St.
Westminster, CA 92683
Phone: 714-893-8529
Fax: 714-894-9492
www.bluwhite.com

BOC Gases
575 Mountain Ave.
Murray Hill, NJ 07974
Phone: 908-464-8100
Fax: 888-262-3298
www.boc.com

Boliden Intertrade, Inc.
3379 Peachtree, Ste. 300
Atlanta, GA 30326
Phone: 404-239-6700
Fax: 404-239-6701

Brackett Geiger
P.O. Box 210163
76151 Karlsruhe, Germany
Phone: 49-721-5001-355
Fax: 49-721-5001-344
www.brackettgeiger.de

Brackett Geiger Ltd.
Severalls Lane
Colchester Essex, C04 4PD England
Phone: 44-1206-756600
Fax: 44-1206-756500
www.brackettgeiger.co.uk

Brackett Geiger USA, Inc.
1335 Regents Park Dr, Ste. 140
Houston, TX 77058
Phone: 281-480-7955
Fax: 281-480-8225
www.bgusa.com

Bran+Luebbe
1025 Busch Pkwy.
Buffalo Grove, IL 60089
Phone: 847-520-0700
Fax: 847-520-0855
www.bran-luebbe.com

Brandt Co.
P.O. Box 2327
Conroe, TX 77305-2327
Phone: 409-756-4800
Fax: 409-756-8102

Brentwood Industries, Inc.
P.O. Box 605
Reading, PA 19603
Phone: 610-374-5109
Fax: 610-376-6022

Browning-Ferris Industries, Inc.
P.O. Box 3151
Houston, TX 77253
Phone: 281-870-8100
Fax: 281-870-7000
www.bfi.com

Brunel Corp.
1304 Twin Oaks
Wichita Falls, TX 76302
Phone: 940-723-7800
Fax: 940-723-7888

BTG, Inc.
2815 Colonnades Ct.
Norcross, GA 30071
Phone: 770-209-6900
Fax: 770-447-1128

Buffalo Technologies, Inc.
P.O. Box 1041
Buffalo, NY 14240
Phone: 716-895-2100
Fax: 716-895-8263
www.buffalotechnologies.com

Calciquest, Inc.
P.O. Box 1709
Belmont, NC 28012
Phone: 704-394-9868
Fax: 704-394-6784
www.calciquest.com

Calgon Carbon Corp.
P.O. Box 717
Pittsburgh, PA 15230-0717
Phone: 412-787-6700
Fax: 412-787-6713
www.calgoncarbon.com

Calgon Corp.
P.O. Box 1346
Pittsburgh, PA 15230
Phone: 412-777-8000
Fax: 412-777-8154

Capital Controls Co.
P.O. Box 211
Colmar, PA 18915
Phone: 215-997-4000
Fax: 215-997-4062
www.capitalcontrols.com

Carboline Co.
350 Hanley Industrial Ct.
St. Louis, MO 63144
Phone: 314-644-1000
Fax: 314-644-4617
www.carboline.com

Carbonite Filter Corp.
P.O. Box 1
Delano, PA 18220
Phone: 717-467-3350
Fax: 717-467-7272

Carus Chemical Co., Inc.
P.O. Box 599
Peru, IL 61354-0599
Phone: 815-223-1500
Fax: 815-224-6697
www.caruschem.com

CASS Water Engineering, Inc.
17671 Cowan, Ste. 150
Irvine, CA 92614-6031
Phone: 949-474-3933
Fax: 949-955-2021
www.sbrcass.com

Catalytic Combustion Corp.
709 21st Ave.
Bloomer, WI 54724
Phone: 715-568-2882
Fax: 715-568-2884
www.catalyticcombustion.com

Catalytic Products International
980 Ensell Rd.
Lake Zurich, IL 60047-1557
Phone: 847-438-0334
Fax: 847-438-0944
www.cpilink.com

CBI Walker, Inc.
601 W. 143rd St.
Plainfield, IL 60544
Phone: 815-439-3100
Fax: 815-439-4010
www.chicago-bridge.com

Celgard LLC
13800 S. Lakes Dr.
Charlotte, NC 28273
Phone: 704-588-5310
Fax: 704-587-8585
www.liquicel.com

Celite Corp.
P.O. Box 519
Lampoc, CA 93438-0519
Phone: 805-735-7791
Fax: 805-735-5699

Centrisys Corp.
501 North Ave.
Libertyville, IL 60048-0441
Phone: 847-816-8210
Fax: 847-367-1787

Chemineer, Inc.
P.O. Box 1123
Dayton, OH 45401
Phone: 937-454-3200
Fax: 937-454-3379
www.chemineer.com

Chemtrac Systems, Inc.
P.O. Box 921188
Norcross, GA 30092
Phone: 770-449-6233
Fax: 770-447-0889
www.chemtrac.com

Chemunex, Inc.
1 Deer Park Dr. Ste. H2
Monmouth Jct., NJ 08852
Phone: 732-329-1493
Fax: 732-329-1182

Cherrington Corp.
P.O. Box A
Fairfax, MN 55332
Phone: 507-426-7261
Fax: 507-426-7263

Chief Industries, Inc.
Box 2078
3942 W. Old Highway 30
Grand Isle, NE 68802-2078
Phone: 308-389-7296
Fax: 308-381-8475
www.chiefind.com

Chlorinators, Inc.
1044 SE Dixie Cutoff Rd.
Stuart, FL 34994
Phone: 561-288-4854
Fax: 561-287-3238
www.regalchlorinators.com

Church & Dwight Co. Inc.
469 N. Harrison St.
Princeton, NJ 08543-5297
Phone: 609-497-7574
Fax: 609-497-7176
www.churchdwight.com

Ciba Specialty Chemicals
P.O. Box 820
Suffolk, VA 23434-0820
Phone: 757-538-3700
Fax: 757-538-3989
www.cibasc.com

Clack Corp.
P.O. Box 500
Windsor, WI 53598-0500
Phone: 608-846-3010
Fax: 608-846-2586
www.clackcorp.com

Claude Laval Corp.
1365 N. Clovis Ave.
Fresno, CA 93727
Phone: 559-255-1601
Fax: 559-255-8093
www.lakos-laval.com

Clear-Flo International
6224 Lakeland Ave. N. #108
Brooklyn Park, MN 55428-2937
Phone: 612-504-4554
Fax: 612-504-4553
www.clean-flo.com

Cleaver-Brooks
P.O. Box 421
Milwaukee, WI 53201
Phone: 414-359-0600
Fax: 414-577-3185
www.aqua-chem.com

ClorTec
1077 Dell Ave., Ste. A
Campbell, CA 95008
Phone: 408-871-1300
Fax: 408-871-1314
www.clortec.com

CMI-Schneible Co.
P.O. Box 100
Holly, MI 48442
Phone: 248-634-8211
Fax: 248-634-2240

CMS Group, Inc.
140 Snow Blvd., Ste. 200
Concord Ontario, Canada L4K 4N9
Phone: 905-660-7580
Fax: 905-660-0243
www.rotordisk.com

Cochrane Inc.
P.O. Box 60191
King of Prussia, PA 19406
Phone: 610-265-5050
Fax: 610-265-5432
www.cochrane.com

Colloid Environmental Technologies Co.
1500 West Shure Dr.
Arlington Heights, IL 60004-1434
Phone: 847-392-5800
Fax: 847-577-5571
www.cetco.com

Composite Structures
P.O. Box 737
Aromas, CA 95004
Phone: 831-726-2644
Fax: 408-726-2609
www.accuraflo.com

Conservatek Industries, Inc.
498 Loop 336 E,
Conroe, TX 77305
Phone: 409-539-1747
Fax: 409-539-5355
www.conservatek.com

Conversion Systems, Inc.
1155 Business Center Dr.
Horsham, PA 19044
Phone: 215-956-5500
Fax: 215-956-5433

Cook Screen Technologies, Inc.
460 Hoover St. NE
Minneapolis, MN 55413
Phone: 612-617-4470
Fax: 612-379-3149
www.pramfiltration

Coors Ceramics Co.
1100 Commerce Park Dr.
Oak Ridge, TN 37830
Phone: 423-481-8021
Fax: 423-481-0090

Copa Group
Crest Industrial Estate, Marden
Tonebridge, Kent, TN12 9QJ England
Phone: 162-2-832444
Fax: 162-2-831466

Corning, Inc.
45 Nagog Park
Acton, MA 01720
Phone: 978-635-2200
Fax: 978-635-2476
www.scienceproducts.corning.com

Corrpro Waterworks, Inc.
2421 Iorio St.
Union, NJ 07083-8105
Phone: 908-686-1770
Fax: 908-686-1704

Coster Engineering
P.O. Box 3407
Mankato, MN 56002
Phone: 507-625-6621
Fax: 507-625-5883

Crane Pumps & Systems
P.O. Box 603
Piqua, OH 45356
Phone: 937-615-3595
Fax: 937-773-7157

Cromaglass Corp.
P.O. Box 3215
Williamsport, PA 17701
Phone: 570-326-3396
Fax: 570-326-6426

Crompton & Knowles Colors, Inc.
P.O. Box 33188
Charlotte, NC 28233
Phone: 704-372-5890
Fax: 704-332-8785

Cross Machine, Inc.
167 Glen Ave.
Berlin, NH 03570
Phone: 603-752-6111
Fax: 603-752-3825

CSM Worldwide/RaySolv
P.O. Box 207
Bound Brook, NJ 08805
Phone: 732-981-0500
Fax: 732-356-3629
www.csmworldwide.com

Culligan International Corp.
One Culligan Pkwy.
Northbrook, IL 60062
Phone: 847-205-6000
Fax: 847-205-6005
www.culligan.com

Cytec Industries, Inc.
Five Garret Mountain Plaza
West Paterson, NJ 07424
Phone: 973-357-3100
Fax: 973-357-3065
www.cytec.com

D.R. Sperry & Co.
112 North Grant St.
North Aurora, IL 60542
Phone: 630-892-4361
Fax: 630-892-1664

D&A Instrument Co.
40 Seton Rd.
Port Townsend, WA 98368
Phone: 360-385-0272
Fax: 360-385-0460

Dacar Chemical Co.
1007 McCartney St.
Pittsburgh, PA 15220
Phone: 412-921-3620
Fax: 412-921-4478

Danfoss/Instrumark
1124 Wrigley Way
Milpitas, CA 95035
Phone: 408-262-0717
Fax: 408-262-3610

David Brown Union Pumps Co.
4600 W. Dickman Rd.
Battle Creek, MI 49015-1098
Phone: 616-966-4600
Fax: 616-962-3534
www.unionpump.com

DBS Manufacturing, Inc.
5421 Hillside Dr.
Forest Park, GA 30297
Phone: 404-768-2131
Fax: 404-761-6360
www.dbsmfg.com

Dedert Corp.
2000 Governors Dr.
Olympia Fields, IL 60461-1074
Phone: 708-747-7000
Fax: 708-755-8815
www.dedert.com

Delta Chemical Corp.
2601 Cannery Ave.
Baltimore, MD 21226
Phone: 410-354-0100
Fax: 410-354-1021

Delta Cooling Towers, Inc.
P.O. Box 952
Fairfield, NJ 07004-2970
Phone: 973-227-0300
Fax: 973-227-0458
www.deltacooling.com

Derrick Corp.
588 Duke Rd.
Buffalo, NY 14225
Phone: 716-683-9010
Fax: 716-683-4991

DesalCo Ltd
48 Par-la-Ville, Ste. 381
Hamilton HM 11, Bermuda
Phone: 441-292-2060
Fax: 441-292-2024
www.desalco.bm

Dexsil Corp.
One Hamden Park Dr.
Hamden, CT 06517
Phone: 203-288-3509
Fax: 203-248-6523

DHV Water BV
P.O. Box 484
3800 AL Amersfoort, The Netherlands
Phone: 31-33682200
Fax: 31-33682301

Diagenex, Inc.
288 Lindbergh Ave.
Livermore, CA 94550
Phone: 925-606-5600
www.diagenex.com

Dieterich Standard
P.O. Box 9000
Boulder, CO 80301
Phone: 303-530-9600
Fax: 303-530-7064

Diversified Remediation Controls, Inc.
21801 Industrial Blvd.
Rogers, MN 55374
Phone: 612-428-3000
Fax: 612-428-3660

Dontech, Inc.
76 Center Dr.
Gilberts, IL 60136
Phone: 847-428-8222
Fax: 847-428-6855

Douglas Engineering
1015 Shary Circle
Concord, CA 94518
Phone: 925-827-4100
Fax: 925-827-4999
www.douglaseng.com

Dow Chemical Co.
P.O. Box 1206
Midland, MI 48641-9940
Phone: 517-636-1000
Fax: 517-638-9783
www.dow.com/liquidseps

Dresser Industries/Roots Division
900 West Mount St.
Connersville, IN 47331
Phone: 765-827-9200
Fax: 765-825-7669
www.rootsblower.com

Duall Division, Met-Pro Corp.
1550 Industrial Dr.
Owosso, MI 48867
Phone: 517-725-8184
Fax: 517-725-8188
www.met-pro.com

Dürr Environmental, Inc.
P.O. Box 930459
Wixom, MI 48393-0459
Phone: 248-668-5200
Fax: 248-926-6570
www.durrenvironmental.com

Dustex Corp.
P.O. Box 7368
Charlotte, NC 28241-7368
Phone: 704-588-2030
Fax: 704-588-2032
www.dustex.com

DWT-Engineering Oy
Raviraitti 3
23800 Laitila, Finland
Phone: 358-(0)2-461-800
Fax: 358-(0)2-461-8400

Dynal, Inc.
5 Delaware Dr.
Lake Success, NY 11042
Phone: 516-326-3270
Fax: 516-326-3298

Dynaphore, Inc.
2709 Willard Rd.
Richmond, VA 23294
Phone: 804-672-3464
Fax: 804-282-1325

E. Beaudrey & Co.
14 Boulevard Orano
75018 Paris, France
Phone: 42-57-14-35
Fax: 42-64-74-62

E.I. Dupont De Nemours, Inc.
P.O. Box 6101
Newark, DE 19714-6101
Phone: 302-774-1011
Fax: 302-892-1705

E.S.G. Manufacturing, LLC
P.O. Box 431
Holden, LA 70744
Phone: 225-567-9200
Fax: 225-567-3091
www.pumpit.com

E-Cell Corp.
52 Royal Rd.
Guelph, Ontario, N1H 1G3
Phone: 519-836-2260
Fax: 519-836-0982
www.ecell.com

Eagle-Picher Minerals, Inc.
P.O. Box 12130
Reno, NV 89510
Phone: 775-824-7600
Fax: 775-824-7694
www.epcorp.com

Eaglebrook, Inc.
4801 Southwick Dr.
Matteson, IL 60443
Phone: 708-747-5038
Fax: 708-747-3278
www.eaglebrook.net

Earth Science Laboratories, Inc.
P.O. Box 5007
Bella Vista, AR 72714
Phone: 501-855-5800
Fax: 501-855-5806
www.earthsciencelabs.com

Eco Purification Systems USA, Inc.
1450 South Rolling Rd.
Baltimore, MD 212227
Phone: 410-455-5770
Fax: 410-455-5777
www.eps.com

Ecolab Inc.
1345 Taney
North Kansas City, MO 64116
Phone: 816-842-0560
Fax: 816-842-6388

Ecolochem, Inc.
P.O. Box 12775
Norfolk, VA 23502
Phone: 757-855-9000
Fax: 757-855-1478
www.ecolochem.com

Ecoloquip Inc.
1657 Oak Tree Dr.
Houston, TX 77080
Phone: 713-849-1984
Fax: 713-827-9396
www.ecoloquip.com

Ecolotech Corp.
P.O. Box 1268
Flat Rock, NC 28731-1268
Phone: 828-692-2276
Fax: 828-692-2229

Eden Equipment Co.
17552 Griffin
Huntington Beach, CA 92647
Phone: 714-842-8181
Fax: 714-842-1284

EET Inc.
4710 Bellaire Blvd., Ste. 300
Houston, TX 77401
Phone: 713-662-0727
Fax: 713-662-2322

Eichrom Industries, Inc.
8205 South Cass, Ste. 111
Darien, IL 60561
Phone: 630-963-0320
Fax: 630-963-0381

Ejector Systems, Inc.
232 Westgate Dr.
Carol Stream, IL 60188
Phone: 630-668-5150
Fax: 630-668-6270

EKA Chemicals
1775 West Oak Commons Ct.
Marietta, GA 30062
Phone: 770-578-0858
Fax: 770-321-5865
www.ekachem.com

Ekokan
P.O. Box 5354
Cary, NC 27512
Phone: 919-469-3727
Fax: 919-467-0294

Elf Atochem North America, Inc.
2000 Market St.
Philadelphia, PA 19103
Phone: 215-419-7000
Fax: 215-419-5230
www.elf-atochem.com

EMI, Inc.
P.O. Box 912
Clinton, CT 06413
Phone: 860-669-1199
Fax: 860-669-7461

EncyNova International, Inc.
557 C Burbank St.
Broomfield, CO 80020
Phone: 303-404-3583
Fax: 303-466-7105

Endress+Hauser
2350 Endress Pl.
Greenwood, IN 46143
Phone: 317-535-7138
Fax: 317-535-8498
www.endress

Enduro Composite Systems
1005 Bluemound Rd.
Fort Worth, TX 76131
Phone: 817-232-1127
Fax: 817-232-1582

Energy Recovery, Inc.
820 Greenbriar Circle, Ste. 24
Chesapeake, VA 23320
Phone: 757-420-8149
Fax: 757-523-7349

Engelhard Corp.
554 Engelhard Dr.
Seneca, SC 29678
Phone: 864-882-9841
Fax: 864-885-1374

Engineering Specialties, Inc.
P.O. Box 2960
Covington, LA 70434
Phone: 504-892-0071
Fax: 504-892-0474

Entoleter, Inc.
251 Welton St.
Hamden, CT 06517
Phone: 203-787-3575
Fax: 203-787-1492

Enviro-Care Co.
1214 Shappert Dr.
Rockford, IL 61115
Phone: 815-282-9064
Fax: 815-282-4121
www.enviro-care.com

Enviroflow, Inc.
12181 Balls Ford Rd.
Manassas, VA 22110
Phone: 703-368-9067
Fax: 703-368-7336

Envirogen
4100 Quakerbridge Rd.
Lawrenceville, NJ 08648
Phone: 609-936-9300
Fax: 609-936-9221

Environetics, Inc.
1201 Commerce St.
Lockport, IL 60441
Phone: 815-838-8331
Fax: 815-838-8336
www.environeticsinc.com

Environment One Corp.
2773 Balltown Rd.
Niskayuna, NY 12309-1090
Phone: 518-346-6161
Fax: 518-346-6188
www.eone.com

Environmental Devices Corp.
80 Essex Hill
Haverhill, MA 01832
Phone: 978-521-1514
Fax: 978-521-1628

Environmental Dynamics Inc.
5601 Paris Rd.
Columbia, MO 65202-9399
Phone: 573-474-9456
Fax: 573-474-6988
www.wastewater.com

Enviropax, Inc.
P.O. Box 65039
Salt Lake City, UT 84165
Phone: 801-263-8880
Fax: 801-263-8898

Enviroquip, Inc.
P.O. Box 9069
Austin, TX 78728-8519
Phone: 512-834-6000
Fax: 512-834-6039
www.enviroquip.com

EnviroSystems Supply
735 Commerce Circle
Longwood, FL 32750
Phone: 407-767-8007
Fax: 407-767-2474
www.alenco.com

Envirotech Pumpsystems
P.O. Box 209
Salt Lake City, UT 84110-0209
Phone: 801-359-8731
Fax: 801-530-7531

EPG Companies, Inc.
P.O. Box 427
Rogers, MN 55374
Phone: 612-424-2613
Fax: 612-493-4812

Equipment Manufacturing Corp.
2615 Pacific Park Dr.
Whittier, CA 90601
Phone: 562-908-7696
Fax: 562-908-7698

Etus, Inc.
1511 Kastner Pl.
Sanford, FL 32771
Phone: 407-321-7910
Fax: 407-321-3098
www.env-sol.com/etus

Eutek Systems, Inc.
El Camino Ave., #100
Carmichael, CA 95608-4733
Phone: 916-972-9272
Fax: 916-972-9357

Everfilt Corp.
3167 Progress Circle
Mira Loma, CA 91752
Phone: 909-360-8380
Fax: 909-360-8384

Exceltec International Corp.
1110 Industrial Blvd.
Sugar Land, TX 77478
Phone: 281-240-6770
Fax: 281-240-6762
www.sanilec.com

F.B. Leopold Co., Inc.
227 S. Division St.
Zelienople, PA 16063
Phone: 724-452-6300
Fax: 724-452-1377
www.fbleopold.com

F.E. Myers Co.
1101 Myers Pkwy.
Ashland, OH 44805
Phone: 419-289-1144
Fax: 419-289-6658
www.pentairpump.com

Fairbanks Morse Pump Corp.
3601 Fairbanks Ave.
Kansas City, KS 66109-0999
Phone: 913-371-5000
Fax: 913-371-2272
www.pentairpump.com

Fairfield Service Co.
240 Boone Ave.
Marion, OH 43302
Phone: 740-387-3335
Fax: 740-387-4869
www.fairfieldengineering.com

Falmouth Products
Box 541
Falmouth, MA 02541
Phone: 508-548-6686
Fax: 508-548-8144

Farmer Automatic of America, Inc.
P.O. Box 39
Register, GA 30452
Phone: 912-681-2763
Fax: 912-681-1096
www.farmerautomaticusa.com

Farr Co.
3501 Airport Rd.
Jonesboro, AR 72401
Phone: 870-933-8048
Fax: 800-222-6891
www.farrco.com

FE3, Inc.
P.O. Box 808
Celina, TX 75009
Phone: 972-382-2381
Fax: 972-382-3211
www.fe3.com

Fen-Tech Environmental, Inc.
4306 South Hwy. 377
Brownwood, TX 76801
Phone: 800-777-4512
Fax: 830-393-4092

Ferro Corp.
P.O. Box 389
East Rochester, NY 14445
Phone: 716-586-8770
Fax: 716-586-7154

Filter Products
25 Grants Mill Rd.
Wrentham, MA 02093
Phone: 508-384-0233
Fax: 508-384-0234

Filter Specialists, Inc.
P.O. Box 735
Michigan City, IN 46361
Phone: 219-879-3307
Fax: 219-877-0632
www.fsifilters.com

Filtra-Systems Hydromation
4000 Town Center, Ste. 100
Southfield, MI 48075
Phone: 248-356-9090
Fax: 248-356-2818

Filtronics, Inc.
4000 Leaverton Ct.
Anaheim, CA 92807
Phone: 714-630-5040
Fax: 714-630-1160

Fisher Controls International, Inc.
P.O. Box 190
Marshalltown, IA 50158
Phone: 515-754-3000
Fax: 515-754-2830

Flo Trend Systems, Inc.
707 Lehman
Houston, TX 77018-1513
Phone: 713-699-0152
Fax: 713-699-8054
www.flotrend.com

Flow Process Technologies, Inc.
11917 Windfern Rd.
Houston, TX 77064
Phone: 281-469-2777
Fax: 281-469-2232

Floway Pumps, Inc.
2494 S. Railroad Ave.
Fresno, CA 93706
Phone: 209-442-4000
Fax: 209-442-3098

Fluid Dynamics, Inc.
6595 Odell Place, Ste. E
Boulder, CO 80301-3316
Phone: 303-530-7300
Fax: 303-530-7754

Fluidyne Corp.
2816 West First St.
Cedar Falls, IA 50613
Phone: 319-266-9967
Fax: 319-277-6034
www.fluidynecorp.com

FMC Corp., Hydrogen Peroxide
 Division
US Highway #1
Princeton, NJ 08543
Phone: 609-951-3122
Fax: 609-951-3668

Force Flow Equipment
1150-D Burnett Ave.
Concord, CA 94520
Phone: 925-686-6700
Fax: 925-686-6713
www.forceflow.com

Ford Hall Co., Inc.
P.O. Box 54312
Lexington, KY 40555
Phone: 606-624-3320
Fax: 606-624-3320
www.fordhall.com

Formulabs, Inc.
1710 Commerce Dr.
Piqua, OH 45356
Phone: 937-773-8933
Fax: 937-773-3055

Franklin Miller, Inc.
60 Okner Pkwy.
Livingston, NJ 07039
Phone: 973-535-9200
Fax: 973-535-6269
www.franklinmiller.com

Futura Coatings, Inc.
9200 Latty Ave.
Hazelwood, MO 63042-2805
Phone: 314-521-4100
Fax: 314-521-7255

G.E.T. Industries, Inc.
P.O. Box 640
Brampton, Ontario, Canada L6V 2L6
Phone: 905-451-9900
Fax: 905-451-5376
www.grindhog.com

GA Industries, Inc.
9025 Marshall Rd.
Cranberry Township, PA 16046-3696
Phone: 724-776-1020
Fax: 724-776-1254
www.gaindustries.com

Gardner Denver Blower Division
100 Gardner Park
Peachtree City, GA 30209
Phone: 770-486-5655
Fax: 770-486-5629
www.gardnerdenver.com

Garland Manufacturing Co.
P.O. Box 538
Saco, ME 04072-0538
Phone: 207-283-3693
Fax: 207-283-4834

Gast Manufacturing, Inc.
P.O. Box 97
Benton Harbor, MI 49023
Phone: 616-926-6171
Fax: 616-925-8288
www.gastmfg.com

Gauld Equipment Sales Co.
P.O. Box 1129
Theodore, AL 36582
Phone: 334-653-8558
Fax: 334-653-0533
www.gauld.com

GDT Corp.
20805 N. 19th St.
Phoenix, AZ 85027
Phone: 623-587-8858
Fax: 623-587-1511
www.gdt-h2o.com

GE Reuter-Stokes
Edison Park, 8499 Darrow Rd.
Twinsburg, OH 44087-2398
Phone: 330-435-3755
Fax: 330-425-4045

Gebr. Bellmer GmbH
P.O. Box 1369
Niefern-Oschelbronn, D-75220
Germany
Phone: 072-33-740
Fax: 072-74-215

General Chemical Corp.
90 E. Halsey Rd.
Parsippany, NJ 07054
Phone: 973-515-0900
Fax: 973-515-4461
www.genchem.com

Geo-Chem Technologies, Inc.
57436 Gearharts Landing Rd.
Three Rivers, MI 49093
Phone: 616-244-5373
Fax: 616-244-5373

Geoenergy International Corp.
7617 South 180th St.
Kent, WA 98032
Phone: 425-251-0407
Fax: 425-251-0414
www.geoenergy.com

Geotech/ORS Environmental Systems
1441 West 46th Ave. #17
Denver, CO 80211
Phone: 303-433-7101
Fax: 303-477-1230
www.geotechenv.com

Geschmay GmbH
P.O Box 460
D-73035 Goppingen, Germany
Phone: 07161-604-0
Fax: 07161-604-105
www.geschmay.com

Girard Industries, Inc.
6531 N. Eldridge Pkwy.
Houston, TX 77041
Phone: 713-466-3100
Fax: 713-466-8050

GL&V/Dorr-Oliver, Inc.
612 Wheelers Farm Rd.
Milford, CT 06460-8719
Phone: 203-876-5400
Fax: 203-876-5432
www.dorr-oliver.com

Glegg Water Conditioning Co.
29 Royal Rd.
Guelph, Ontario, Canada N1H 1G2
Phone: 519-836-0500
Fax: 519-836-9373
www.glegg.com

GLI International
9020 W. Dean Rd.
Milwaukee, WI 53224
Phone: 414-355-3601
Fax: 414-355-8346
www.gliint.com

Global Technologies
8855 N. 55th St.
Milwaukee, WI 53223-2358
Phone: 414-365-6430
Fax: 414-365-6410
www.anguil.com

GNESYS, Inc.
3259 Whitebrook Dr., Ste. D100
Memphis, TN 38118
Phone: 901-794-2665
Fax: 901-794-1960
www.hydrasep.com

Goal Line Environmental Technologies
11141 Outlet Dr.
Knoxville, TN 37932
Phone: 423-671-4045
Fax: 423-671-4047
www.glet.com

Goodnature Products, Inc.
P.O. Box 866
Buffalo, NY 14240
Phone: 716-855-3325
Fax: 716-855-3328

Goodtech ASA
P.O. Box 429
N-1301 Sandvika, Norway
Phone: 47-6755-1999
Fax: 47-6755-1990
www.goodtech.no

Goodwin Pumps of America
One Floodgate Rd.
Bridgeport, CT 08014
Phone: 609-467-3636
Fax: 609-467-4841

Gorman-Rupp Co.
305 Bowman St.
Mansfield, OH 44901
Phone: 419-755-1011
Fax: 419-755-1251
www.gormanrupp.com

Graham Manufacturing Co.
20 Florence Ave.
Batavia, NY 14020
Phone: 716-343-2216
Fax: 716-343-1097
www.graham-mfg.com

Grande, Novac & Associates, Inc.
3532 Ashby
Montreal, Quebec, Canada H4R 2C1
Phone: 514-339-1131
Fax: 514-339-9720
www.gnainc.com

Graver Co.
750 Walnut Ave.
Crawford, NJ 07016-3348
Phone: 908-653-4200
Fax: 908-653-4300
www.graver.com

Gravity Flow Systems, Inc.
9542 Hardpan Rd.
Angola, NY 14006
Phone: 716-549-2500
Fax: 716-549-6250
www.aquacarefsd.com

Great Lakes International, Inc.
1905 Kearny Ave.
Racine, WI 53403
Phone: 262-634-2386
Fax: 262-634-6259

Gundle Lining Systems, Inc.
19103 Gundle Rd.
Houston, TX 77073
Phone: 281-443-8564
Fax: 281-875-6010
www.gseworld.com

Guzzler Manufacturing, Inc.
575 N. 37th St.
Birmingham, AL 35222
Phone: 205-591-2477
Fax: 205-591-2495

Gyulavari Consulting Kft.
Post Office Box 62
H-1399 Budapest, Hungary
Phone: 36-1-318-5688
Fax: 36-1-137-1526

H.I.L. Technology, Inc.
94 Hutchins Dr.
Portland, ME 04102
Phone: 207-756-6200
Fax: 207-756-6212
www.hil-tech.com

H&H Eco Systems, Inc.
505 Evergreen Dr.
North Bonneville, WA 98639
Phone: 509-427-7353
Fax: 509-427-3627

Hach Co.
P.O. Box 389
Loveland, CO 80539-0389
Phone: 970-669-3050
Fax: 970-669-2932
www.hach.com

Hadley Industries
5900 West Fourth St.
Ludington, MI 49431
Phone: 231-845-0537
Fax: 231-843-3882

Hammonds
15760 W. Hardy #400
Houston, TX 77060
Phone: 281-820-5674
Fax: 281-847-1857
www.hammondscos.com

Hankin Environmental Services, Inc.
P.O. Box 935
Somerville, NJ 08876
Phone: 908-722-9595
Fax: 908-722-9514

Hans Huber GmbH
Postfach 63
D-92332 Berching, Germany
Phone: 49-8462-2010
Fax: 49-8462-27103
www.huber.de

Hans Huber/Huber Technology
5825 Glenridge Dr. #3-101G
Atlanta, GA 30328
Phone: 404-250-3582
Fax: 404-250-3583
www.huber-technology.com

Harmsco Filtration Products
P.O. Box 14066
North Palm Beach, FL 33408
Phone: 561-848-9628
Fax: 561-845-2474
www.harmsco.com

Harris Waste Management Group, Inc.
200 Clover Reach Dr.
Peachtree City, GA 30269
Phone: 770-631-7290
Fax: 770-631-7299

Haynes International, Inc.
1020 W. Park Ave.
Kokomo, IN 46904
Phone: 765-456-6000
Fax: 765-456-6905

Hayward Industrial Products, Inc.
Box 18
Elizabeth, NJ 07207
Phone: 908-351-5400
Fax: 908-351-7893

Hazleton Environmental, Inc.
125 Butler Dr.
Hazleton, PA 18201
Phone: 570-454-7515
Fax: 570-454-7520
www.hazletonenvr.com

Headworks, Inc.
9601 Katy Freeway, Ste. 470
Houston, TX 77024
Phone: 713-647-6667
Fax: 713-647-0999

Healy-Ruff Co.
2485 North Fairview Ave.
St. Paul, MN 55113
Phone: 651-633-7522
Fax: 651-633-2671

Heinkel Filtering Systems, Inc.
520 Sharptown Rd.
Swedesboro, NJ 08085
Phone: 856-467-3399
Fax: 856-467-1010

Heyl & Patterson, Inc.
P.O. Box 36
Pittsburgh, PA 15230-0036
Phone: 412-788-6900
Fax: 412-788-6913

HF Scientific, Inc.
3170 Metro Pkwy.
Ft. Myers, FL 33916-7597
Phone: 941-337-2116
Fax: 941-332-7643
www.hfscientific.com

Hi-Tech Environmental, Inc.
P.O. Box 360597
Birmingham, AL 35236
Phone: 205-987-8976
Fax: 205-987-8996
www.hi-techenv.com

Highland Tank & Manufacturing
1 Highland Rd.
Stoystown, PA 15563
Phone: 814-893-5701
Fax: 814-893-6126
www.highlandtank.com

Hindon Corp.
2055 Bee's Ferry Rd.
Charleston, NC 29414
Phone: 843-763-6616
Fax: 843-763-2338

Hinsilblon Laboratories
516-1 SE 47th Terrace
Cape Coral, FL 33904
Phone: 941-540-7766
Fax: 941-540-0076

Hitachi Maxco, Ltd.
1630 Cobb International Rd.
Kennesaw, GA 30152
Phone: 770-424-9350
Fax: 770-424-9145
www.hitmax.com

Hitachi Metals America, Ltd.
2400 Westchester Ave.
Purchase, NY 10577
Phone: 914-694-9200
Fax: 914-694-9279

Hoffland Environmental Inc.
10391 Silver Springs Rd.
Conroe, TX 77303
Phone: 409-856-4515
Fax: 409-856-4589

Homa Pump Technology
18 Elmcroft
Stamford, CT 06902
Phone: 203-327-6365
Fax: 203-356-1064

Hosokawa Bepex Corp.
333 Taft St. NE
Minneapolis, MN 55413
Phone: 612-331-4370
Fax: 612-331-1046
www.bepex.com

Howe-Baker Engineers, Inc.
P.O. Box 956
Tyler, TX 75710
Phone: 903-597-0311
Fax: 903-597-8670

HRI-Biotek
203 S. First St., Ste. 100
Lufkin, TX 75901
Phone: 409-632-9945
Fax: 409-632-9948
www.hri-rig.com

Huber+Suhner AG
CH-8330 Pfaffikon, Switzerland
Phone: 41-1952-2211
Fax: 41-1952-2552
www.hubersuhner.com

Hungerford & Terry, Inc.
P.O. Box 650
Clayton, NJ 08312
Phone: 609-881-3200
Fax: 609-881-6859

Huntington Environmental Systems, Inc.
707C West Algonquin Rd.
Schaumburg, IL 60005
Phone: 847-545-8800
Fax: 847-545-1946
www.Huntington1.com

Huron Tech Corp.
Box 189
Delco, NC 28436
Phone: 910-655-3845
Fax: 910-655-3892

Hychem, Inc.
10014 N. Dale Maybry Hwy.
Tampa, FL 33618
Phone: 813-963-6214
Fax: 813-960-0175
www.hychem.com

Hyde Marine, Inc.
28045 Ranney Pkwy.
Westlake, OH 44145
Phone: 440-871-8000
Fax: 440-871-8104

Hydranautics
401 Jones Rd.
Oceanside, CA 92054
Phone: 760-901-2500
Fax: 760-901-2578
www.membranes.com

Hydro Gate Corp.
6101 North Dexter St.
Commerce City, CO 80022
Phone: 303-288-7873
Fax: 303-287-8531
www.hydrogate.com

Hydro-Flo Technologies, Inc.
205 E. Kehoe Blvd.
Carol Stream, IL 60188
Phone: 630-462-7550
Fax: 630-462-7728
www.hydroflotech.com

HydroCal, Inc.
22732 Granite Way, Ste. A
Laguna Hills, CA 92653
Phone: 949-455-0765
Fax: 949-455-0764
www.hydrocal.com

Hydrolab Corp.
8700 Cameron Rd., Ste. 100
Austin, TX 78754
Phone: 512-255-8841
Fax: 512-832-8838

Hydropress Wallander & Co., AB
P.O. Box 125
Lindome, SE-437 22, Sweden
Phone: 46-31-995050
Fax: 46-31-995133

IDEXX Laboratories, Inc.
1 IDEXX Dr.
Westbrook, ME 04092
Phone: 207-856-0300
Fax: 207-856-0346

Idreco USA, Ltd.
3494 Progress Dr., Ste. B
Bensalem, PA 19020
Phone: 215-638-2111
Fax: 215-638-2114
www.idreco.com

ILC Dover, Inc.
One Moonwalker Rd.
Frederica, DE 19946
Phone: 302-335-3911
Fax: 302-335-1320
www.ilcdover.com

IN USA, Inc.
100 Crescent Rd., Unit 1B
Needham, MA 02194
Phone: 781-444-2929
Fax: 781-444-9229
www.inusaozone.com

In-Situ, Inc.
210 S. Third St.
Laramie, WY 82073
Phone: 307-742-8213
Fax: 307-721-7598
www.in-situ.com

Inchen USA, Inc.
1920 E. Hallandale Beach Blvd.,
    Ste. 607
Hallandale, FL 33009
Phone: 954-456-7165
Fax: 954-455-8856

Industrial Analytics Corp.
1 Orchard Park Rd.
Madison, CT 06443
Phone: 203-245-0380
Fax: 203-245-3698

Industrial Fabrics Corp.
7160 Northland Circle
Minneapolis, MN 55428
Phone: 612-535-3220
www.ifcfabrics.com

Industrial Filter & Pump Mfg. Co.
5900 Ogden Ave.
Cicero, IL 60650-3888
Phone: 708-656-7800
Fax: 708-656-7816

Infilco Degremont, Inc.
P.O. Box 1390
Richmond, VA 23255-1390
Phone: 804-756-7600
Fax: 804-756-7645
www.infilcodegremont.com

Integrated Environmental Solutions, Inc.
3787 Old Middleburg Rd., Ste. 3
Jacksonville, FL 33210
Phone: 904-778-1188
Fax: 904-778-0201

International Dioxide, Inc.
554 Ten Rod Rd.
N. Kingston, RI 02852-4220
Phone: 401-294-9575
Fax: 401-295-7108

International Filter Media
P.O. Box 216
Hazelton, PA 18201-0458
Phone: 570-459-1491
Fax: 570-455-7510

Invincible AirFlow Systems
P.O. Box 380
Baltic, OH 43804
Phone: 330-897-3200
Fax: 330-897-3400
www.invincibleair.com

Ionics, Inc.
65 Grove St.
Watertown, MA 02172
Phone: 617-926-2500
Fax: 617-926-4303
www.ionics.com

Ionics RCC
3006 Northup Way
Bellevue, WA 98004-1407
Phone: 425-828-2400
Fax: 425-828-0526
www.ionicsrcc.com

Isco, Inc.
P.O. Box 82531
Lincoln, NE 68501-2531
Phone: 402-464-0231
Fax: 402-465-3064
www.isco.com

ITT A-C Pump
1150 Tennessee Ave.
Cincinnati, OH 45229
Phone: 513-482-2500
Fax: 513-482-2569
www.ittacpump.com

ITT Flygt Corp.
P.O. Box 1004
Trumball, CT 06611-0943
Phone: 203-380-4700
Fax: 203-380-4705
www.flygt.com

IX Services Co.
1102 Holly St.
Las Cruces, NM 88005
Phone: 505-526-2838
Fax: 505-526-2838

Jaeger Products, Inc.
1611 Peachleaf
Houston, TX 77039
Phone: 281-449-9500
Fax: 281-449-9400
www.jaeger.com

Jay R. Smith Mfg. Co.
P.O. Box 3237
Montgomery, AL 36109-0237
Phone: 334-277-8520
Fax: 334-272-7396
www.jrsmith.com

JBF Environmental Technology
18 Beach St.
Seymour, CT 06583
Phone: 203-888-7700
Fax: 203-888-7720
www.jbfenv.com

JBS Instruments
311 D St.
West Sacramento, CA 95605
Phone: 916-372-0534
Fax: 916-372-1624

JDV Equipment Corp.
216 Little Falls Rd., Unit 1
Cedar Falls, NJ 07009
Phone: 973-571-7110
Fax: 973-571-7112

Jeffrey Chain Corp.
2307 Maden Dr.
Morristown, TN 37813-2898
Phone: 423-586-1951
Fax: 423-581-2399
www.jeffreychain.com

Jet, Inc.
750 Alpha Dr.
Cleveland, OH 44143
Phone: 440-461-2000
Fax: 440-442-9008

John Meunier, Inc.
6290 Perinault
Montreal Quebec, Canada H4K 1K5
Phone: 514-334-7230
Fax: 514-334-5010
www.johnmeunier.com

John Zink Co.
P.O. Box 21220
Tulsa, OK 74121-1220
Phone: 918-234-1800
Fax: 918-234-1975

Jones MacCrea, Inc.
P.O. Box 6030
Syracuse, NY 13217-6030
Phone: 315-478-3119
Fax: 315-478-0802

JWC Environmental
290 Paularino Ave.
Costa Mesa, CA 92626
Phone: 949-833-3888
Fax: 949-833-8858
www.jwce.com

K-Tron North America
P.O. Box 888
Pitman, NJ 08071
Phone: 856-589-0500
Fax: 856-589-8113
www.ktron.com

Kady International
P.O. Box 847
Scarborough, ME 04070-0847
Phone: 207-883-4141
Fax: 207-883-8241
www.kadyinternational.com

Kason Corp.
67-71 East Willow St.
Milburn, NJ 07041-1416
Phone: 973-467-8140
Fax: 973-258-9533
www.kason.com

KCC Corrosion Control Co.
4010 Trey Rd.
Houston, TX 77084
Phone: 281-550-1199
Fax: 281-550-9097
www.kcccontrol.com

Kem-Tron
10404 Cash Rd.
Stafford, TX 77477
Phone: 281-261-5778
Fax: 281-499-4080
www.kemtron.com

Kemco Systems, Inc.
11500 47th St. N.
Clearwater, FL 33762
Phone: 727-573-2323
Fax: 727-573-2346

Kemiron
316 Bartow Airport
Bartow, FL 33830
Phone: 863-533-5990
Fax: 863-533-7077

Kimre Inc.
P.O. Box 570846
Perrine, FL 33257-0846
Phone: 305-233-4249
Fax: 305-233-8687

Kinetico Engineered Systems, Inc.
P.O. Box 127
Newbury, OH 44065
Phone: 440-564-0111
Fax: 440-564-7696
www.kinetico.com

King Lee Technologies
8949 Kenamar Dr., Bldg. 107
San Diego, CA 92121-2453
Phone: 619-693-4062
Fax: 619-693-4917
www.kingleetech.com

Klenzoid, Inc.
P.O. Box 444
Wayne, PA 19087
Phone: 610-825-0218
Fax: 610-825-5390

Knapp Polly Pig, Inc.
1209 Hardy
Houston, TX 77020
Phone: 713-222-0146
Fax: 713-222-7403
www.pollypig.com

Knight Manufacturing Corp.
P.O. Box 167
Brodhead, WI 53520
Phone: 608-897-2131
Fax: 608-897-2561
www.knightmfg.com

Koch Membrane Systems, Inc.
850 Main St.
Wilmington, MA 01887-3388
Phone: 978-657-4250
Fax: 978-657-5208
www.kochmembrane.com

Koch-Otto York
P.O. Box 3100
Parsippany, NJ 07054
Phone: 973-299-9200
Fax: 973-299-9401

Koflo Corp.
309 Cary Point Rd.
Cary, IL 60013
Phone: 847-516-3700
Fax: 847-516-3724

Komax Systems, Inc.
P.O. Box 1323
Wilmington, CA 90748-1323
Phone: 310-830-4320
Fax: 310-830-9826
www.komax.com

Komline-Sanderson Engineering Corp.
12 Holland Ave.
Peapack, NJ 07977-0257
Phone: 908-234-1000
Fax: 908-234-9487
www.komline.com

Krauss Maffei Corp.
7095 Industrial Rd.
Florence, KY 41042-6270
Phone: 606-283-0200
Fax: 606-283-1878
www.krauss-maffei.com

Krebs Engineers
5505 West Gillette Rd.
Tucson, AZ 85743
Phone: 520-744-8200
Fax: 520-744-8300
www.krebs.com

KriStar Enterprises
P.O. Box 7352
Santa Rosa, CA 95407
Phone: 707-524-2424
Fax: 707-524-8186
www.kristar.com

Krofta Engineering Corp.
P.O. Box 972
Lenox, MA 01240
Phone: 413-637-0740
Fax: 413-637-0768
www.krofta.com

Krüger A/S
Gladsaxevej 363
DK-2860 Soborg, Denmark
Phone: 45-39-690222
Fax: 45-39-690806
www.kruger.dk

Krupp Thyssen Nirosta GmbH
Oberschlesienstr. 16
D-47807 Krefeld, Germany
Phone: 49-21-5183-01
Fax: 49-21-5183-2022
www.nirosta.ce

Kvaerner Chemetics
1818 Cornwall Ave.
Vancouver, BC, Canada V6J 1C7
Phone: 604-734-1200
Fax: 604-734-0340

Kvaerner Eureka USA
17 Sawmill Way
Georgetown, MA 01833
Phone: 508-352-4487
Fax: 508-352-4417

Lakeside Equipment Corp.
P.O. Box 8448
Bartlett, IL 60103
Phone: 630-837-5640
Fax: 630-837-5647
www.lakeside-equipment.com

Lakeview Engineered Products, Inc.
2010 Lakeview Dr.
Fort Wayne, IN 46802
Phone: 219-432-3479
Fax: 219-432-6239

Land Combustion
2525-B Pearl Buck Rd.
Bristol, PA 19007
Phone: 215-781-0810
Fax: 215-781-0798
www.landinst.com

Landa, Inc.
4275 NW Pacific Rim Blvd.
Camas, WA 98607
Phone: 360-833-9100
Fax: 360-833-9200
www.landa-inc.com

Landustrie Sneek BV
P.O. Box 199
8600 AD Sneek, The Netherlands
Phone: 31-515-486888
Fax: 31-515-412398
www.landustrie.nl

Lang Filter Media Co.
910 Sheraton Dr. #100
Mars, PA 16046
Phone: 724-779-3990
Fax: 724-779-3993
www.langfiltermedia.com

Lantec Products, Inc.
5308 Derry Ave., Unit E
Agoura Hills, CA 91301
Phone: 818-707-2285
Fax: 818-707-9367
www.lantecp.com

Larox Inc.
9730 Patuxent Woods Dr.
Columbia, MD 21046
Phone: 410-381-3314
Fax: 410-381-4490
www.larox.fi

LAS International
3811 Lockport St.
Bismark, ND 58501
Phone: 701-222-8331
Fax: 701-222-2773
www.lasinternational.com

Layne Christensen, Ranney Division
801 W. Cherry St.
Sudbury, OH 43074
Phone: 740-965-2833
Fax: 740-965-2834
www.laynechristensen.com

LCI Corp.
P.O. Box 16348
Charlotte, NC 28297-9984
Phone: 704-394-8341
Fax: 704-393-8590
www.lcicorp.com

LEEM Filtration Products, Inc.
25 Arrow Rd.
Ramsey, NJ 07446
Phone: 201-236-4833
Fax: 201-236-2004

Lemna Corp.
1408 Northland Dr., Ste. 310
St. Paul, MN 55120-1013
Phone: 612-688-0836
Fax: 612-688-8813
www.lemna.com

Lighthouse Separation Systems, Inc.
P.O. Box 370
Dahlonega, GA 30533
Phone: 706-864-8644
Fax: 706-864-8677
www.lighthousefilters.com

Lightnin
P.O. Box 1370
Rochester, NY 14603
Phone: 716-527-1623
Fax: 716-527-1720

Linde-KCA-Dresden GmbH
Schumannstrabe 21
D-39264 Dobritz, Germany
Phone: 41-351-4560-207
Fax: 41-351-4560-272
www.linde.com

Liquid Dynamics Corp.
36W897 Dean St.
St. Charles, IL 60174
Phone: 630-513-8366
Fax: 630-513-6447
www.jetmix.com

Liquid Metronics, Inc.
8 Post Office Sq.
Acton, MA 01720-3948
Phone: 978-263-9800
Fax: 978-264-9172
www.lmipumps.com

Liquid Waste Technology, Inc.
Box 250
Somerset, WI 54025
Phone: 715-247-5464
Fax: 715-247-3934
www.lwtpithog.com

Liquid-Solids Separation Corp.
P.O. Box 9
Northvale, NJ 07647
Phone: 201-784-1570
Fax: 201-784-1575

LIST, Inc.
42 Nagog Park
Acton, MA 01720
Phone: 978-635-9521
Fax: 978-635-0570

Longwood Engineering Co., Ltd.
Parkwood Mills, Longwood,
      Huddersfield
West Yorkshire, HD3 4TP England
Phone: 0484-642011
Fax: 0484-642935

Lotepro Corp.
115 Stevens Ave.
Valhalla, NY 10595
Phone: 914-749-5228
Fax: 914-747-3422
www.linde.com

Lowry Aeration Systems, Inc.
P.O. Box 1239
Blue Hill, ME 04614
Phone: 207-374-3502
Fax: 207-374-3503
www.lowryh2o.com

LSR Technologies, Inc.
898 Main St.
Acton, MA 01720
Phone: 978-635-0123
Fax: 978-635-0058
www.concentric.net

Lurgi Bamag GmbH
Wetzlarer Strabe 136
D-635510 Butzbach, Germany
Phone: 49-60-33-839
Fax: 49-60-33-83-506
www.lurgi.com

M&W Industries, Inc.
16792 Talisman Lane #210
Huntington Beach, NC 27045
Phone: 910-969-9526
Fax: 910-969-2156

Markland Specialty Engineering Ltd.
48 Shaft Rd.
Rexdale Toronto Ontario,
    Canada M9W 4M2
Phone: 416-244-4980
Fax: 416-244-2287
www.sludgecontrols.com

Marolf, Inc.
4430 Erie Dr.
New Port Richey, FL 34652
Phone: 727-843-0681
Fax: 727-849-3272

Marsh-McBirney, Inc.
4539 Metropolitan Ct.
Frederick, MD 21701-8364
Phone: 301-874-5599
Fax: 301-874-2172
www.marsh-mcbirney.com

Martin Marietta Specialties, Inc.
P.O. Box 15470
Baltimore, MD 21220-0470
Phone: 410-780-5500
Fax: 410-780-5555
www.magspecialties.com

Matheson Gas Products
166 Keystone Dr.
Montgomeryville, PA 18936
Phone: 215-641-2700
Fax: 215-641-2714
www.mathesongas.com

Matrix Desalination, Inc.
3295 SW 11th Ave.
Fort Lauderdale, FL 33315
Phone: 954-524-5120
Fax: 954-524-5216

Matt-Son, Inc.
28W005 Industrial Ave.
Barrington, IL 60010
Phone: 847-382-7810
Fax: 847-382-5814
www.matt-son.com

Mazzei Injector Corp.
500 Rooster Dr.
Bakersfield, CA 93307-9555
Phone: 661-363-6500
Fax: 661-363-7500
www.mazzei-injector.com

McCrometer, Inc.
3255 West Stetson Ave.
Hemet, CA 92545-7799
Phone: 909-652-6811
Fax: 909-652-3078
www.mccrometer.com

McGill Airclean Corp.
1779 Refugee Rd.
Columbus, OH 43207-2119
Phone: 614-443-0192
Fax: 614-445-8759
www.mcgillairclean.com

McLanahan Corp.
200 Wall St.
Hollidaysburg, PA 16648
Phone: 814-695-9807
Fax: 814-695-6684

McTighe Industries, Inc.
3405 S. Westport Ave.
Sioux Falls, SD 57106
Phone: 605-363-3407
Fax: 605-362-3494
www.mctighe.com

Mechanical Equipment Co., Inc.
861 Carondelet St.
New Orleans, LA 70130
Phone: 504-599-4000
Fax: 504-599-4100
www.meco.com

Medina Products Bioremediation
  Division
P.O. Box 309
Hondo, TX 78861
Phone: 830-426-3011
Fax: 830-426-2288

Megator Corp.
562 Alpha Dr.
Pittsburgh, PA 15238
Phone: 412-963-9200
Fax: 412-963-9214

Megtec Systems, Inc.
P.O. Box 5030
DePere, WI 54115-0030
Phone: 920-336-5715
Fax: 920-336-3404
www.megtec.com

Membrex, Inc.
155 Route 146 West
Fairfield, NY 07004
Phone: 973-575-8388
Fax: 973-575-7011
www.membrex.com

Mer-Made Filter, Inc.
185 Le Grand Ave.
Northvale, NJ 07647
Phone: 201-784-3523
Fax: 201-784-1575

Mercer International, Inc.
P.O. Box 540
Mendham, NJ 07945
Phone: 973-543-9000
Fax: 973-543-4343

Meridian Diagnostics, Inc.
3741 River Hills Dr.
Cincinnati, OH 45244
Phone: 513-271-3700
Fax: 513-272-5432

Merrick Industries, Inc.
10 Arthur Dr.
Lynn Haven, FL 32444
Phone: 850-265-3611
Fax: 850-265-9768

Met-Pro Corp.
P.O. Box 144
Harleysville, PA 19438
Phone: 215-723-6751
Fax: 215-723-6758
www.met-pro.com

Meurer Industries, Inc.
15611 West 6th Ave.
Golden, CO 80401
Phone: 303-279-8431
Fax: 303-279-8429

Micronair LLC
11259 Phillips Pkwy. Dr. E.
Jacksonville, FL 32256
Phone: 904-268-5457
Fax: 904-268-5597
www.micronairusa.com

MIE, Inc.
7 Oak Park
Bedford, MA 01730
Phone: 781-275-1919
Fax: 781-275-2121
www.mieinc.com

Millipore Corp.
80 Ashby Rd.
Bedford, MA 01730
Phone: 781-275-9200
Fax: 781-533-8878

Milltronics, Inc.
709 Stadium Dr. E.
Arlington, TX 76011-9870
Phone: 817-277-3543
Fax: 817-277-3894
www.milltronics.com

Milton Roy Co.
201 Ivyland Rd.
Ivyland, PA 18974-1706
Phone: 215-441-0800
Fax: 215-441-8620
www.miltonroy.com

Minntech Fibercor
14605 28th Ave. N.
Minneapolis, MN 55447
Phone: 612-553-3300
Fax: 612-553-3387
www.minntech.com

Miox Corp.
5500 Midway Park Pl. NE
Albuquerque, NM 87109
Phone: 505-343-0900
Fax: 505-343-0093
www.miox.com

Misonix, Inc.
1938 New Highway
Farmingdale, NY 11735
Phone: 631-694-9555
Fax: 631-694-9412
www.misonix.com

MixAir Technologies, Inc.
4712 68th Ave.
Kenosha, WI 53144
Phone: 262-657-6788
Fax: 262-657-8383
www.mixairtech.com

ModuTank, Inc.
41-04 35th Ave.
Long Island City, NY 11101
Phone: 718-392-1112
Fax: 718-786-1008

Molecular Analytics LLC
25 Loveton Circle, Box 1123
Sparks, MD 21152-1123
Phone: 410-472-2146
Fax: 410-472-2156

Monitek Technologies, Inc.
2021 Las Positas Ct., Ste. 145
Livermore, CA 94550
Phone: 925-243-0050
Fax: 925-243-0051
www.monitek.com

Monoflo
16503 Park Row
Houston, TX 77084
Phone: 281-599-4700
Fax: 281-599-4733

Monsanto Enviro-Chem Systems, Inc.
P.O. Box 14547
St. Louis, MO 63178
Phone: 314-275-5700
Fax: 314-275-5701
www.enviro-chem.com

Mother Environmental Systems, Inc.
1124 Purina Dr.
Gainesville, GA 30501
Phone: 770-534-3118
Fax: 770-534-3117
www.mycelx.com

Moyno Industrial Products
P.O. Box 960
Springfield, OH 45501
Phone: 937-327-3510
Fax: 937-327-3064
www.moyno.com

MPR Services, Inc.
1201 FM646
Dickinson, TX 77539-3022
Phone: 281-337-7424
Fax: 281-337-6534
www.mprserv.com

Mt. Fury Co., Inc.
1460 19th Ave., NW
Issaquah, WA 98027
Phone: 425-391-0747
Fax: 425-391-9708

Munters
P.O. Box 6428
Fort Myers, FL 33911
Phone: 941-936-1555
Fax: 941-936-6582
www.munters.com

Munters Zeol Corp.
P.O. Box 600
Amesbury, MA 01913
Phone: 978-241-1103
Fax: 978-241-1220
www.munterszeol.com

MWD Technologies Ltd.
111440 West Bernado
San Diego, CA 92127
Phone: 858-674-6902
Fax: 858-674-6903

N-Viro International Corp.
3450 West Central Ave., Ste. 328
Toledo, OH 43606
Phone: 419-535-6374
Fax: 419-535-7008
www.nviro.com

Nalco Chemical Co.
One Nalco Center
Naperville, IL 60566-1024
Phone: 630-305-1000
Fax: 630-305-2900
www.nalco.com

NAO Inc.
1284 E. Sedgley Ave.
Philadelphia, PA 19134
Phone: 215-743-5300
Fax: 215-743-3018
www.nao.com

National Fluid Separators, Inc.
827 Hanley Industrial Ct.
St. Louis, MO 63144-1402
Phone: 314-968-2838
Fax: 314-968-4773
www.mjind.com

National Seal Co.
1245 Corporate Blvd., Ste. 300
Aurora, IL 60504
Phone: 630-898-1161
Fax: 630-898-3461

Nature Plus, Inc.
52 Lakeview Ave.
New Canaan, CT 06840
Phone: 203-972-1100
Fax: 203-966-2200

NEFCO, Inc.
P.O. Box 30493
Palm Beach Gardens, FL 33420-0493
Phone: 561-775-9303
Fax: 561-775-6043
www.nefcoinc.com

Neotronics of North America
P.O. Box 2100
Flowery Branch, GA 30542-2100
Phone: 770-967-2196
Fax: 770-967-1854

Neptune Chemical Pump Co.
P.O. Box 247
Lansdale, PA 19446
Phone: 215-699-8701
Fax: 215-699-0370
www.neptune1.com

Netzsch, Inc.
119 Pickering Way
Exton, PA 19341-1393
Phone: 610-363-8010
Fax: 610-363-0971
www.netzschusa.com

Neutraman, Inc.
2701 S. Coliseum Blvd., Ste. 1000
Fort Wayne, IN 46803
Phone: 219-422-5953
Fax: 219-422-5160

New Logic International
1295 67th St.
Emeryville, CA 94608
Phone: 510-655-7305
Fax: 510-655-7307
www.vsep.com

Newport Electronics
2229 S. Yale St.
Santa Ana, CA 92704
Phone: 714-540-4914
Fax: 714-546-3022
www.newportus.com

Niro, Inc.
9165 Rumsey Rd.
Colombia, MD 21045
Phone: 410-997-8700
Fax: 410-997-5021
www.niro.com

Nitrate Removal Technologies, LLC
1667 Cole Blvd., Ste. 400
Golden, CO 80401
Phone: 303-274-1426
Fax: 303-237-1103

Noggerath GmbH
Feldstrabe 2
D-31708 Ahmsen, Germany
Phone: 49-5722-882-0
Fax: 49-5722-882-282
www.noggerath.de

Nopon Oy
Turvekuja 6
Helsinki, Finland 00700
Phone: 358-9-351-5700
Fax: 358-9-351-5620
www.nopon.fi

Norair Engineering Corp.
337 Brightseat Rd., Ste. 200
Landover, MD 20785
Phone: 301-499-2202
Fax: 301-499-1342

Norchem Industries
18651 Graphic Ct.
Tinley Park, IL 60477
Phone: 708-802-9700
Fax: 708-802-9775
www.norchemindustries.com

Norit Americas, Inc.
1050 Crown Point Pkwy., Ste. 1500
Atlanta, GA 30338
Phone: 770-512-4610
Fax: 770-512-4622
www.norit.com

North East Environmental Products, Inc.
17 Technology Dr.
West Lebanon, NH 03784
Phone: 603-298-7061
Fax: 603-298-7063
www.neepsystems.com

Northwest Cascade, Inc.
P.O. Box 73399
Puyallup, WA 98373
Phone: 253-848-2371
Fax: 253-848-2545
www.nwcascade.com

Norton Co.
P.O. Box 350
Akron, OH 44309-0350
Phone: 330-673-5860
Fax: 330-677-7245

Norton Performance Plastics
150 Dey Rd.
Wayne, NJ 07470
Phone: 973-696-4700
Fax: 973-628-5550

NRG, Inc.
P.O. Box 306
Ardmore, PA 19003-9998
Phone: 610-896-6850
Fax: 610-649-5083

NSF International
P.O. Box 130140
Ann Arbor, MI 48113-0140
Phone: 734-769-8010
Fax: 734-760-0109
www.nsf.org

NSW Corp.
530 Gregory Ave.
Roanoke, VA 24016
Phone: 540-981-0362
Fax: 540-345-8421
www.nswcorp.com

Nuove Energie s.r.l.
Via della Meccanica
23/25 Vicenza, Italy
Phone: 0444-963453
Fax: 0444-960959

NuTech Environmental Corp.
5350 N. Washington St.
Denver, CO 80216-1951
Phone: 303-295-3702
Fax: 303-295-6145
www.nutechenvironmental.com

Odor Management, Inc.
18-4 E. Dundee Rd., Ste. 200
Barrington, IL 55427-9707
Phone: 847-304-9111
Fax: 847-304-0989

Olds Filtration Engineering, Inc.
Highway 98, 907 Halls Lane
Daphne, AL 36526-0970
Phone: 334-626-9492
Fax: 334-626-7988
www.oldsfiltration.com

Omega Engineering, Inc.
P.O. Box 4047
Stamford, CT 06907-0047
Phone: 203-359-1660
Fax: 203-359-7700
www.omega.com

On-Demand Environmental Systems,
   Inc.
761 Coleman Ave.
San Jose, CA 95110
Phone: 408-287-7012
Fax: 408-287-7147

Oritex Corp.
5530 Ferguson Dr.
City of Commerce, CA 90022
Phone: 323-890-1588
Fax: 323-890-1591

Orival, Inc.
40 N. Van Brunt St.
Englewood, NJ 07631
Phone: 201-568-3311
Fax: 201-568-1916
www.orival.com

Osmonics Desal
760 Shadowridge Dr.
Vista, CA 92083-7986
Phone: 760-598-3334
Fax: 760-598-3335
www.osmonics.com

Osmonics, Inc.
5951 Clearwater Dr.
Minnetonka, MN 55343-8995
Phone: 612-933-2277
Fax: 612-933-0141
www.osmonics.com

Osmosis Technology, Inc.
6900 Hemosa Circle
Buena Park, CA 90620
Phone: 714-670-9303
Fax: 714-670-9323
www.osmotik.com

Osprey Biotechnics
2530 B. Trailmate Dr.
Sarasota, FL 34243
Phone: 941-755-7770
Fax: 941-755-0626

Ozone Pure Water, Inc.
5330 Ashton Ct.
Sarasota, FL 34233
Phone: 941-923-8528
Fax: 941-923-8231
www.ozonepure.com

Ozonia North America
P.O. Box 455
Elmwood Park, NJ 07407
Phone: 201-794-3100
Fax: 201-794-3358
www.ozonia.com

Pacific Keystone Technologies
P.O. Box 360
Black Diamond, WA 98010-0491
Phone: 360-886-1396
Fax: 360-886-2480
www.clearwaterworld.com

Pacific Ozone Technology, Inc.
730 Concord Ave.
Brentwood, CA 94513
Phone: 925-634-7252
Fax: 925-634-7291
www.pacificozone.com

PacTec, Inc.
P.O. Box 8069
Clinton, LA 70722
Phone: 225-683-8062
Fax: 225-683-8711
www.pactecinc.com

Pall Corp.
2200 Northern Blvd.
East Hills, NY 11548-1289
Phone: 516-484-5400
Fax: 516-484-6164
www.pall.com

Parkson Corp.
P.O. Box 408399
Fort Lauderdale, FL 33340-8399
Phone: 954-974-6610
Fax: 954-974-6182
www.parkson.com

Particle Measuring Systems, Inc.
5475 Airport Blvd.
Boulder, CO 80301
Phone: 303-443-7100
Fax: 303-449-6870
www.pmeasuring.com

Passavant-Roediger GmbH
Aarbergen
D-65322, Germany
Phone: 49-6120-282791
Fax: 49-6120-282672
www.passavant-roediger.de

Patterson Candy International, Ltd.
21 The Mall,
Ealing London, W5 2PU England
Phone: 01-579-1311
Fax: 01-840-6180

Patterson Pump Co.
P.O. Box 790
Toccoa, GA 30577
Phone: 706-886-2101
Fax: 706-886-0023
www.pattersonpumps.com

Paul Mueller Co.
P.O. Box 828
Springfield, MO 65801-0828
Phone: 417-831-3000
Fax: 417-831-3528
www.muel.com

PBR Industries
143 Cortland St.
Lindenhurst, NY 11757
Phone: 516-226-2930
Fax: 516-226-3125
www.pbrind.com

PCI-Wedeco Environmental
  Technologies, Inc.
One Fairfield Crescent
West Caldwell, NJ 07006
Phone: 973-575-7052
Fax: 973-575-8941
www.pci-wedeco.com

PennProcess Technologies, Inc.
P.O. Box 427
Plumsteadville, PA 18949-0427
Phone: 215-766-7766
Fax: 215-766-8290
www.pennprocess.com

Pepcon Systems, Inc.
3770 Howard Hughes Pkwy., Ste. 340
Las Vegas, NV 89109
Phone: 702-735-2324
Fax: 702-735-9456

PerkinElmer Instruments
761 Main Ave.
Norwalk, CT 06859
Phone: 203-762-4003
Fax: 203-761-5330
www.perkinelmer.com

Phelps Dodge Refining Co.
P.O. Box 20001
El Paso, TX 79998
Phone: 915-775-8826
Fax: 915-775-8350
www.pdsales.com

Philadelphia Mixers
1221 E. Main
Palmyra, PA 17078
Phone: 717-838-1341
Fax: 717-832-8802
www.philamixers.com

Phipps & Bird
P.O. Box 27324
Richmond, VA 23261
Phone: 804-254-2737
Fax: 804-254-2955
www.phippsbird.com

Phoenix Process Equipment Co.
2402 Watterson Trail
Louisville, KY 40299
Phone: 502-499-6198
Fax: 502-499-1079
www.dewater.com

Pica USA, Inc.
432 McCormick Blvd.
Columbus, OH 43213
Phone: 614-864-8100
Fax: 614-864-9914
www.picausa.com

Pipeline Pigging Products, Inc.
P.O. Box 692005
Houston, TX 77269
Phone: 281-351-6688
Fax: 281-255-2385
www.pipepigs.com

Pitt-Des Moines, Inc.
3400 Grand Ave.
Pittsburgh, PA 15225
Phone: 412-331-3000
Fax: 412-331-3188
www.pdm.com

Plasti-Fab, Inc.
P.O. Box 100
Tualatin, OR 97062-0100
Phone: 503-692-5460
Fax: 503-692-1145
www.plasti-fab.com

Pollution Control Systems, Inc.
5827 Happy Hollow Rd.
Milford, OH 45150
Phone: 513-831-1165
Fax: 513-965-4812

Polybac Corp.
3894 Courtney St.
Bethlehem, PA 18017
Phone: 610-867-7338
Fax: 610-861-0991

Polychem Corp.
P.O. Box 527
Phoenixville, PA 19460-0527
Phone: 610-935-0225
Fax: 610-935-7151
www.polychemcorp.com

Polydyne, Inc.
P.O. Box 351420
Toledo, OH 43635
Phone: 419-843-8066
Fax: 419-843-3081
www.polydynemso.com

PPG Industries, Inc.
1 PPG Pl.
Pittsburgh, PA 15272
Phone: 412-434-3131
Fax: 412-434-4578
www.ppg.com

PQ Corp.
P.O. Box 840
Valley Forge, PA 19482
Phone: 610-651-4200
Fax: 610-251-5249

Praxair, Inc.
810 Jorie Blvd.
Oak Brook, IL 60521-2216
Phone: 708-572-7500
Fax: 708-572-7935
www.praxair.com

Praxair-Trailigaz Ozone Co.
11501 Goldcoast Dr.
Cincinnati, OH 45249-1623
Phone: 513-530-7702
Fax: 513-530-7711
www.ptoc.com

Premier Chemicals
7521 Engle Rd., Ste. 415
Middleburg Heights, OH 44130
Phone: 440-234-4600
Fax: 440-234-5772
www.premierchemicals.com

Pro Products Corp.
502 Incentive Dr.
Fort Wayne, IN 46825
Phone: 219-490-5970
Fax: 219-490-9431
www.redbgone.com

Pro-Ent, Inc.
P.O. Box 23611
Jacksonville, FL 32241
Phone: 904-737-3536
Fax: 904-737-3537

Pro-Equipment, Inc.
237 Wisconsin Ave.
Waukesha, WI 53186
Phone: 262-513-8801
Fax: 262-513-8897
www.proequipment.com

Probiotic Solutions
201 S. Roosevelt
Chandler, AZ 85226
Phone: 480-961-1220
Fax: 480-961-3501

Process Combustion Corp.
5460 Horning Rd.
Pittsburgh, PA 15241
Phone: 412-655-0955
Fax: 412-650-5560

Product Level Control, Inc.
11929 Portland Ave. S.
Burnsville, MN 55337
Phone: 612-707-9101
Fax: 612-707-1075
www.productlevel.com

Professional Water Technologies, Inc.
1145 Industrial Ave., Ste. I
Escondido, CA 92029
Phone: 760-741-7404
Fax: 760-741-5645
www.pwtinc.com

ProGuard Filtration Systems
P.O. Box 678
Nowata, OK 74048
Phone: 918-273-2208
Fax: 918-273-2101

ProMinent Fluid Controls, Inc.
136 Industry Dr.
Pittsburgh, PA 15275
Phone: 412-787-2484
Fax: 412-787-0704
www.pfc-amer.com

PTI Advanced Filtration
2340 Eastman Ave.
Oxnard, CA 93030
Phone: 805-604-3400
Fax: 805-604-3401
www.pti-afi.com

Pulsafeeder, Inc.
2883 Brighton Henrietta TL Rd.
Rochester, NY 14623
Phone: 716-292-8000
Fax: 716-424-5619
www.pulsa.com

Pump Engineering, Inc.
1004 W. Hurd Rd.
Monroe, MI 48161
Phone: 734-242-1772
Fax: 734-242-9777

Pumps Unlimited
19 Affonso Dr.
Carson City, NV 89706
Phone: 775-246-0800
Fax: 775-246-0847

Purafil, Inc.
2654 Weaver Way
Doraville, GA 30340
Phone: 770-662-8545
Fax: 770-263-6922
www.purafil.com

Purestream, Inc.
P.O. Box 68
Florence, KY 41042-0068
Phone: 606-371-9898
Fax: 606-371-3577
www.purestreaminc.com

Purolite Co.
150 Monument Rd., Bala Cynwyd
Philadelphia, PA 19004
Phone: 610-668-9090
Fax: 610-668-8139
www.purolite.com

QED Environmental Systems, Inc.
P.O. Box 3726
Ann Arbor, MI 48106-9897
Phone: 734-995-2547
Fax: 734-995-1170
www.qedenv.com

Quantum Technologies, Inc.
1632 Enterprise Pkwy.
Twinsburg, OH 44087
Phone: 330-425-7880
Fax: 330-425-0955

R.P. Adams Co., Inc.
P.O. Box 963
Buffalo, NY 14240-0963
Phone: 716-877-2608
Fax: 716-877-9385
www.rpadams.com

RDP Technologies, Inc.
2495 Blvd. of the Generals
Norristown, PA 19403-5236
Phone: 610-650-9900
Fax: 610-650-9070
www.rdptech.com

Recra Environmental
1576 Sweet Home Rd.
Amherst, NY 14228
Phone: 716-636-1550
Fax: 716-636-1598

Red Fox Environmental, Inc.
P.O. Box 53809
Lafayette, LA 70505
Phone: 337-235-2499
Fax: 337-235-2999

Red Valve Co., Inc.
700 N. Bell Ave.
Carnegie, PA 15106
Phone: 412-279-0044
Fax: 412-279-7878
www.redvalve.com

Refinite Water Conditioning Co.
P.O. Box 11676
Rock Hill, SC 29731
Phone: 803-324-7600
Fax: 803-324-1116

Refractron Technologies Corp.
5750 Stuart Ave.
Newark, NJ 14513
Phone: 315-331-6222
Fax: 315-331-7254
www.refractron.com

Regenesis
1011 Calle Sombra
San Clemente, CA 92672
Phone: 949-366-8000
Fax: 949-366-8090
www.regenesis.com

Reheis, Inc.
235 Snyder Ave.
Berkely Heights, NJ 07922
Phone: 908-464-1500
Fax: 908-464-7726

REKO Industrial Equipment B.V.
Delta Industrieweg 36
3251 LX Stellendam, The Netherlands
Phone: 31-187-492988
Fax: 31-187-492781

RGF O3 Systems, Inc.
3875 Fiscal Ct.
West Palm Beach, FL 33404
Phone: 561-848-3187
Fax: 561-848-2170
www.rgf.com

Ringlace Products, Inc.
9902 N.E. Glison St.
Portland, OR 97220
Phone: 503-251-1295
Fax: 503-256-7325

Roberts Filter Group
P.O. Box 167
Darby, PA 19023
Phone: 610-583-3131
Fax: 610-583-0117
www.robertsfiltergroup.com

Rochem Environmental, Inc.
610 North Milby St.
Houston, TX 77003
Phone: 713-224-7626
Fax: 713-224-7627
www.rochem.com

Rochester Midland
333 Hollenbeck St.
Rochester, NY 14621
Phone: 716-336-2200
Fax: 716-266-1606
www.rochestermidland.com

Rodney Hunt Co.
46 Mill St.
Orange, MA 01364-1268
Phone: 978-544-2511
Fax: 978-544-7204
www.rodneyhunt.com

Roediger Pittsburgh, Inc.
3812 Route 8
Allison Park, PA 15101
Phone: 412-487-6010
Fax: 412-487-6005
www.roediger.com

Rohm & Haas, Co.
5000 Richmond St.
Philadelphia, PA 19105
Phone: 215-537-4000
Fax: 215-537-4219
www.rohmhaas.com

Ronningen-Petter
9551 Shaver Rd.
Portage, MI 49081
Phone: 616-323-1313
Fax: 616-323-0065

Ropur AG
4142 Munchebstein 1
Switzerland
Phone: 41-61-415-8710
Fax: 41-61-415-8720

Rosemount Analytical, Inc.
2400 Barranca Pkwy.
Irvine, CA 92606
Phone: 949-863-1181
Fax: 949-474-7250

Rosenmund
St. James Ct., Wilderspool Causeway
Warrington, Chesire, WA4 6PS England
Phone: 0925-52621
Fax: 0925-416790

Rosenmund, Inc.
9110 Forsyth Park Dr.
Charlotte, NC 28273
Phone: 704-587-0440
Fax: 704-588-6866

Roto-Sieve AB
Hjorthagsgatan 10
S-413 17, Goteborg, Sweden
Phone: 031-427890
Fax: 031-422070

Rubber Millers, Inc.
709 S. Caton
Baltimore, MD 21229
Phone: 410-947-8400
Fax: 410-233-6537

Rupprecht & Patashnick Co., Inc.
25 Corporate Circle
Albany, NY 12203
Phone: 518-452-0065
Fax: 518-452-0067
www.rpco.com

S.P. Kinney Engineers, Inc.
P.O. Box 445
Carnegie, PA 15106-0445
Phone: 412-276-4600
Fax: 412-276-6890

S&G Enterprises, Inc.
N115 W1900 Edison Dr.
Germantown, WI 53022
Phone: 262-251-8300
Fax: 262-251-1616
www.ramflat.com

S&N Airoflo, Inc.
P.O. Drawer 1139
Greenwood, MS 38935-1139
Phone: 601-453-2588
Fax: 601-453-1991
www.airoflo.com

SAMI
940 Kulp Rd.
Pottstown, PA 19465
Phone: 610-495-6858
Fax: 610-495-0560
www.sami1.com

Sanitaire Corp.
9333 N. 49th St.
Brown Deer, WI 53223
Phone: 414-365-2200
Fax: 414-365-2210
www.sanitaire.com

Schlicher & Schuell
P.O. Box 2012
Keene, NH 03431
Phone: 603-352-3810
Fax: 603-357-3627

Schloss Engineered Equipment
10555 E. Dartmouth #230
Aurora, CO 80014
Phone: 303-695-4500
Fax: 303-695-4507

Schlueter Co.
P.O. Box 548
Janesville, WI 53547
Phone: 608-755-5455
Fax: 608-755-5450

Schreiber Corp.
100 Schreiber Dr.
Trussville, AL 35173
Phone: 205-655-7466
Fax: 205-655-7669
www.schreiber-water.com

Schreiber-Klaranlagen
Postfach 1580
30853 Langenhagen, Germany
Phone: 0511-77990
Fax: 0511-7799220

SciCorp Systems, Inc.
274 Burton Ave. Ste. 203B
Barrie, Ontario, Canada L4N 5W4
Phone: 705-733-2626
Fax: 705-733-2618
www.scicorpbiologic.com

Science Application International Corp.
3240 Schoolhouse Rd.
Middletown, PA 17057
Phone: 717-944-5501
Fax: 717-944-4551

Scienco/FAST Systems
3240 N. Broadway
St. Louis, MO 63147-3515
Phone: 314-621-2536
Fax: 314-621-1952

Screening Systems International
P.O. Box 760
Slaughter, LA 70777
Phone: 225-654-3900
Fax: 225-654-3966
www.screeningsystems.com

Seaman Corp.
1000 Venture Blvd.
Wooster, OH 44691
Phone: 330-262-1111
Fax: 330-263-6950
www.seamancorp.com

Sefar America, Inc.
333 S. Highland Ave.
Briarcliff Manor, NY 10510
Phone: 914-941-7767
Fax: 914-762-8599
www.sefaramerica.com

Seghers Better Technology USA
3114 Emery Circle
Austell, GA 30168
Phone: 770-739-4205
Fax: 770-739-0117
www.bettertechnology.com

Selecto, Inc.
5933 Peachtree Industrial Blvd.
Norcross, GA 30092
Phone: 770-448-2433
Fax: 770-448-5214

Semblex, Inc.
1635 W. Walnut
Springfield, MO 65806-1643
Phone: 417-866-1035
Fax: 417-866-0235

Sentex Systems, Inc.
553 Broad Ave.
Ridgefield, NJ 07657
Phone: 201-945-3694
Fax: 201-941-6064

Serfilco, Ltd.
1777 Shermer Rd.
Northbrook, IL 60062-5360
Phone: 847-559-1777
Fax: 847-559-1995
www.serfilco.com

Sernagiotto Technologies
1903 Corona Ave.
Jasper, AL 35501
Phone: 205-221-3709
Fax: 205-221-5237
www.sernagiotto.it

Serpentix Conveyor Corp.
9085 Marshall Ct.
Westminster, CO 80030
Phone: 303-430-8427
Fax: 303-430-7337
www.serpentix.com

Servomex Co.
90 Kerry Pl.
Norwood, MA 02062
Phone: 781-769-7710
Fax: 781-769-2834
www.servomex.com

Shannon Chemical Corp.
P.O. Box 376
Malvern, PA 19355
Phone: 610-363-9090
Fax: 610-524-6050
www.shannonchem.com

Silbrico Corp.
6300 River Rd.
Hodgkins, IL 60525-4257
Phone: 708-354-3350
Fax: 708-354-6698

Simon-Hartley, Ltd.
Stoke-On-Trent
Staffordshire, ST4 7BH England
Phone: 0782-202300
Fax: 0782-260534

Smith & Loveless, Inc.
14040 Sante Fe Trail Dr.
Lenexa, KS 66215-1284
Phone: 913-888-5201
Fax: 913-888-2173
www.smithandloveless.com

Solidur Plastics Co.
200 Industrial Dr.
Delmont, PA 15626
Phone: 724-468-6868
Fax: 724-468-4044

Solinst Canada Ltd.
35 Todd Rd.
Georgetown, Ontario, L7G 4R8
Phone: 905-873-2255
Fax: 905-873-1992
www.solinst.com

Solucorp Industries Corp.
250 W. Nyack Rd.
West Nyack, NY 10994
Phone: 914-623-2333
Fax: 914-623-4987
www.solucorpltd.com

Solvay America
3333 Richmond
Houston, TX 77098
Phone: 713-525-6000
Fax: 713-524-7887
www.solvay.com

Somat Corp.
855 Fox Chase
Coatesville, PA 19320
Phone: 610-384-7000
Fax: 610-380-8500
www.somatcorp.com

Spaulding Composites Co.
1300 S. Seventh St.
Dekalb, IL 60115
Phone: 815-758-8181
Fax: 815-758-1900
www.spauldingcom.com

Spencer Turbine Co.
600 Day Hill Rd.
Windsor, CT 06095
Phone: 860-688-8361
Fax: 860-688-0098

Sper Chemical Corp.
14770 62nd St.
Clearwater, FL 33760-2331
Phone: 727-535-9033
Fax: 727-530-0741

Spirac AB
Box 30033
200 61 Malmo, Sweden
Phone: 46-(0)40-162020
Fax: 46-(0)40-153650

Spirac USA, Inc.
P.O. Box 3137
Peachtree City, GA 30269
Phone: 770-632-9833
Fax: 770-632-9838
www.spirac.se

SRE, Inc.
510 Franklin Ave.
Nutley, NJ 07110
Phone: 973-661-5192
Fax: 973-661-3713
www.sreinc.com

SRS Crisafulli, Inc.
P.O. Box 1051
Glendive, MT 59330-9985
Phone: 406-365-3393
Fax: 406-365-8088
www.crisafulli.com

SRS Industrial Engineering
362 S. Main
Clearfield, UT 84015
Phone: 801-773-1311
Fax: 801-773-8608

Stahler GmbH
Muhlenhof 1
D-65589 Hadamar, Germany
Phone: 49-64-33-93000
Fax: 49-64-33-5960

Stancor Pump, Inc.
515 Fan Hill Rd.
Monroe, CT 06468
Phone: 203-268-7513
Fax: 203-268-7958

Stanley Pump & Equipment, Inc.
2525 South Clearbrook Dr.
Arlington Heights, IL 60005
Phone: 847-439-9200
Fax: 847-439-9388
www.stanleypump.com

Star Systems, Inc.
P.O. Box 518
Timmonsville, SC 29161
Phone: 843-346-3101
Fax: 843-346-3736
www.hilliardcorp.com

Steel Tank Institute
570 Oakwood Rd.
Lake Zurich, IL 60047
Phone: 847-438-8265
Fax: 847-438-4509
www.steeltank.com

Sterling Fluid Systems (USA)
P.O. Box 460
Grand Island, NY 14072
Phone: 716-773-6450
Fax: 716-773-2330
www.sterlingfluidsystems.com

Sternson Ltd.
P.O. Box 1540
Brantford, Ontario, Canada N3T 5V6
Phone: 519-759-7570
Fax: 519-759-8962

Stevens Water Monitoring Systems
P.O. Box 40
Beaverton, OR 97075-0040
Phone: 503-646-9171
Fax: 503-526-1471
www.stevenswater.com

Stiles-Kem Divison, Met Pro Corp.
1570 Lakeside Dr.
Waukegan, IL 60085-8309
Phone: 847-689-1100
Fax: 847-689-9289

Stockhausen, Inc.
3408 Doyle St.
Greensboro, NC 27406
Phone: 910-333-3500
Fax: 910-333-3518

Stormceptor Corp.
600 Jefferson Plaza, Ste. 304
Rockville, MD 20852
Phone: 301-762-8361
Fax: 301-762-4190
www.csrstormceptor.com

StormTreat Systems, Inc.
3408 Doyle St.
Barnstable, MA 02630
Phone: 508-778-4449
Fax: 508-362-5335

Stormwater Management
2035 N.E. Colombia Blvd.
Portland, OR 97211
Phone: 503-240-9553
Fax: 503-240-9553
www.stormwatermgt.com

Strategic Diagnostics, Inc.
111 Pencader Dr.
Newark, DE 19702
Phone: 302-456-6789
Fax: 302-456-6782
www.sdix.com

Süd-Chemie Prototech Inc.
32 Fremont St.
Needham, MA 02194
Phone: 781-444-5188
Fax: 781-444-0130
www.prototechco.com

SulfaTreat Co.
17998 Chesterfield Airport Rd., Ste. 215
Chesterfield, MO 63005
Phone: 314-532-2189
Fax: 314-532-2764
www.sulfatreat.com

Sumitomo Machinery Corp.
4200 Holland Blvd.
Chesapeake, VA 23323
Phone: 757-485-3355
Fax: 757-487-3193
www.smcyclo.com

Svedala Industries, Inc.
P.O. Box 15312
York, PA 17405-7312
Phone: 717-843-8671
Fax: 717-845-5154

SWECO Engineering Corp.
7120 New Buffington Rd.
Florence, KY 41042
Phone: 859-727-5147
Fax: 859-727-5122
www.sweco.com

Sybron Chemicals, Biochemical
  Division
111 Kesler Mill Rd.
Salem, VA 24153
Phone: 540-389-9361
Fax: 540-389-9364
www.sybronchemicals.com

Sybron Chemicals, Inc.
P.O. Box 66
Birmingham, NJ 08011
Phone: 609-893-1100
Fax: 609-894-8641
www.sybronchemicals.com

Synetix
2 Transam Plaza Dr., Ste. 230
Oak Brook Terrace, IL 60181
Phone: 630-268-6300
Fax: 630-268-9797
www.synetix.com

TAH Industries, Inc.
107 N. Gold Dr.
Robbinsville, NJ 08691
Phone: 609-259-9222
Fax: 609-259-0957

Tate Andale, Inc.
1941 Lansdowne Rd.
Baltimore, MD 21227-1789
Phone: 410-247-8700
Fax: 410-247-9672

Techniflo Systems
12300 Perry Highway
Wexford, PA 15090
Phone: 412-749-0600
Fax: 412-935-0777

Tecnetics Industries, Inc.
1811 Buerkle Rd.
St. Paul, MN 55110
Phone: 651-777-4780
Fax: 651-777-5582
www.tecweigh.com

Tellkamp Systems, Inc.
15523 Carmenita Rd.
Sante Fe Springs, CA 90670
Phone: 562-802-1621
Fax: 562-802-1303
www.tellkamp.com

Tenco Hydro, Inc.
4620 Forest Ave.
Brookfield, IL 60513
Phone: 708-387-0700
Fax: 708-387-0732

Tetra Process Technologies
Park West One, Ste. 600
Pittsburgh, PA 15275
Phone: 412-788-8300
Fax: 412-788-8304

Tetratec
1741 Loretta Ave.
Feasterville, PA 19053
Phone: 215-355-7111
Fax: 215-355-6745
www.tetratex.com

Thermaco, Inc.
646 Greensboro St.
Asheboro, NC 27204
Phone: 336-629-4651
Fax: 336-626-5739
www.big-dipper.com

Thermacon Enviro Systems, Inc.
1983 Marcus Ave.
Lake Success, NY 11042
Phone: 516-328-6600
Fax: 516-328-7988
www.thermacon.com

ThermaFab, Inc.
200 Rich Lex Dr.
Lexington, SC 29072
Phone: 803-794-2543
Fax: 803-796-0999

Thermal Black Clawson
605 Clark St.
Middletown, OH 45042
Phone: 513-424-7400
Fax: 513-424-1168

Thermatrix, Inc.
308 N. Peteres Rd.
Knoxville, TN 37922
Phone: 423-539-9603
Fax: 423-539-9643
www.thermatrix.com

Thermax, Ltd.
40440 Grand River
Novi, MI 48050
Phone: 248-474-3050
Fax: 248-474-5790

TIGG Corp.
Box 11661
Pittsburgh, PA 15228
Phone: 412-563-4300
Fax: 412-563-6155

TN Technologies, Inc.
P.O. Box 800
Round Rock, TX 78680-0800
Phone: 512-388-9100
Fax: 512-388-9200
www.tnksi.com

Tonka Equipment Co.
13305 Watertower Circle
Plymouth, MN 55441
Phone: 612-559-2837
Fax: 612-559-1979
www.tankwater.com

Toray Industries, Inc.
2-2, Nichonbashi-Muromachi, Chuo-Ku
Tokyo, 103 Japan
Phone: 03-245-5607
Fax: 03-245-5555

Toyobo Co., Ltd.
2-8 Dojima Hama 2-chrome, Kita-ku
Osaka 530-8230, Japan
Phone: 81-6-6348-3360
Fax: 81-6-6348-3418
www.toyobo.co.jp

Tri-Mer Corp.
P.O. Box 730
Owosso, MI 48867
Phone: 517-723-7838
Fax: 517-723-7844
www.tri-mer.com

TriSep Corp.
93 S. La Patera Ln.
Goleta, CA 93117
Phone: 805-964-8003
Fax: 805-964-1235
www.trisep.com

Triton Technologies
11917 FM 529
Houston, TX 77041
Phone: 713-937-0101
Fax: 713-937-1979

Trojan Technologies, Inc.
3020 Gore Rd.
London, Ontario, Canada N5V 4T7
Phone: 519-457-3400
Fax: 519-457-3030
www.trojanuv.com

Trusty Cook, Inc.
10530 E. 59th St.
Indianapolis, IN 46236
Phone: 317-823-6821
Fax: 317-823-6822
www.trusty-cook.com

Turner Designs
845 W. Maude Ave.
Sunnyvale, CA 94086
Phone: 408-749-0994
Fax: 408-749-0998
www.turnerdesigns.com

Turnkey Solutions, Inc.
103 Godwin Ave.
Midland Park, NJ 07432-1813
Phone: 201-848-7676
Fax: 201-848-1643

Tuthill Pneumatics Group
P.O. Box 2877
Springfield, MO 65801-2877
Phone: 417-865-8715
Fax: 417-865-2950
www.mdpneumatics.com

Tytronics, Inc.
25 Wiggins Ave.
Bedford, MA 01730-2323
Phone: 781-275-9660
Fax: 781-275-9665
www.tytronics.com

Ultra Additives, Inc.
460 Straight St.
Paterson, NJ 07501
Phone: 973-279-1306
Fax: 973-279-0602

Ultraflote Corp.
8558 Katy Freeway, Ste. 100
Houston, TX 77024
Phone: 713-461-2100
Fax: 713-461-2213
www.ultraflote.com

Underwriters Laboratories, Inc.
333 Pfingsten Rd.
Northbrook, IL 60062
Phone: 847-272-8800
Fax: 847-509-6219
www.ul.com

Unicel, Inc.
P.O. Box 15203
Baton Rouge, LA 70895
Phone: 225-753-7129
Fax: 225-753-7125
www.unicelinc.com

Unifilt Corp.
P.O. Box 389
Zelienople, PA 16063-0389
Phone: 724-758-3833
Fax: 724-758-3870
www.unifilt.com

Unimin Corp.
258 Elm St.
New Canaan, CT 06840
Phone: 203-966-8880
Fax: 203-966-3453

Unipure Corp.
1440 North Harbor, Ste. 125
Fullerton, CA 92835
Phone: 714-870-1578
Fax: 714-870-4576
www.unipure.com

Unisol
1810 W. Drake Dr. #103
Tempe, AZ 85283
Phone: 480-491-7145
Fax: 480-491-7185
www.unisol-biologics.com

United Industries, Inc.
P.O. Box 3838
Baton Rouge, LA 70821-3838
Phone: 225-292-5527
Fax: 225-293-1655
www.ui-inc.com

Universal Process Equipment Co.
P.O. Box 338
Roosevelt, NJ 08555-0338
Phone: 609-443-4545
Fax: 609-259-0644
www.upe.com

USFilter Recovery Services
2430 Rose Pl.
Roseville, MN 55113
Phone: 651-638-1300
Fax: 651-633-5074
www.usfilter.com

USFilter/Aerator Products
11765 Main St.
Roscoe, IL 61073
Phone: 815-623-2111
Fax: 815-623-6416
www.usfilter.com

USFilter/Asdor
250 Royal Crest Ct.
Markham, Ontario, Canada L3R 3S1
Phone: 905-944-2828-836-7700
Fax: 905-474-1334
www.usfilter.com

USFilter/Bekox
Pol.Ind. Santa Ana, C/El Electrodo, 52
Rivas Vaciamadrid, 28529
Phone: 34-91-660-4000
Fax: 34-91-666-7716
www.usfilter.com

USFilter/Contra-Shear
CPO Box 1611
Auckland, New Zealand
Phone: 64-9-818-6108
Fax: 64-9-818-6599
www.contra-shear.co.nz

USFilter/Control Systems
1239 Willow Lake Blvd.
Vadnais, MN 55110
Phone: 651-766-2700
Fax: 651-766-2701
www.usfilter.com

USFilter/CPC
441 Main St.
Sturbridge, MA 01566
Phone: 508-347-7344
Fax: 508-347-7049
www.usfilter.com

USFilter/Davco
P.O. Box 1419
Thomasville, GA 31792
Phone: 912-226-5733
Fax: 912-228-0312
www.usfilter.com

USFilter/Davis Process
2650 Tallevast Rd.
Sarasota, FL 34243
Phone: 941-355-2971
Fax: 941-351-4756
www.usfilter.com

USFilter/Dewatering Systems
2155 112th Ave.
Holland, MI 49424-9604
Phone: 616-772-9011
Fax: 616-772-4516
www.usfilter.com

USFilter/Diffused Air Products Group
8506 Beechmont Ave.
Cincinnati, OH 45255
Phone: 513-388-4100
Fax: 513-388-4111
www.usfilter.com

USFilter/Electrocatalytic
2 Milltown Ct.
Union, NJ 07083
Phone: 908-851-6952
Fax: 908-851-6906
www.usfilter.com

USFilter/Envirex
P.O. Box 1604
Waukesha, WI 53187
Phone: 262-541-0141
Fax: 262-541-4120
www.usfilter.com

USFilter/Filtration & Separation
2118 Greenspring Dr.
Timonium, MD 21093
Phone: 410-252-0800
Fax: 410-560-2857
www.usfilter.com

USFilter/Gas Technologies
1501 E, Woodfield Rd.
Schaumburg, IL 60173
Phone: 847-706-6900
Fax: 847-706-6996
www.usfilter.com

USFilter/General Filter
600 Arrasmith Trail
Ames, IA 50010
Phone: 515-232-4121
Fax: 515-232-2571
www.usfilter.com

USFilter/Gutling
Postfach 1445
Fellbach, Germany 70704
Phone: 07-11-518550-0
Fax: 07-11-518550-100
www.usfilter.com

USFilter/Headworks Products
100 Highpoint Dr.
Chalfont, PA 18914
Phone: 215-712-0280
Fax: 215-996-1136
www.usfilter.com

USFilter/HPD
2 West Main St.
Plainfield, IL 60544
Phone: 815-436-3013
Fax: 815-436-3010
www.usfilter.com

USFilter/Hubert
P.O. Box 29
8715 ZH Stavoren, The Netherlands
Phone: 31-514-684444
Fax: 31-514-682198
www.usfilter.com

USFilter/Industrial Wastewater Systems
181 Thorn Hill Rd.
Warrendale, PA 15086-7257
Phone: 724-772-0044
Fax: 724-772-1202
www.usfilter.com

USFilter/Jet Tech
P.O. Box 13306
Edwardsville, KS 66113-0306
Phone: 913-422-7600
Fax: 913-422-7667
www.usfilter.com

USFilter/Johnson Screens
P.O. Box 64118
St. Paul, MN 55164
Phone: 651-636-3900
Fax: 651-638-3132
www.usfilter.com

USFilter/Krüger
401 Harrison Oaks Blvd., Ste. 100
Cary, NC 27513
Phone: 919-677-8310
Fax: 919-677-0082
http://KrugerWorld.com/

USFilter/Lowell
10 Technology Dr.
Lowell, MA 01581
Phone: 978-934-9349
Fax: 978-441-6025
www.usfilter.com

USFilter/Memcor
2118 Greenspring Dr.
Timonium, MD 21093
Phone: 410-308-2947
Fax: 410-561-3017
www.usfilter.com

USFilter/Microfloc
44 Main St.
Sturbridge, MA 01566
Phone: 508-347-7344
Fax: 508-347-7049
www.usfilter.com

USFilter/RJ Environmental
13100 Gregg St.
Poway, CA 92064
Phone: 858-486-8500
Fax: 858-486-8501
www.usfilter.com

USFilter/Rockford
4669 Shepherd Trail
Rockford, IL 61103
Phone: 815-877-3041
Fax: 815-877-0172
www.usfilter.com

USFilter/Rossmark
P.O. Box 109
7600 AC Almelo, The Netherlands
Phone: 31-546-838000
Fax: 31-546-814141
www.usfilter.com

USFilter/Schumacher Filters
P.O. Box 8040
Asheville, NC 28814
Phone: 828-252-9000
Fax: 828-253-7773
www.usfilter.com

USFilter/Smogless
Via L. Mascheroni, 29
20145 Milan, Italy
Phone: 02-48595-1
Fax: 02-48008417

USFilter/Stranco
P.O. Box 389
Bradley, IL 60915
Phone: 815-939-1265
Fax: 815-932-0674
www.usfilter.com

USFilter/Wallace & Tiernan
1901 West Garden Rd.
Vineland, NJ 08360
Phone: 856-507-9000
Fax: 856-507-4125
www.usfilter.com

USFilter/Whittier
12442 East Putnam St.
Whittier, CA 90602
Phone: 562-698-9414
Fax: 562-698-1960
www.usfilter.com

USFilter/Zimpro
301 W. Military Rd.
Rothschild, WI 54474
Phone: 715-359-7211
Fax: 715-355-3219
www.usfilter.com

UV Systems Technology, Inc.
2800 Ingleton Ave.
Burnaby, BC V5C 6G7
Phone: 604-451-1069
Fax: 604-451-1072
www.ultraguard.com

Val-Matic Valve & Manufacturing Corp.
905 Riverside Dr.
Elmhurst, IL 60126
Phone: 630-941-7600
Fax: 630-941-8042
www.valmatic.com

Vanton Pump & Equipment Corp.
201 Sweetland Ave.
Hillside, NJ 07205
Phone: 908-688-4216
Fax: 908-686-9314
www.vanton.com

Vara International
1201 19th Pl.
Vero Beach, FL 32960
Phone: 561-567-1320
Fax: 561-567-4108

Vaughan Co., Inc.
364 Monte Elma Rd.
Montesano, WA 98563
Phone: 360-249-4042
Fax: 360-249-6155
www.chopperpumps.com

Verder, Inc.
P.O. Box 1329
West Chester, PA 19380
Phone: 610-429-4200
Fax: 610-429-1139
www.verderflex.com

Vibra Screw, Inc.
755 Union Blvd.
Totowa, NJ 07511
Phone: 973-256-7410
Fax: 973-256-7567

Vikoma International Ltd.
Propect Rd., Cowes
Isle of Wight PO31 7AD, England
Phone: 44 (0)983 296021
Fax: 44 (0)983 299035

Vincent Corp.
P.O. Box 5747
Tampa, FL 33675
Phone: 813-248-2650
Fax: 813-247-7557
www.vincentcorp.com

VMI Inc.
1125 N. Maitlen Dr.
Cushing, OK 74023
Phone: 918-225-7000
Fax: 918-225-0333
www.vmi-dredges.com

Voith Sulzer
P.O. Box 688
Neenah, WI 54957
Phone: 920-722-7713
Fax: 920-725-8615

Vortechnics, Inc.
41 Evergreen Dr.
Portland, ME 04103-1067
Phone: 207-878-3662
Fax: 207-878-8507
www.vortechnics.com

Vortex Ventures
6611 Portwest Dr.
Houston, TX 77024
Phone: 713-869-2593
Fax: 713-869-2596
www.vortexventures.com

Vulcan Industries, Inc.
P.O. Box 390
Missouri Valley, IA 51555
Phone: 712-642-2755
Fax: 712-642-4256
www.vulcanindustries.com

Vulcan Performance Chemicals
P.O. 38015
Birmingham, AL 35238-5015
Phone: 205-298-3000
Fax: 205-298-2955
www.vul.com

W.L. Gore & Associates, Inc.
P.O. Box 1100
Elkton, MD 21922-1100
Phone: 410-392-3300
Fax: 410-398-6624
www.wlgore.com

Walker Process Equipment
840 N. Russell Ave.
Aurora, IL 60506
Phone: 630-892-7921
Fax: 630-892-7951
www.walker-process

Warren Rupp, Inc.
P.O. Box 1568
Mansfield, OH 44901-1568
Phone: 419-524-8388
Fax: 419-522-7867
www.warrenrupp.com

Waste Water Systems, Inc.
4386 Lilburn Industrial Way
Lilburn, GA 30047
Phone: 770-921-0022
Fax: 770-564-0409

Waste-Tech, Inc.
1931 Industrial Dr.
Libertyville, IL 60048-9738
Phone: 847-367-5150
Fax: 847-367-1787

Water and Power Technologies
P.O. Box 27836
Salt Lake City, UT 84127-0836
Phone: 801-974-5500
Fax: 801-973-9733
www.wpt.com

Water Resources Group, Inc.
P.O. Box 470
Auburn, AL 36831
Phone: 334-821-1135
Fax: 334-821-2451

Waterlink Biological Systems
630 Currant Rd.
Fall River, MA 02720-4732
Phone: 508-679-6770
Fax: 508-672-5779
www.waterlink.com

Waterlink Inc.
4100 Holiday St. N.W.
Canton, OH 44718-2532
Phone: 330-649-4000
Fax: 330-649-4008
www.waterlink.com

Waterlink Separations, Inc.
29850 N. Skokie Hwy.
Lake Bluff, IL 60044
Phone: 847-473-3700
Fax: 847-473-0477
www.waterlink.com

Waterlink/Aero-Mod Systems
7927 US Highway 24
Manhattan, KS 66502-4995
Phone: 785-537-4995
Fax: 785-537-0813
www.waterlink.com

Waterlink/Barnebey Sutcliffe
835 N. Cassady Ave.
Columbus, OH 43219
Phone: 614-258-9501
Fax: 614-258-3464
www.waterlink.com

Watson Marlow, Inc.
220 Ballardvale
Wilmington, MA 01887
Phone: 978-658-6168
Fax: 978-658-5558
www.watson-marlow.com

Weatherly
1100 Spring St., Ste. 800
Atlanta, GA 30309
Phone: 404-873-5030
Fax: 404-873-1303
www.weatherlyinc.com

Wescor, Inc.
459 S. Main St.
Logan, UT 84321
Phone: 435-753-7760
Fax: 435-753-6756
www.wescor.com

WesTech Engineering Inc.
P.O. Box 65068
Salt Lake City, UT 84115-0068
Phone: 801-265-1000
Fax: 801-265-1080
www.westech-inc.com

Western States Machine Co.
P.O. Box 327
Hamilton, OH 45012
Phone: 513-863-4758
Fax: 513-863-3846
www.westernstates.com

Westfalia Separator, Inc.
100 Fairway Ct.
Northvale, NJ 07647
Phone: 201-767-3900
Fax: 201-767-4399
www.westfaliaseparatorus.com

Westport Environmental Systems
251 Forge Rd.
Westport, MA 02790-0217
Phone: 508-636-8811
Fax: 508-636-2088
www.wesenvsys.com

Westwood Chemical Corp.
46 Tower Dr.
Middletown, NY 10940
Phone: 914-692-6721
Fax: 914-695-1906

Wheelabrator Air Pollution Control, Inc.
441 Smithfield St.
Pittsburgh, PA 15222-2292
Phone: 412-562-7300
Fax: 412-562-7254
www.wapc.com

Wheelabrator Water Technologies, Inc.
1110 Benfield Blvd., Ste. B
Millersville, MD 21108
Phone: 410-729-1440
Fax: 410-729-0854
www.bio-gro.com

Wilfley Weber, Inc.
P.O. Box 2330
Denver, CO 80201
Phone: 303-779-1777
Fax: 303-779-1277

WRc Process Engineering
Aynho Rd., Adderbury, Banbury
Oxan, OX17 3NL England
Phone: 0295-812282
Fax: 0295-812283

WTW Measurement Systems, Inc.
3170 Metro Pkwy.
Fort Myers, FL 33916-7597
Phone: 941-337-7112
Fax: 941-337-2045

Wyssmont Co., Inc.
P.O. Box 1397
Fort Lee, NJ 07024
Phone: 201-947-4600
Fax: 201-947-0324
www.wyssmont.com

Yeomans Chicago Corp.
3905 Enterprise Ct.
Aurora, IL 60504
Phone: 630-236-5500
Fax: 630-235-5511
www.yccpump.com

Zellweger Analytics, Inc.
100 Park Ave.
League City, TX 77573
Phone: 281-316-7700
Fax: 281-316-7800
www.zelma.com

Zenon Environmental, Inc.
3239 Dundas St. W.
Oakville, Ontario, Canada L6M 4B3
Phone: 905-465-3030
Fax: 905-465-3050
www.zenonenv.com

ZMI/Portec Chemical Processing
P.O. Box 274
Sibley, IA 41249
Phone: 712-754-4661
Fax: 712-754-3607
www.zmichemical.com